Thre

Studies in Environmental Science 39

AQUATIC BIOENVIRONMENTAL STUDIES: THE HANFORD EXPERIENCE 1944–84

C.D. Becker

Geosciences Department, Pacific Northwest Laboratory,
Richland, WA 99352, U.S.A.

''Onward ever,
Lovely River,
Softly calling to the sea,
Time that scars us,
Maims and mars us,
Leaves no track or trench on thee.''

From ''Beautiful Willamette''
By Samuel S. Simpson

ELSEVIER
Amsterdam — Oxford — New York — Tokyo 1990

ELSEVIER SCIENCE PUBLISHERS B.V.
Sara Burgerhartstraat 25
P.O. Box 211, 1000 AE Amsterdam, The Netherlands

Distributors for the United States and Canada:

ELSEVIER SCIENCE PUBLISHING COMPANY INC.
655, Avenue of the Americas
New York, NY 10010, U.S.A.

Library of Congress Cataloging-in-Publication Data

Becker, C. Dale (Clarence Dale)
Aquatic bioenvironmental studies : the Hanford experience, 1944-84 / C. Dale Becker.
p. cm. -- (Studies in environmental science ; 39)
Includes bibliographical references and index.
ISBN 0-444-88653-2
1. Nuclear reactors--Environmental aspects--Washington (State)--Hanford Reach. 2. Nuclear power plants--Environmental aspects--Washington (State)--Hanford Reach. 3. Aquatic organisms--Washington (State)--Hanford Reach--Effect of radiation on. 4. Aquatic organisms--Washington (State)--Hanford Reach--Effect of water pollution on. 5. Environmental monitoring--Washington (State)--Hanford Reach. 6. Indicators (Biology)--Washington (State)--Hanford Reach. 7. Hanford Works (Wash.) I. Title. II. Series.
QH545.N8B43 1990
628.1'685--dc20 90-41371
CIP

ISBN 0-444-88653-2

This book is printed on acid-free paper.

Printed in The Netherlands

PREFACE

From 1944 to 1971, the Hanford Reach of the Columbia River in Washington State was used to provide cooling water for up to eight single-purpose reactors producing a new and unique fissionable material: plutonium. The release of more than 60 radionuclides, vast amounts of waste heat, and some process chemicals in the effluent to the Hanford Reach during that era resulted in no apparent impairment of ecological functions in the Columbia River. The knowledge that releases of this type and amount can be mitigated by a river ecosystem is important to present and future generations. Today, only traces of radioactivity from the plutonium-production era at the Hanford Site remain in the form of a few long-lived radionuclides in sediment deposits and river organisms.

From 1971 through 1984, water from the Hanford Reach was used in Hanford Site operations for closed-cycle cooling of one dual-purpose facility (plutonium and electricity), dilution of process effluents, one nuclear power generating plant, and a variety of offsite purposes. Releases of radioactivity, heat, or chemicals to the Hanford Reach declined to relatively small amounts. All releases became subject to applicable federal regulations designed to protect water quality and public health. Long-term environmental monitoring programs were expanded onsite and offsite to demonstrate compliance.

The quality of water in the Hanford Reach remains high in the 1980s, in a near-pristine condition, almost representative of the years before the first dams were constructed on the mainstem Columbia River. The river's quality complies well with all state standards for drinking water. Activities and events taking place upstream now influence changes in the quality of water flowing through the Hanford Reach to a greater extent than do site activities.

My objective is to review bioenvironmental studies related to the Hanford Reach of the Columbia River on the Hanford Site from 1944 to 1984. These studies dealt, in large part, with the potential effects of specific Hanford Site activities on aquatic organisms and the physicochemical properties of the river ecosystem. The studies encompassed an extended series of interrelated field and laboratory investigations.

This book covers early experiments at the University of Washington on radiological effects and aquatic organisms. It details laboratory and field studies associated with operation of single-purpose, plutonium-production reactors at Hanford from 1944 to 1971. It covers subsequent investigations to identify any effects on the Columbia River and its aquatic organisms from Hanford Site energy production activities. A historical framework is used to help explain not only why certain studies were conducted but their contribution to radioecology, aquatic ecology, and decisions to guide onsite operations.

We now know that initial bioenvironmental studies at Hanford were primitive by today's more exacting standards. Yet the initial studies retain value, and they provide important environmental lessons. In 1971, John R. Totter, then Director of DOE's Division of Biology and Medicine, drew attention to "the importance of looking back as we move ahead in science...(because)... things have a way of being rediscovered periodically; sometimes this must be pointed out to those who cannot remember the past." Then, as now, progress in science builds on foundations built by others.

In style, I have emphasized results of findings rather than methods. A complete reference list for each chapter provides further insight to readers. The material used has been drawn extensively from publications and periodic reports issued by government, industrial, and institutional scientists who have conducted research related to the Hanford Reach since 1944 and, to a lesser extent, on operational documents and annual progress reports that may be difficult to locate today. However, the "grey" literature produced by Hanford contractors often presents incomplete findings that were subsequently reexamined, analyzed, consolidated, and published in open-literature journals and symposia. Whenever a study was reported in the open literature, it was used as a principal reference source. Open-literature publications describing long-term, coordinated research efforts in the Hanford Reach were particularly valuable as references.

In retrospect, perhaps nowhere in the world were bioenvironmental studies conducted in a flowing river ecosystem with the same intensity and thoroughness as they were in the Hanford Reach from 1944 to 1984. One conclusion is inescapable: these studies advanced scientific knowledge and provided lasting benefit to mankind. Review of stresses imposed on the Columbia River ecosystem by activities at Hanford, why and how various studies were undertaken, and the significance of research findings, are best understood when they are placed in a frame of historical events.

This I have attempted to do.

C.D. Becker
August 1990

CONTENTS

ACKNOWLEDGMENTS

Completion of this book rests on the broad shoulders of many individuals. Some are currently employed by Pacific Northwest Laboratory (PNL) on the Hanford Site, while others are now retired or working elsewhere. Contributions were made in one form or another based on an individual's research and administrative backgrounds, technical expertise, depth of environmental concern, and willingness to contribute. Among the many, the assistance of the following was especially helpful.

From the ranks of PNL research scientists and administrators: Carl D. Corbitt, Colbert E. Cushing, Dennis D. Dauble, Richard F. Foster (retired), Robert H. Gray, Frank P. Hungate, Duane A. Neitzel, Iral C. Nelson, Thomas L. Page, Keith R. Price, Roy C. Thompson, Jr. (retired), and Donald G. Watson (retired). Former PNL employees include Charles C. Coutant, Oak Ridge National Laboratory; Mark J. Schneider, Bonneville Power Administration; and Roy E. Nakatani (retired), Fisheries Research Institute, University of Washington. From the former Department of Radiation Biology at the College of Fisheries, University of Washington: Lauren R. Donaldson (retired), Allyn H. Seymour (retired), and Ahmad E. Nevissi.

The final drive to publication was made possible by the editorial effort of Julie M. Gephart, Publication and Administration Department, PNL, and by funding from the U.S. Department of Energy, Richland Operations Office under Contract DE-ACO6-76RLO 1830.

As an author attempting to synthesize historical records, I must point out the significance of long-term support from the U.S. Department of Energy and its predecessors on the Hanford Site. These administrators recognized the need for bioenvironmental studies related to the Hanford Reach of the Columbia River long before current environmental protection laws became effective, and they provided for the continuity of these studies over 40 consecutive years.

GLOSSARY OF SCIENTIFIC AND COMMON NAMES

Common Name	Scientific Name
FISH	
white sturgeon	*Acipenser transmontanus*
largescale sucker	*Catostomus macrocheilus*
bridgelip sucker	*C. columbianus*
pumpkinseed	*Lepomis gibbosus*
largemouth bass	*Micropterus salmoides*
smallmouth bass	*M. dolomieui*
torrent sculpin	*Cottus rhotheus*
common carp	*Cyprinus carpio*
northern squawfish	*Ptychocheilus oregonensis*
redside shiner	*Richardsonius balteatus*
black bullhead	*Ictalurus melas*
brown bullhead	*I. nebulosus*
channel catfish	*I. punctatus*
mountain whitefish	*Prosopium williamsoni*
chinook salmon	*Oncorhynchus tshawytscha*
coho salmon	*O. kisutch*
sockeye salmon	*O. nerka*
cutthroat trout	*O. clarki (formerly Salmo clarki)*
rainbow trout or steelhead (anadromous form)	*O. mykiss (formerly Salmo gairdneri)*
brook trout	*Salvelinus fontinalis*
AQUATIC INVERTEBRATES	
freshwater mussel	*Anodonta californiensis*
snail	*Stagnicola spp.*
giant Columbia River limpet	*Fisherola nuttalli*
great Columbia River spire snail	*Lythoglyphus columbiana*
crayfish	*Pacifasticus leniusculus*

sponge	*Spongilla lacustris*
caddis fly	*Hydropsyche cockerelli*
black fly	*Simulium vittatum*
mayfly	*Paraleptophlebia bicornata*
midge fly	Family *Chironomidae*

AQUATIC PLANTS

reed canarygrass	*Scirpus spp.*
common wheatgrass	*Agropyron spicatum*
barnyard grass	*Echinochla crusgallii*
willows	*Salix spp.*
yellow cress	*Rorippa columbiae*

TERRESTRIAL ANIMALS

American bald eagle	*Haliaeetus leurocephalus*
mule deer	*Odocoileus hemionus*
coyote	*Canis latrans*
Great Basin Canada goose	*Branta canadensis moffitti*
American osprey	*Pandion haliaetus*

Chapter 1

INTRODUCTION

The Hanford Site

The Hanford Site is a semiarid expanse in southcentral Washington covering about 1460 square kilometers (560 square miles) (Figure 1.1). The site is administered by the U.S. Department of Energy (DOE) for research and development (R&D) activities in the areas of defense, energy, and environmental studies. Hanford was established during World War II after a nationwide search for a remote, sparsely settled location to produce a unique fissionable material of great importance to mankind. While activities at Hanford have diversified over a span of 40 years (Anonymous 1984), the site remained a key location for formulating and implementing government policy in defense, energy, and the environment.

The Columbia River has always played a key role at Hanford. The river flows through the northern part and along the eastern border of the site. Columbia River water was initially used for the once-through cooling of up to eight single-purpose plutonium-production reactors and for the chemical recovery of isotopes from irradiated fuels. From 1944 to 1971, these reactors discharged cooling water to the Columbia River, thus adding large amounts of radioactivity, heat, and chemicals to the river environment (Figure 1.2).

More recently, water from the Hanford Reach has been used to cool a dual-purpose reactor that produces plutonium and steam, an adjacent power plant that converts the steam to electricity, and a commercial power-generating plant that uses nuclear fuel. This water also serves to dilute some process effluent after their release, and is used to meet a variety of offsite regional needs. Today, nearly all use and return of water to the Columbia River from the Hanford Site is controlled by federal and state regulations to limit impairment of water quality.

The Hanford Reach of the Columbia River extends from Vernita (below Priest Rapids Dam) for 94 kilometers (58 miles) downstream to the city of Richland (see Figure 1.1). The Hanford Reach remains flowing

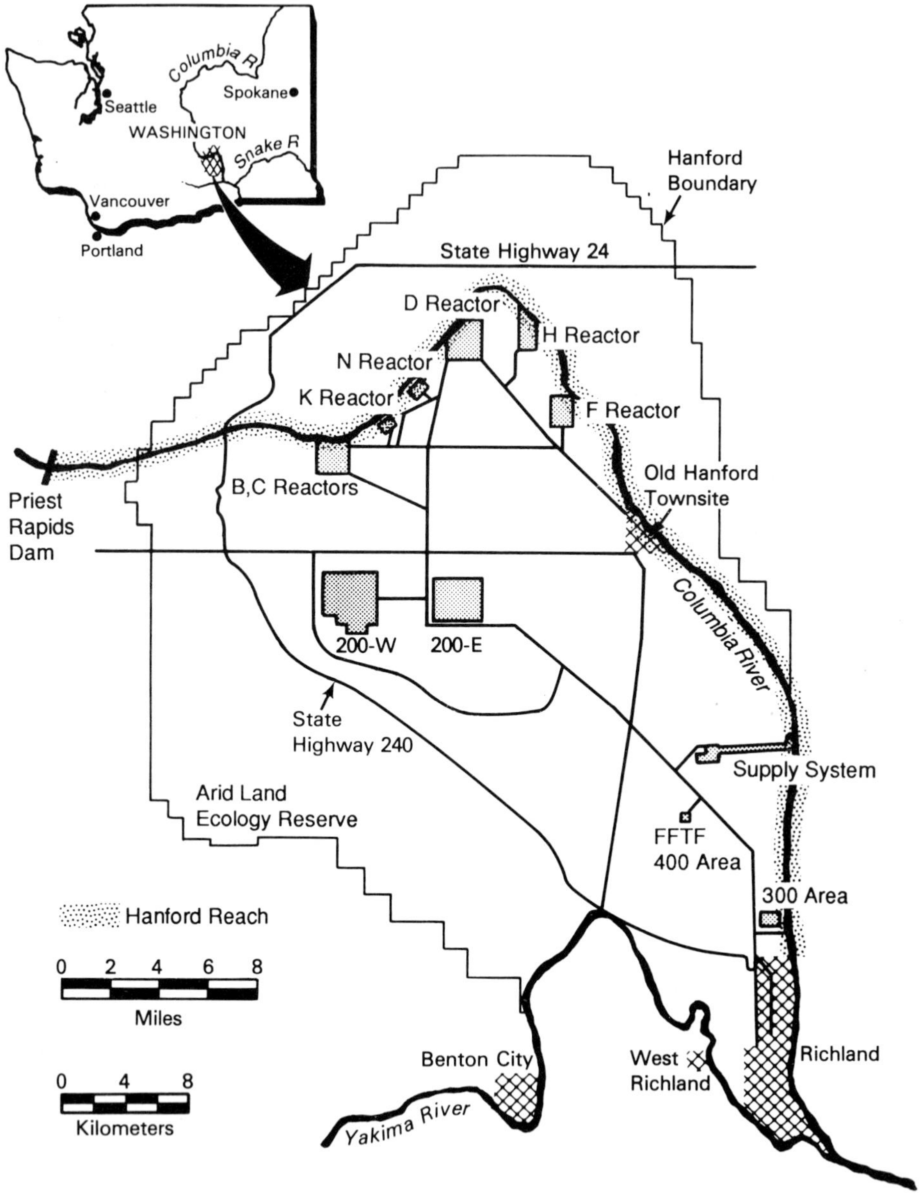

Fig. 1.1. Relative isolation and the availability of large amounts of water were two reasons that Hanford was sited on the mainstem Columbia River in southeastern Washington in 1943. Today, the Hanford Reach is the only portion of the mainstem Columbia River below the international border that remains flowing.

Fig. 1.2. The 100-B Reactor was the first single-purpose unit built to produce plutonium at Hanford. It operated from September 1944 to February 1968, providing nearly 24 years of service.

today, and thus has unique historical and ecological value. The rest of the mainstem Columbia River below the Canada/United States border has been impounded (PNWRBC 1979). Furthermore, over the past 25 years, it has evolved into an important mainstem spawning area for fall chinook salmon. Because it still flows, the Hanford Reach has become essential to maintaining populations of valued resources, such as the anadromous salmon and steelhead trout, in the mid-Columbia Basin (Becker 1985). A study of the Hanford Reach was authorized in 1988 to determine its eligibility for designation and protection under the Wild and Scenic Rivers Act.

This book was inspired by the realization that information from 40 years of laboratory and field investigations related to the Hanford Reach needed to be compiled for use by present and future generations. Fortunately, field studies in the Hanford Reach also involved controlled laboratory studies that were closely correlated with direct field investigations. Further, the continuity of these studies for 40 years allowed thorough evaluation of ecosystem responses. Such a situation may never occur again anywhere.

It is noteworthy that Hanford studies from 1944 to 1984 were conducted in a river environment subject to many external stresses. Some scientists now maintain that the most meaningful ecological research is conducted, or has been conducted, in environments directly impacted by human activities or managed for mankind's specific purposes (Hinds 1979).

Environmental awareness

Scientists and engineers organizing the unique undertaking at Hanford in the early 1940s recognized that use of large amounts of water from the Columbia River for reactor cooling might impair its quality and create environmental problems. In fact, General Leslie R. Groves, director of the Manhattan Project, U.S. Army Corps of Engineers, was acutely aware that Columbia River salmon might be affected (Groueff 1967). And top officials of E. I. du Pont de Nemours and Company, Inc., which constructed and operated the original Hanford Engineering Works, realized that preventing harm to fish in the Columbia River was a predominant issue (Carter 1987).

Concerns about possible adverse effects from discharging water with radioactive materials from the prototype plutonium-production units, which then existed only on paper, materialized in the form of radioecological studies with aquatic organisms – the first of their type anywhere. Subsequently, environmental studies at Hanford became the model for other studies involving aquatic ecosystems, particularly after passage of the National Environmental Policy Act in 1969.

Under pressure of World War II, producing a new fissionable material at Hanford under the veil of secrecy received top priority (Groves 1962). Few environmental regulations then existed, and knowledge on the effects of radioactivity, particularly penetrating radiation, was limited. In the decades after World War II, the United States grew aware of potential environmental and health problems associated with advancing energy technologies and expanding industries. If these problems were ignored, they would only be compounded by future population growth. As a result, federal and state regulations were drafted to maintain the quality of the nation's air, surface water, and groundwater (Table 1.1).

Laboratory and field studies on radioactivity at Hanford contributed to two significant developments. The first was to strengthen radiation standards for the protection of human health. The second, more recently, was the principle of limiting radiation exposure to as low as possible.

Table 1.1.
Environmental regulations with major influence on Hanford Site Operations, 1941–1984. [a]

Regulation	Enactment date	Scope
Federal Water Pollution Control Act ("Clean Water Act")	1948	Enables control of water pollution through U.S. Environmental Protection Water Agency (EPA); sets minimum quality criteria for state regulations; controls point-source pollution via National Pollutant Discharge Elimination System permits; sets limits for release of toxic substances.
Atomic Energy Act	1954	Requires compliance with criteria for radioactive emissions and other rules and regulations set by the U.S. Nuclear Regulatory Commission. Defense-related activities are regulated by Department of Energy Orders.
Clean Air Act	1963	Enables control of air pollution through EPA; sets minimum quality standards for state regulations; sets limits for release to air of hazardous substances.
National Environmental Policy Act	1969	Requires environmental consideration for federal actions via impact statement process; creates Council on Environmental Quality.
Endangered Species Act	1973	Identifies plants and animals that are threatened or endangered; provides for their protection and preservation of their habitat.
Toxic Substances Control Act	1976	Requires EPA to obtain information on chemical substances and control those identified as hazardous to public health or the environment.
Resource Conservation and Recovery Act	1976	Provides basis for regulating solid wastes; authorizes EPA to provide criteria to assist states in safe disposal of solid wastes.
Nuclear Waste Policy Act	1982	Provides for establishment of nuclear waste repositories; authorizes related research and development activities.

[a] Because Hanford was a government-controlled facility, thousands of studies were conducted onsite to comply with environmental regulations. In many cases, studies later required by law were under way long before applicable regulations were enacted.

The concept of maintaining radiation exposure "As Low As Reasonably Achievable," or ALARA, was formally introduced by the National Council on Radiation Protection and Measurements in 1954. Major energy contractors for the U.S. Atomic Energy Commission (AEC), then the nation's lead agency for atomic energy programs, incorporated this philosophy into their radiation safety guidelines. Requirements for limiting radiation exposures to "as low as practical" were introduced by the AEC's successor, the U.S. Energy Research and Development Administration (ERDA), in Chapter 0524 of its operating manual in 1975. Subsequently in 1980, DOE published "A Guide to Reducing Radiation Exposure to As Low As Reasonably Achievable (ALARA)" as DOE/EV/1830-T5. This guide was drafted at Hanford by staff at Pacific Northwest Laboratory, a major onsite contractor. It represented, in large part, a description of the laboratory's current radiation safety program (Highby and Denovan 1982).

Environmental monitoring has been required at all DOE sites, on the basis of DOE Order 5484.1, and the results are reported annually. DOE's policy is to operate its facilities so that radiation doses to members of the public are ALARA, consistent with technical feasibility, costs, and applicable dose standards. This policy, issued through DOE Orders, is the basis of environmental monitoring at Hanford today (Price 1987).

On a broader scale, the only radiation dose limit for protecting the public by the early 1980s was 170 millirems per year, which was the limit allowed for exposure of large numbers of the public (Hall 1984). The AEC, which controlled operation of the single-purpose reactors at Hanford until 1975, technically could have irradiated every man, woman, and child to a dose equal to this amount. However, a commission appointed by the AEC established a much more conservative limit--that the radiation dose at the boundary fence of a nuclear power reactor should not exceed 5 millirems per year.

The conservative nature of this standard can be demonstrated by a few comparisons. Five millirems are equal to the radiation dose from

- one transatlantic trip by jet
- moving from a wood to a concrete house for a few weeks
- a 7-day vacation in Denver for a person living in Seattle
- the average exposure of a person living near the Three Mile Island Unit 2 nuclear plant in late March and early April 1979 multiplied by five.

The AEC and its successors, ERDA and DOE, have been criticized because they paid little attention to ecology and the problem of radiation in the environment, while doing a fairly respectable job of protecting workers at their plants. In reality, many programs under way at Hanford over its initial 20 years were mainly concerned with environmental problems and, therefore, completely refute this charge (Parker 1972). Environmental aspects have required a sizable portion of each year's operating budget at Hanford for more than four decades.

This book reviews bioenvironmental studies related to the Hanford Reach, Columbia River, from a historical perspective. Perspective helps explain activities at Hanford in relation to past and present uses of the Columbia River. It also helps us to understand current political, institutional, and philosophical restraints on past and present activities on the Hanford Site.

References

Anonymous. 1984. *Hanford*. Richland Operations Office, U.S. Department of Energy, Richland, Washington.

Becker, C.D. 1985. *Anadromous Salmonids of the Hanford Reach, Columbia River: 1984 Status*. PNL-5371, Pacific Northwest Laboratory, Richland, Washington.

Carter, L.J. 1987. *Nuclear Imperatives and Public Trust. Dealing with Radioactive Waste*. Resources for the Future, Washington, D.C.

Groueff, S. 1967. *Manhattan Project. The Untold Story of the Making of the Atomic Bomb*. Little, Brown, and Colorado, Boston.

Groves, L.R. 1962. *Now It Can Be Told. The Story of the Manhattan Project*. DaCapo Press, Inc., New York. 464 p.

Hall, E.J. 1984. *Radiation and Life*. 2nd Ed. Pergamon Press, New York.

Highby, D.P., and J. T. Denovan. 1982. *Pacific Northwest Laboratory Plan to Maintain Radiation Exposure As Low As Reasonably Achievable (ALARA)*. PNL-4560, Pacific Northwest Laboratory, Richland, Washington.

Hinds, W.T. 1979. "The Cesspool Hypothesis Versus Natural Areas for Research in the United States." *Environ. Cons*. 6:11–20.

Pacific Northwest River Basins Commission (PNWRBC). 1979. *Water Today and Tomorrow, Vol. II*. The Region. PNWRBC, Vancouver, Washington.

Parker, H.M. 1972. Remarks at the dedication of the Life Sciences Laboratory. In: *Annual Report for 1971 to USAEC Division of Biology and Medicine, Vol. I Life Sciences, Part 2 Ecological Sciences*. BNWL-1650,Pacific Northwest Laboratory, Richland, Washington.

Price, K.E. 1987. "Environmental Monitoring." In: *Environmental Monitoring at Hanford for 1986*, pp. 2.1–2.14. PNL-6120, Pacific Northwest Laboratory, Richland, Washington.

Totter, J.R. 1972. Remarks at the dedication of the Life Sciences Laboratory. In: *Annual Report for 1971 to USAEC Division of Biology and Medicine, Vol. I Life Sciences, Part 2 Ecological Sciences*. BNWL-1650, Pacific Northwest Laboratory, Richland, Washington.

Chapter 2

HISTORICAL INFLUENCES ON HANFORD OPERATIONS

Genesis of the Hanford site

Hanford was born in haste as a result of World War II. The site was selected in January 1943, and onsite activities were directed by the Manhattan District, U.S. Army Corps of Engineers (Corps) under what was called the Manhattan Project, a top-secret wartime undertaking. Hanford's specific purpose was to produce plutonium, a newly discovered element capable of releasing tremendous amounts of energy in a spontaneous nuclear reaction. All obstacles were removed to make the Hanford project successful to help bring an abrupt end to the war.

Work on the first prototype reactor at Hanford began on June 7, 1943, and operation began 18 months later in September 1944, an extraordinary achievement by today's standards. A few months later, two more prototype reactors became operational.

Hanford was one phase in the expansive effort by the Corps for the Manhattan Project. The concept emerged in 1939, when scientists working primarily in Germany and the United States theorized that splitting the nucleus of the U-235 atom would produce energy. Theory soon turned to reality. In December 1942, nuclear physicists in the United States, led by Enrico Fermi, produced a sustained and controlled nuclear reaction within a crude, graphite-moderated "pile," or reactor, at the University of Chicago (Dawson 1976). The experiment proved that an atomic bomb was possible. Wartime intelligence hinted that physicists in Nazi Germany were nearing the same breakthrough, although how near was entirely speculative. A race with high stakes began.

Several problems had to be resolved in progressing from the crude Fermi pile to a workable, fission bomb. One problem was to obtain enough fissionable material to produce such a weapon. Only two different elements could be used to produce a chain reaction, either U-235 or Pu-239. Less than 1% of all uranium in the earth's crust was fissionable

U-235, while the rest was primarily nonfissionable U-238 (Kathren 1984). The two isotopes have identical chemical properties and can be separated only by slight differences in their atomic mass. Plutonium 239, on the other hand, could be created by irradiating U-238 in a nuclear reactor, then extracting it chemically from the irradiated fuel.

World War II imparted urgency to the undertaking. Both methods of obtaining fissionable material were pursued under the Manhattan Project to reduce the risk of failure. The task of producing U-235 was assigned to the Oak Ridge National Laboratory in Tennessee. Plutonium production was assigned to the E. I. du Pont de Nemours and Company, Inc. (du Pont) at Hanford (Groves 1962; Groueff 1967). At the same time, the Los Alamos National Laboratory was established in New Mexico under the auspices of the University of California. Scientists at Los Alamos were to explore more fully the theory of nuclear fission and, on the basis of this knowledge, develop workable atomic bombs from both U-235 and Pu-239

Fig. 2.1. Isolation of the Hanford Site in a remote desert location was one of the factors in its selection by the U.S. Army Corps of Engineers under the Manhattan Project. Rattlesnake Mountain provides a backdrop along the western margin of the site.

Fig. 2.2. Banking services were provided on the Hanford Site to thousands of immigrant workers in 1944. (Photograph contributed by the U.S. Department of Energy.)

Fig. 2.3 Housing quarters constructed for Hanford workers in 1944. The women's barracks were surrounded by a barbed-wire fence to keep out unauthorized males. (Photograph contributed by the U.S. Department of Energy.)

(Kunetka 1979; Szasz 1984). The tasks facing each laboratory were formidable.

Hanford, located at a remote spot in semiarid southeastern Washington (Figure 2.1), was chosen for plutonium production because the rural population was small, and it was located far from major cities, two desirable features for ensuring security and safety. Furthermore, large amounts of good quality water were available from the Columbia River, and large amounts of electricity were available from Grand Coulee Dam, completed in 1941. The site also had rail transportation, a substrate that could support the foundations of heavy industrial plants, sand and gravel for making concrete, and a climate generally favorable for construction (Groves 1962; Foster 1972). More than 150,000 people worked at the Hanford Site during World War II, all involved on a project whose significance they could only guess at (Figures 2.2 and 2.3).

On August 6, 1945, the first atomic bomb was dropped on Hiroshima, Japan. That bomb contained enriched uranium from Oak Ridge. On August 9, a second bomb was dropped on Nagasaki, Japan. That bomb

Fig. 2.4. Hanford has always been administered as a secured area. Personal articles endangering onsite activities are controlled, the public has limited access, and information exchange is based on "need to know."

contained plutonium from Hanford. Within 24 hours, the Japanese government surrendered, and the war ended.

Hanford expanded its defense role after World War II. At first the site provided plutonium for several United States weapons laboratories in a national effort to improve the performance of fission bombs – the first generation of nuclear weapons. Technical development surged in the 1950s, in part because of the "cold war" with the Soviet Union. The national laboratories received massive funding and a mandate to pursue new weapons development. This effort led to the second generation of nuclear weapons, the hydrogen, or fusion, bomb and its derivatives.

The nature of activities at Hanford has always required administration of the site as a secured area (Figure 2.4) to protect the interests of the U.S. Government and its defense activities.

Milestones

Forty years of bioenvironmental studies related to the Hanford Reach of the Columbia River were shaped and forged by significant historical events (Table 2.1).

The War Years (1940s)

The first of three prototype reactors (B Reactor) built at Hanford became operational in September 1944, the second (D Reactor) in December 1944, and the third (F Reactor) in February 1945. By mid-1945, the toxicity of cooling water effluent was being examined in a special laboratory built near F Reactor. The first bioassays with fish indicated the effluent would not harm trout and salmon when diluted in the Columbia River below the discharge outfalls (Foster 1972). Because of the war, initial bioassays were reported in classified documents available only to involved personnel (e.g., Foster 1946; Olson 1948).

Field work with fish and other aquatic biota in the Hanford Reach began in 1946. These studies (e.g., Herde 1947) focused on the fate of radionuclides released to the river with the cooling water effluents, which was the most immediate concern. Almost immediately, certain radionuclides were found to be concentrated in river biota. This led to more comprehensive radiological surveys that included aquatic invertebrates as well as fish (Coopey 1948; Davis and Cooper 1951).

Table 2.1.
Chronology of Major Historical Milestones Related to Operations and Environmental Studies on the Hanford Site, 1941 to 1984.

THE WAR YEARS AND AFTERMATH: 1941–1950

October 9, 1941. President Franklin D. Roosevelt, informed about the potential of atomic energy, initiated policy steps that led to the development of an atomic bomb. The U.S. Army would be responsible for the development program.

December 7, 1941. Japanese airplanes bombed Pearl Harbor at Honolulu, Hawaii, drawing the United States into World War II.

June 25, 1942. Officials of the Army and the Office of Scientific Research and Development held a meeting in Washington, D.C., to plan policy, site selection, contracting, and priorities in the fledgling atomic program.

September 23, 1942. General Leslie R. Groves took official charge of the United States atomic program – soon to be named "The Manhattan Project."

November 1942. E. I. du Pont de Nemours and Company, Inc. (du Pont), accepted responsibility for the design, construction, and operation of a large-scale plutonium plant at an unselected site. The techniques to be applied were largely unknown.

December 2, 1942. The world's first self-sustaining nuclear chain reaction was initiated and stopped in a 400-ton pile of graphite blocks at the University of Chicago.

December 1942. Colonel Franklin T. Matthias, under the Manhattan Project, toured the western United States searching for a suitable site for plutonium production facilities.

January 16, 1943. General Groves inspected and approved the Hanford Site for construction of plutonium-producing "piles" under strict conditions of wartime secrecy.

February 23, 1943. An order of expropriation was issued by a federal court under the War Powers Act for the condemnation of more than 600 square miles of land encompassing the towns of Richland, Hanford, and White Bluffs. More than 1500 residents were ordered to leave within 30 days.

March 1943. Construction started on the first plutonium-production piles in the 100 Areas and on chemical processing facilities in the 200 Areas at Hanford. More than 150,000 people were to work at the Hanford Site during World War II.

April 6, 1943. Ground was broken for a base construction camp at the tiny town of Hanford, 30 miles north of Richland.

May 20, 1943. The possibility of contaminating the Columbia River by plutonium-production facilities at Hanford was discussed in Chicago by 20 key persons involved in developing an atomic bomb.

September 13, 1944. Testing of B Reactor, Hanford's first prototype pile, commenced 18 months after construction began.

December 28, 1944. Full-scale plutonium production began at B Reactor. It was the first of three prototype reactors built and operated at full capacity until the end of World War II.

January 1, 1945. The first irradiated fuel was dissolved at T Plant, 200 West Area.

February 1945. Construction of D and F prototype reactors was completed, and production of weapons-grade plutonium began at the "Hanford Works."

February 1945. The first shipment of plutonium 239 was delivered from Hanford to Los Alamos, New Mexico. About 340,000 curies of radioactivity were released to the air at Hanford during the year.

July 16, 1945. The world's first atomic bomb was exploded in the New Mexico desert near Alamogordo under the code name "Trinity." The bomb was made from plutonium 239 produced at Hanford.

August 6, 1945. A uranium-235 bomb was dropped on Hiroshima, Japan. Three days later, a plutonium bomb was exploded over Nagasaki.

August 10, 1945. Japan agreed to accept surrender terms stipulated by the United States.

August 14, 1945. World War II ended. The atomic bombs allowed the Japanese government to surrender "with honor."

Year 1946. More plutonium was produced at the Hanford Works. Onsite studies were conducted on radiation effects in plants, livestock, wildlife, and aquatic organisms. Environmental monitoring was first described in quarterly reports.

July 1946. The United States exploded two atomic bombs at Bikini atoll in the Marshall Islands, initiating "Operation Crossroads."

August 1, 1946. President Harry S. Truman signed the Atomic Energy Act. The measure established control of nuclear weapons and technology under the U.S. Atomic Energy Commission (AEC).

September 1, 1946. Responsibility for management of the Hanford Works was transferred from du Pont to General Electric Company.

January 1, 1947. The AEC replaced the Manhattan District, U.S. Army Corps of Engineers, as the agency with authority for controlling research and development activities at Hanford.

September 1947. Plutonium production was increased at Hanford, and construction of "replacement reactors" received government priority. Operating experience led to improved design of all reactors built after World War II.

April-May 1948. The United States exploded three atomic weapons at Enewetok atoll. One bomb had six times the power of the atomic bomb dropped on Nagasaki.

Year 1948. Hanford environmental scientists, working with more sophisticated equipment, begin to detect radioactive isotopes with extended half-lives on onsite vegetation.

October 1948. A dike at a retention basin gave way, releasing 14.5 million gallons of low-level radioactive liquids to the Columbia River. Levels of radioactivity outside of the Richland area remained so low they were undetectable.

THE COLD WAR: 1950–1960

July 1949. Government officials reported that the United States now had a sizable arsenal of nuclear weapons, forming an "atomic shield" against the Soviet Union.

Year 1949. The AEC established the Columbia River Advisory Group (CRAG) to review studies on the Columbia River dealing with the fate and uptake of radionuclides released in reactor effluent.

August 1949. Sources indicated that the Soviet Union had exploded its first atomic bomb, 3 to 5 years ahead of forecasts by the western world.

September 1949. Radiation detectors aboard a U.S. Air Force weather plane on routine patrol provided evidence that the Soviet Union had, in fact, detonated its first atomic bomb. The AEC decided to accelerate its expansion program.

September 1949. A fourth plutonium-production reactor (H) started up at Hanford. The military increased its demands for plutonium. The government announced that eight reactors would be built at Hanford by the mid-1950s.

Table 2.1. (continued)

December 1–3, 1949. An estimated 5000 to 7000 curies of radioiodine were released to the atmosphere at Hanford to examine the United States' ability to detect atomic bomb tests in the Soviet Union.

June 1950. The United States entered the Korean War. Late 1950. Onsite officials reported that emissions of radioiodine to the air at Hanford had been cut 99.9%.

Year 1951. Air filters failed at Hanford plants, inadvertently releasing about 19,000 curies of radioiodine to the atmosphere.

Year 1953. By the end of 1953, the AEC had exploded more than 30 test devices in the atmosphere, either over the Pacific Ocean or at the Nevada Test Site.

Fall 1954. Congress passed the Atomic Energy Act of 1954. The Act included provisions to implement President Dwight D. Eisenhower's "Atoms for Peace" Program.

Year 1955. The first international conference on the Peaceful Uses of Atomic Energy was held in Geneva, Switzerland, under the auspices of the United Nations. Many papers were presented by Hanford scientists.

Year 1956. The Hanford Laboratories were created, primarily for research and development activities related to peaceful uses of nuclear energy. Engineering, fuel fabrication, and operational facilities in use since 1944 were renovated and enlarged.

October 1, 1957. The International Atomic Energy Agency was inaugurated in Vienna, Austria.

Year 1958. The AEC initiated research to eventually solidify high-level radioactive wastes at Hanford. Results of environmental monitoring at Hanford were first released to the public as annual reports.

October 31, 1958. The United States and the Soviet Union agreed to refrain from all tests of nuclear weapons because of international concern about adverse effects of radiation from atmospheric fallout.

May 13, 1959. Construction began on the New Production Reactor (N Reactor), a dual-purpose facility at Hanford. It was authorized by Congress to produce both plutonium and steam for electric power.

PERIOD OF DIVERSIFICATION: 1960–1970

August 31, 1961. The Soviet Union announced it would resume tests of nuclear weapons and detonate several high-yield weapons in the atmosphere during the fall. The United States resumed atmospheric tests in 1962.

April 7, 1962. A serious accident at Hanford was reported. Plutonium reached criticality at the Plutonium Finishing Plant, exposing three workers to large doses of radiation. (All three workers were alive and well when this list was compiled in 1985.)

August 5, 1963. The United States and the Soviet Union signed a new 'limited' test-ban agreement. It prohibited testing in the atmosphere, outer space, or under water, but allowed underground tests.

September 23, 1963. Hanford's N Reactor began operation. It was dedicated by President John F. Kennedy on September 26, and attained full power in 1964.

September 26, 1963. Construction started on the Hanford Generating Plant (HGP) adjacent to N Reactor; it began commercial production of electricity in November 1966.

November 1963. President Lyndon B. Johnson ordered a 25% reduction in production of enriched uranium and a shutdown of four plutonium-production reactors. He challenged other nations to do the same.

January 1964. General Electric Company announced its withdrawal from Hanford; the AEC announced a multicontractor program to segment and diversify the Hanford Works.

November 9, 1964. N Reactor reached design capacity of 4000 megawatts of thermal power.

December 1964. The 100-DR Reactor was shut down, the first of the eight single-purpose reactors that were operated at Hanford.

January 1, 1965. United States Testing Company took over bioassays, processing of film badges, and analysis of environmental samples at Hanford.

January 4, 1965. Battelle Memorial Institute assumed responsibility for managing the Hanford Laboratories, which were renamed the Pacific Northwest Laboratory. A new policy to attract diversified contractors to Hanford was emphasized.

January 1965. The Pacific Northwest Laboratory became the independent monitoring entity for all onsite work and expanded the environmental monitoring program at Hanford.

June 25, 1965. The 100-F Reactor was shut down after 20 years of service.

August 1, 1965. The Hanford Environmental Health Foundation took over industrial medicine and health responsibilities at Hanford. The Federal Building was completed in Richland to house the AEC and certain contractor personnel.

November 1, 1965. Douglas United Nuclear, Inc., took over operation of the remaining Hanford reactors and onsite fabrication of their fuel elements.

December 31, 1965. ISOCHEM, Inc., assumed General Electric Company's role in chemical processing of irradiated fuels and managing radioactive waste at Hanford.

March 1, 1966. ITT Federal Support Services, Inc., took over operation of support services at Hanford.

April 8, 1966. The HGP produced its first electricity with steam from N Reactor.

July 1, 1967. General Electric Company completed 21 years as prime contractor for the AEC, leaving facilities valued at $1.25 billion.

July 1968. Federal and state agencies initiated the Columbia River Thermal Effects Study to examine the effects of heated water from Hanford reactors on the Columbia River.

February 12, 1968. Hanford's first plutonium production reactor, B Reactor, ceased operation after 24 years of service.

September 1, 1969. Atlantic Richfield Hanford Company took over chemical processing and radioactive waste management at Hanford from ISOCHEM, Inc.

March 5, 1970. The United States, United Kingdom, Soviet Union, and 45 other countries signed the Nuclear Nonproliferation Treaty.

ENERGY R&D: 1970–1980

January 1971. The KE Reactor, the last of the single-purpose reactors at Hanford, was shut down.

January 28, 1971. President Richard M. Nixon ordered N Reactor closed. The Office of Management and Budget in its budget for fiscal 1972 cut all funding for weapons material production at Hanford.

April 1971. The Nixon Administration agreed to accept $20 million from the State of Washington to pay for the steam produced by N Reactor for 3 years. Operation resumed in August.

Table 2.1. (continued)

Spring 1973. A major leak was detected in a 30-year-old single-shell, underground storage tank containing high-level radioactive waste in the 200 Areas. Although the leak never reached groundwater, it initiated a new policy in control of liquid radioactive waste at Hanford.

Year 1974. Production of 12% fuels-grade plutonium began at N Reactor.

October 11, 1974. The Energy Reorganization Act of 1974 was signed by President Gerald R. Ford. The Act replaced the AEC with the U.S. Energy Research and Development Administration (ERDA), and created the U.S. Nuclear Regulatory Commission (NRC).

Year 1975. About 75% of Hanford's annual operating budget was spent on energy-related research. (A decade later, in 1985, about 60% of the budget returned to military-related spending.)

January 1975. The ERDA officially replaced the AEC and expanded studies of energy alternatives for the future at Hanford.

August 4, 1977. President Jimmy Carter signed the Department of Energy Organization Act, combining all energy and nuclear programs under the U.S. Department of Energy (DOE). The Federal Energy Office was abolished.

October 1977. The ERDA became part of the cabinet-level DOE, which assumed responsibility for energy- and nuclear-related activities at Hanford.

March 28, 1979. An accident at Three Mile Island, a commercial nuclear power plant in Pennsylvania, resulted in a major shake-up of federal safety programs and gave nuclear power a black eye.

STRATEGIC DETERMENT: 1980 AND BEYOND

Year 1981. Emphasis was shifted, once again, to production of plutonium, fuel processing, and related defense activities at Hanford. A $54-million upgrade program was budgeted for N Reactor.

April 1982. The Fast Flux Test Facility, a sodium-cooled reactor for development of breeder technology, began full operation at Hanford.

October 1983. The Plutonium-Uranium Extraction Plant resumed operation at Hanford.

Spring 1984. President Ronald W. Reagan signed a new, 5-year Nuclear Weapons Stockpile Memorandum that set goals for nuclear weapons production over the next 5 years.

May 1984. WNP-2, a nuclear power plant built by the Washington Public Power Supply System, reached full operation at Hanford. Two other plants were being constructed, but one was later mothballed and the other was terminated.

December 1984. Hanford came under consideration as one of three sites for an underground repository to store radioactive wastes from commercial, nuclear power plants.

POSTSCRIPT

Year 1985. Major alterations totaling $1 billion were planned to extend the useful life of N Reactor into the next century.

Year 1985. The Pacific Northwest Laboratory was officially designated by the DOE as one of its national facilities for energy research.

May 1985. An environmental assessment was released for Hanford as a location for a reference repository for radioactive waste. Intense public opposition followed.

October 1985. The government's annual operating budget for Hanford ($948 million) was 60% for defense purposes. The rest was for research and related work on nuclear power and the nation's energy future.

April 28, 1986. An accident at the Chernobyl nuclear complex in the Soviet Union spread airborne radioactivity over northern Europe and ignited new debate over the safety of the N Reactor at Hanford.

January 1987. N Reactor was shut down for $70 million in safety improvements.

July 1987. DOE operations at Hanford were consolidated under four contractors: Battelle-Northwest, Kaiser Engineers, Westinghouse Hanford Company, and Hanford Environmental Health Foundation.

December 17, 1987. Congress agreed to fund an underground repository for nuclear wastes in Nevada, eliminating all feasibility study effort at Hanford.

February 16, 1988. The DOE announced that N Reactor would be placed on cold standby because less plutonium was needed in defense programs.

Onsite studies on effects and fate of radioactivity in the Columbia River at Hanford were initiated by du Pont. These studies were continued by General Electric Company after they took over operations at Hanford on September 1, 1946, from du Pont. Related studies were conducted in Seattle by the University of Washington.

After World War II ended, the Atomic Energy Act of 1946 was passed, which established two government bodies to control and develop the atom (Dawson 1976). One was administrative, the U.S. Atomic Energy Commission (AEC); the other was legislative, the Joint Committee on Atomic Energy. Control of Hanford Site activities passed from the military to the AEC.

Laboratory and field studies at Hanford soon showed that fish in the Columbia River were not threatened by discharges from the prototype reactors. Furthermore, people who used river water or who caught and ate its fish faced little harm. However, the concentration of some radionuclides in aquatic organisms was recognized as a mechanism by which radioactive material could be transferred to higher organisms, including humans. It was also learned that heat and process chemicals (such as sodium dichromate) in the cooling water effluent could adversely affect aquatic life if the quantities were increased by perhaps an order of magnitude (Foster 1972). Plans for constructing more single-purpose reactors in the 1950s stimulated expansion of environmental studies in the Hanford Reach.

The Post-War Years (1950s)

An improved laboratory for aquatic studies was constructed in the 100-F Area in 1952. The AEC sponsored this effort through the general contractor, General Electric Company. In 1950, the U.S. Public Health Service sent a team to Hanford to examine radiological features and water quality in the Columbia River for 2 years. Their report (Robeck et al. 1954) was a comprehensive review of available information, and it remains a valuable reference today. Five additional single-purpose reactors at Hanford were authorized for construction in 1952 and 1953, underlining the need for continued studies related to environmental conditions. In effect, a long-term bioenvironmental monitoring program related to the Hanford Reach was initiated.

During the early 1950s, the tremendous potential for producing electric power from a controlled nuclear reaction received much attention. The first International Conference on the Peaceful Uses of Atomic Energy was held in 1955 in Geneva, Switzerland, sponsored by the young United Nations. Bioenvironmental studies in the Columbia River at Hanford were described for the first time to an international audience (Foster and Davis 1956; Hanson and Kornberg 1956; Parker 1956). Additional data were presented at the second and third Geneva conferences in 1958 and 1964 (Davis et al. 1958; Parker et al. 1965).

At this time, the United States public grew more concerned about potential effects of radionuclides and other manufactured materials released to the environment. Awareness grew partly from extensive testing of new nuclear devices in the atmosphere by the United States and the Soviet Union, and partly from projections of future use of nuclear power in a civilian economy. Several significant events occurred in the 1950s. One was passage of the Atomic Energy Act of 1954, which allowed private ownership of nuclear power reactors (Dawson 1976). Another was the formation of committees within the National Academy of Sciences (NAS) to evaluate the biological effects of atomic radiation. The committees' findings were presented in two reports. One (NAS-NRC 1957) covered the effects of atomic radiation on oceanography and fisheries. The other (NAS-NRC 1962) covered the disposal of radioactive wastes in oceans.

Era of Environmental Awakening (1960s)

A dual-purpose reactor, N Reactor, was completed at Hanford in 1963 by Kaiser Engineers at a cost of $199.7 million (Figure 2.5). One reason it

Fig. 2.5 The N Reactor lower center, a plutonium producer and the Hanford Generating Project No. 2 (left), an electric power producer.

was built was because officials realized that no new reactor should use direct once-through cooling (Dawson 1976). Also in 1963, a contract to obtain steam from the N Reactor was signed with Washington Public Power Supply System (Supply System), a consortium of 76 utilities. A steam-operated power plant, now known as Hanford Generating Project (HGP), was constructed nearby to use steam from N Reactor. The HGP first reached full power in December 1966. The Supply System negotiated further contracts with the U.S. Government for steam from N Reactor after January 1971, when the last single-purpose reactor at Hanford shut down. The HGP proved to be a safe and reliable source of electricity for consumers in the Pacific Northwest for more than 25 years.

In the early 1960s, the AEC announced a substantial reduction in production of special nuclear materials. Four single-purpose reactors at Hanford were shut down on December 30, 1964, and chemical reprocessing to recover plutonium from irradiated fuel was reduced. More reactors at Hanford were shut down later, and the last single-purpose reactor ceased operation on January 28, 1971. This left only N Reactor in

Fig. 2.6. The research operations center of Battelle Northwest is located in private buildings just north of Richland.

operation. The direct discharge of activation products and heated water to the Columbia River from once-through reactor cooling at Hanford ceased.

On January 4, 1965, Battelle Memorial Institute (Battelle) contracted to manage government research and development work at Hanford, replacing General Electric Company (Figure 2.6). The agreement included a mandate to diversify onsite activities. Battelle also assumed responsibility for environmental monitoring programs at Hanford. Research to develop energy sources other than nuclear was emphasized after the U.S. Energy Research and Development Administration (ERDA) replaced the AEC in January 1975. However, ERDA became part of the cabinet-level U.S. Department of Energy (DOE) in October 1977. The DOE assumed responsibility for government-related defense work at Hanford, a position it still holds today.

The federal government strongly supported the development and commercialization of nuclear power throughout the United States in the

1960s. As the public became aware that the discharge of heated water from once-through reactor cooling might damage aquatic environments, "thermal pollution" became an emotional issue (Parker and Krenkel 1969; Krenkel and Parker 1969; Clark and Brownell 1973). Citizens also discovered the words "environment" and "ecology" in the 1960s and took their meaning to heart. This led to passage of the National Environmental Policy Act (NEPA) in 1969 and creation of the Presidential Council on Environmental Protection, both after startup of N Reactor and HGP.

The NEPA provided a means to examine adverse effects from thermal pollution by nuclear and fossil-fueled power plants, and the related issues of impingement and entrainment of aquatic organisms. It required detailed environmental impact statements before new power plants could be licensed; therefore, possible impacts on aquatic environments received close scrutiny during the planning, licensing, construction, and operational stages. Potential impacts from heated water discharged from the single-purpose reactors still operating at Hanford were examined more closely. After these reactors were shut down, the possibility of impacts on the Hanford Reach from both N Reactor and HGP received scrutiny. The same requirements were applied to commercial power plants proposed for construction at Hanford.

Rise and Decline of Nuclear Power (1970s)

Environmental studies to assess the consequences of all new power plants, nuclear or fossil fueled, occurred in full force throughout the United States in the early 1970s. This effort was one result of the NEPA. At Hanford, release of once-through cooling water to the Hanford Reach ceased when KE Reactor shut down in January 1971. However, radiological monitoring of the Columbia River and its aquatic organisms was continued to quantify the rapid decline of short-lived radionuclides. The Columbia River Thermal Effects Study, conducted from 1969 through 1970, led to recommendations by regional water pollution control agencies that all future power plants in the Columbia River basin must use offstream cooling (EPA 1971).

In 1972, construction began on three commercial power plants on the Hanford Site: Supply System Nuclear Projects (called WNP) 1, 2, and 4. WNP-2 was under way in 1972, entered the testing stage in 1983, and reached full operation May 1984. Construction on WNP-1 and -4 was stopped in 1982 before the plants were completed because of escalating costs and reduced projections of regional power needs. Much information on aquatic organisms and ecology of the Hanford Reach was obtained from pre-operational environmental studies associated with these projects.

In the early to mid-1970s, widespread movements to challenge nuclear power took shape. The industry ran headfirst into many roadblocks, including governmental regulations, lack of public confidence, self-inflicted injuries such as poor management and huge cost overruns, and opposition from environmentalists (Stoler 1985). Orders for new nuclear power plants ceased in the late 1970s, and some previous orders were canceled. In 1979, an accident at the Three Mile Island Unit 2 Plant in Pennsylvania capped the industry's problems and gave new impetus to challenges by environmentalists.

Despite these problems within the nuclear industry, environmental research at Hanford continued. And 30 years of expertise in examining environmental issues at Hanford provided a broad scientific and technical base for assessing the impact of developing energy technologies elsewhere.

By the end of the decade, the scientific community had reached an important conclusion: release of relatively large amounts of radioactive materials and heat to the Hanford Reach from 1943 to 1971 had not impaired any ecological function in the Columbia River. Instead, aquatic communities not only adapted and survived, but upriver runs of fall chinook salmon flourished.

Under Siege (Early 1980s)

Activities at Hanford came under withering crossfire in the early 1980s, first from antinuclear activists, and later from state agencies, politicians, special interest groups, the press, and the concerned public. There were four main issues, all fueled primarily by sociopolitical rather than scientific factors. Opponents frequently did not, would not, or could not distinguish between nuclear power and nuclear weapons or between fact and fiction. This frequently confused or misled the general public about onsite activities and their environmental consequences.

One issue involved continued operation of N Reactor. The Reagan Administration and Congress mandated increased production of plutonium to update the nation's nuclear arsenal. Numerous relevant and irrelevant questions were raised about the safety of operating N Reactor and, indeed, the nation's need for more nuclear weapons. Many safety improvements were made to N Reactor, but it was placed on cold standby in 1987.

Another issue involved selection of Hanford for exploration and assessment as the nation's first repository for commercial, high-level radioactive wastes (Carter 1987). On April 7, 1977, President Jimmy Carter announced indefinite deferral of fuel reprocessing and commercial reactor development, which soon made storage of spent radioactive fuels a national problem. The Nuclear Waste Policy Act of 1982 required DOE to take title to spent fuel stored at commercial power plants, and to open a permanent repository for these wastes by 1998. Hanford was one of three candidate sites, all located in the western United States, selected for initial study on May 28, 1985. This effort resulted in environmental studies at Hanford under the Basalt Waste Isolation Project (BWIP). However, the Yucca Mountain site in Nevada was later selected for a repository, and the BWIP was terminated in December 1987.

Another issue involved long-term, possibly "permanent," disposal of radioactive waste from defense production activities that had accumulated at Hanford since 1944. High-level liquid wastes were stored in underground tanks, and solid wastes were placed in subsurface cribs and trenches in the 200 Areas. In addition, large volumes of intermediate and low-level radioactive liquids from fuel reprocessing had been released to the soil where tritium, a highly mobile element, nitrates, and a few process chemicals had reached the groundwater.

Another issue involved airborne emissions of radioiodine during the initial three decades of activities at Hanford. The possibility of long-term effects on local residents living downwind (to the east and southeast) was frequently publicized. Claims were made of impaired human health and deformed animals, even though analysis of scientific data from onsite and offsite monitoring programs clearly indicated otherwise. Historical documents released to the public in February 1986 added fuel to "downwinder" concerns.

Relative to these issues, many critics contended that minute quantities of radioactivity from onsite activities might contaminate the Columbia River via the groundwater. Lessons learned from radioecological studies, 1944 to 1971, when much greater amounts of radioactivity were released to the Hanford Reach in reactor effluents, appeared to be forgotten.

Organizations on the Hanford site, early 1980s

By the early 1980s, work on the Hanford Site reflected 15 years of diversification effort. Hanford had gained national stature for production of defense materials and for research and development (R&D) activities. The main pursuits at Hanford included the production of plutonium (N Reactor); irradiated fuel reprocessing; and R&D programs in nuclear waste management, reactor design, and weapons material development. In addition, scientists at Hanford conducted innovative studies to develop new sources of energy, efficiently manage natural resources, and promote human health.

Eight agencies operated under contract to DOE-Richland Operations (DOE-RL) on the Hanford Site through the early 1980s: Battelle, responsible for operation of DOE's Pacific Northwest Laboratory (PNL); Boeing Computer Services Richland, Inc. (BCSR), automatic data processing services; Hanford Environmental Health Foundation (HEHF), health services; Kaiser Engineers Hanford (KEH), onsite architect and engineering services; Rockwell Hanford Operations, chemical processing, waste management, and support services; UNC Nuclear Industries, management of N Reactor and its fuel fabrication; and Westinghouse Hanford Company, management of the Hanford Engineering Development Laboratory and Fast Flux Test Facility.

Two private agencies also operated facilities on the Hanford Site in the early 1980s. U.S. Ecology, licensed by the U.S. Nuclear Regulatory Commission and the State of Washington, operates a 100-acre burial ground for disposal of low-level radioactive wastes generated offsite. The Supply System continued to operate HGP-2, which used by-product steam from N Reactor, and WNP-2.

Postscript: In early 1987, DOE's work force at Hanford was consolidated under four contractors. KEH began work in February as consolidated architect/engineer contractor. Westinghouse Hanford Company assumed responsibility as the consolidated operations-engineering contractor. Battelle-Northwest remained as the R&D contractor. Health care services continued under HEHF.

References

Carter, L.J. 1987. *Nuclear Imperatives and Public Trust. Dealing with Radioactive Wastes*. Resources for the Future, Inc., Washington, D.C.

Clark, J., and W. Brownell. 1973. *Electric Power Plants in the Coastal Zone: Environmental Issues*. American Littoral Society Special Publication No. 7, American Littoral Society, Highlands, New Jersey.

Coopey, R.W. 1948. *The Accumulation of Radioactivity as Shown by a Limnological Study of the Columbia River in the Vicinity of Hanford Works*. U.S. Atomic Energy Commission Report, HW-11662, Hanford Atomic Products Operation, Richland, Washington.

Davis, J.J., and C.L. Cooper. 1951. *Effect of Hanford Pile Effluent upon Aquatic Invertebrates in the Columbia River*. U.S. Atomic Energy Commission Report, HW-20055, Hanford Atomic Products Operation, Richland, Washington.

Davis, J.J., R.W. Perkins, R.F. Palmer, W.C. Hanson, and J.F. Cline. 1958. "Radioactive Materials in Aquatic and Terrestrial Organisms Exposed to Reactor Effluent Water." In: *Proceedings of the Second International Conference on the Peaceful Uses of Atomic Energy*, Geneva, Switzerland, 1958, Vol. 18, pp. 421–428. United Nations, New York.

Dawson, F.G. 1976. *Nuclear Power. Development and Management of a Technology.* University of Washington Press, Seattle, Washington.

EPA. See U.S. Environmental Protection Agency.

Foster, R.F. 1972. "The History of Hanford and Its Contribution of Radionuclides to the Columbia River." In: *The Columbia River Estuary and Adjacent Waters*, eds. A. T. Pruter and D. L. Alverson, pp. 1–18. University of Washington Press, Seattle.

Foster, R.F. 1946. *Some Effects of Pile Area Effluent Water on Young Chinook Salmon and Steelhead Trout.* U.S. Atomic Energy Commission Report, HW-1–4759, Hanford Engineering Works, Richland, Washington.

Foster, R.F., and J.J. Davis. 1956. "The Accumulation of Radioactive Substances in Aquatic Forms." In: *Proceedings of the International Conference on Peaceful Uses of Atomic Energy*, Geneva, Switzerland, 1955, pp. 361–367. United Nations, New York.

Groueff, S. 1967. *Manhattan Project. The Untold Story of the Making of the Atomic Bomb.* Little, Brown, and Company, Boston, Massachusetts.

Groves, L.R. 1962. *Now It Can Be Told. The Story of the Manhattan Project.* Da Capo Press, Inc., New York.

Hanson, W.C., and H.A. Kornberg. 1956. "Radioactivity in Terrestrial Animals Near an Atomic Energy Site. In: *Proceedings of the International Conference on Peaceful Uses of Atomic Energy*, Geneva, Switzerland, 1955, pp. 381–388. United Nations, New York.

Herde, K.E. 1947. *Radioactivity in Various Species of Fish from the Columbia and Yakima Rivers.* U.S. Atomic Energy Commission Report, HW-35501, Hanford Engineering Works, Richland, Washington.

Highby, D.F., and J.T. Denovan. 1982. *Pacific Northwest Laboratory Plan to Maintain Radiation as Low as Reasonably Achievable (ALARA).* PNL-4560, Pacific Northwest Laboratory, Richland, Washington.

Kathren, R.L. 1984. *Radioactivity in the Environment: Sources, Distribution, and Surveillance.* Harwood Academic Publishers, New York.

Krenkel, P.A., and F.L. Parker, eds. 1969. *Biological Aspects of Thermal Pollution.* Vanderbilt University Press, Nashville, Tennessee.

Kunetka, J.W. 1979. *City of Fire. Los Alamos and the Atomic Age, 1943–1945.* University of New Mexico Press, Albuquerque, New Mexico.

National Academy of Sciences – National Research Council (NAS-NRC). 1957. *The Effects of Atomic Radiation on Oceanography and Fisheries.* Publication No. 551, National Academy of Sciences, Washington, D.C.

National Academy of Sciences – National Research Council (NAS-NRC). 1962. *Disposal of Low-Level Radioactive Waste into Pacific Coastal Waters.* Publication No. 985, National Academy of Sciences, Washington, D.C.

Olson, P.A. 1948. *Some Effects of Pile Area Effluent Water on Young Silver Salmon.* U.S. Atomic Energy Commission Report HW-8944, Hanford Engineering Works, Richland, Washington.

Parker, H.M. 1956. "Radiation Exposure from Environmental Hazards." In: *Proceedings of the International Conference on Peaceful Uses of Atomic Energy*, Geneva, Switzerland, 1955, pp. 301–310. United Nations, New York.

Parker, H.M., R.F. Foster, I.L. Ophel, F.L. Parker, and W.C. Reinig. 1965. "North American Experience in the Release of Low-Level Waste into Rivers and Lakes". In: *Proceedings of the Third International Conference on the Peaceful Uses of Atomic Energy*, Geneva, Switzerland, 1964, pp. 61–71. United Nations, New York.

Parker, F.L., and P.A. Krenkel, eds. 1969. *Engineering Aspects of Thermal Pollution*. Vanderbilt University Press, Nashville, Tennessee.

Robeck, G.G., C. Henderson, and R.C. Palange. 1954. *Water Quality Studies on the Columbia River*. Special Report, U.S. Public Health Service, R.A. Taft Sanitary Engineering Center, Cincinnati, Ohio.

Stoler, P. 1985. *Decline and Fall, the Ailing Nuclear Power Industry*. Dodd, Mead and Company, New York.

Szasz, F.M. 1984. *The Day the Sun Rose Twice*. University of New Mexico Press, Albuquerque, New Mexico.

U.S. Environmental Protection Agency (EPA). 1971. *Columbia River Thermal Effects Study Volume I: Biological Effects Study*. The U.S. Environmental Protection Agency in cooperation with the Atomic Energy Commission and National Marine Fisheries Service, Washington, D.C.

Chapter 3

OPERATION AREAS AND LAND USE AT HANFORD

At its beginning, the sole purpose for Hanford's existence was the production of plutonium, an element new to mankind. The first installations at Hanford were designed with only rudimentary knowledge of plutonium's unique properties. In fact, plutonium was not fully understood by anyone (Groueff 1967). Assumptions from examining microscopic quantities of plutonium created in the laboratory were projected to production needs on a vast scale at Hanford.

Essentially, two processes were to be carried out. The first was the transformation of uranium atoms into plutonium atoms by bombardment with neutrons in a graphite "pile," or nuclear reactor. The second was the recovery of plutonium in purified form by chemical and metallurgical techniques. As initially planned, the Hanford Site had four main operational areas devoted to separate production phases (Foster et al. 1961; Foster 1972; Rostenbach 1956):

1. fabrication facilities, where natural uranium could be fashioned into fuel elements suitable for irradiation
2. reactors, where the uranium fuel elements could be irradiated with neutrons, producing plutonium, activation products, and fission products
3. chemical separation plants, where the irradiated fuel could be processed, the plutonium and uranium recovered, and the waste products isolated
4. disposal areas, where low-level radioactive (nontransuranic) liquids could be released to the ground, and high-level radioactive (transuranic) wastes could be stored pending long-term disposal.

Construction activities spread across more than 917 square kilometers (360 square miles) of semiarid steppe covered primarily with sagebrush. Starting with a rapidly growing town, Richland, and a temporary construction camp, the U.S. Army's Camp Hanford, the Hanford Site eventually encompassed 1482 square kilometers (570 square miles). It contained fuel fabrication facilities, three production reactors, two chemical separation plants, administrative headquarters, a plutonium purification plant (231-Z Plant), and innumerable buildings for special manufacturing processes. All were connected by a network of roads.

Original site layout

The original layout of the Hanford Works was planned to isolate the four main operational areas. Thus, any unexpected event, such as enemy attack, in one area would not compromise work in another area. The single-purpose production reactors were aligned along the northern margin of the Hanford Site in the 100 Areas next to the Columbia River. The reactors needed direct access to river water for their cooling systems, and the high banks provided a margin of safety from spring floods (Figure 3.1). The chemical separation plants were placed in the 200 West and 200 East Areas near the center of the Hanford Site on a low plateau. At this location, the separation facilities were well above the subsurface water table, and unsaturated soils favored containment of radioactive liquids. The main disposal and storage areas for radioactive wastes were included

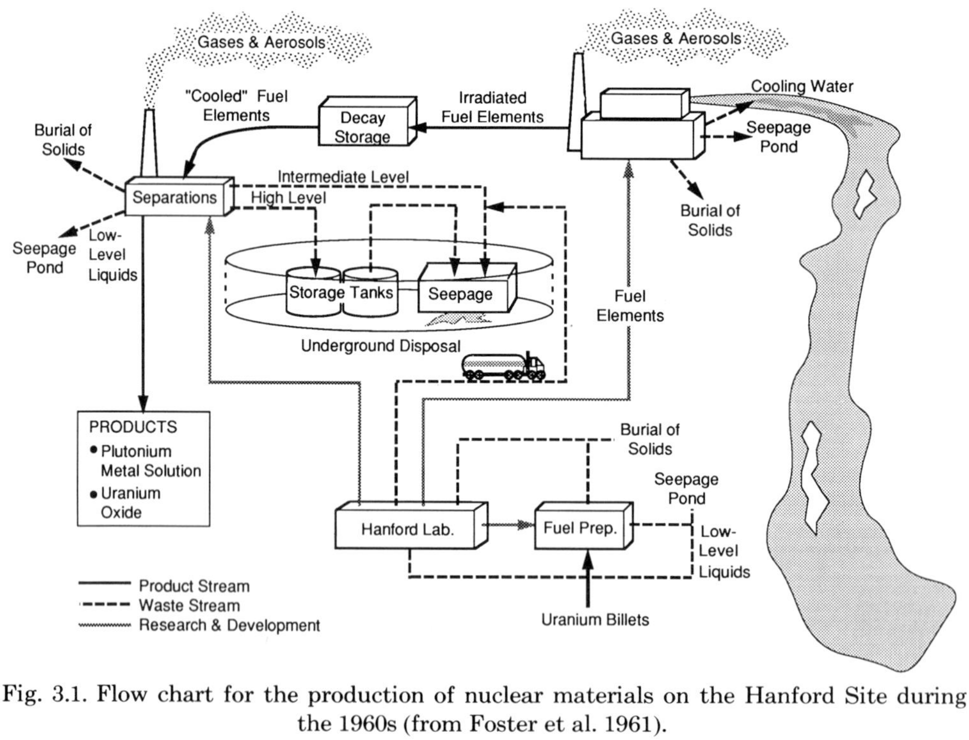

Fig. 3.1. Flow chart for the production of nuclear materials on the Hanford Site during the 1960s (from Foster et al. 1961).

near the chemical separation plants in the 200 Areas. Metal testing and fuel fabrication facilities were placed north of Richland in the 300 Area, along with research and engineering quarters (Jones 1985).

Site layout and activities today

The site layout exists today in modified form (Figure 3.2). Another location, the 400 Area, has been placed between the 200 and 300 Areas. It contains facilities associated with electrical power generation and peaceful uses of atomic energy. Additional areas have been numbered to identify the location of specific activities. The 600 Area represents a buffer zone or space between all operational areas, and includes the Arid Lands Ecology Reserve and the National Environmental Research Park, operated by Pacific Northwest Laboratory. The 700 Area in Richland contains federal administrative offices and quarters for some onsite

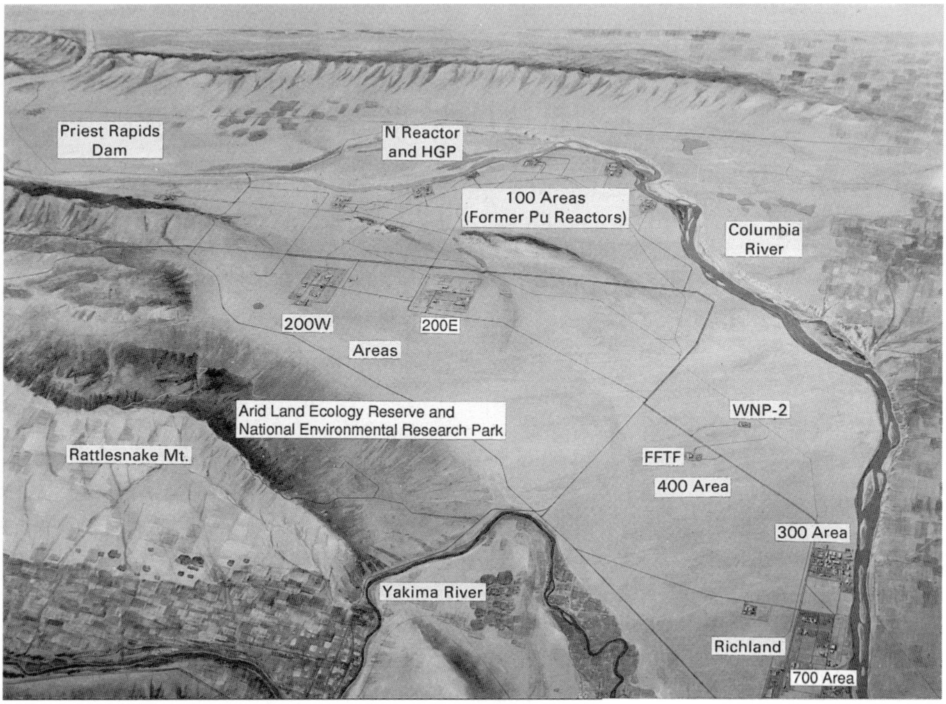

Fig. 3.2. Major operations areas on the Hanford Site. Hanford today is a major asset of the U.S. Department of Energy. Present-day operations are multiprogram and multicontract. In 1985, there were 13,000 federally funded employees.

contractors. The 1100 Area, north of Richland, contains stores, vehicle maintenance shops, and a bus lot. Also north of Richland is the 3000 Area, which contains the headquarters and laboratories of several contractors. Land north of the Columbia River beyond the 100 Areas is now included in the Saddle Mountain National Wildlife Refuge and in a wildlife reserve managed by the State of Washington Department of Game.

100 Areas

The eight single-purpose plutonium-production reactors in the 100 Areas were all designed for once-through cooling by the intake and discharge of Columbia River water. Used coolant or effluent, which had been heated in the reactor core, also contained radionuclides and a small amount of chemicals. Some radioactive wastes from reactor operation, mostly as a result of accidental fuel element ruptures, were retained in trenches and pits near the riverbank. Some leachate containing radioactive materials later entered the groundwater and seeped to the Columbia River through shoreline springs.

Today, the one unit in the 100 Areas still capable of producing plutonium is the N Reactor, part of a unique, dual-purpose facility. The primary cooling system of N Reactor contains demineralized water that recirculates. A secondary cooling system extracts steam from the primary loop and passes it to an affiliate unit, the Hanford Generating Project (HGP) to generate electricity. HGP is cooled with river water by a direct once-through system. Operation of N Reactor in a dual-purpose mode considerably reduces the temperature of each cooling water discharge.

Nearly all radioactivity is confined to water in the primary loop of N Reactor. In the past, bleed-off from this loop was routinely released to a trench near the riverbank. Because some radioactive materials began leaching to the river, a disposal trench farther from the shoreline was constructed for bleed-off, and a special treatment facility was constructed in the mid-1980s.

200 Areas

The chemical separation plants in the 200 Areas (Figure 3.3) released large amounts of cooling water and process liquids containing low-level radioactivity (nontransuranic) to the ground, while high-level radioactive wastes (plutonium and transuranics) were stored in large underground tanks (ERDA 1975). Disposal of liquids containing limited radioactivity

Fig. 3.3. The PUREX (plutonium-uranium-extraction) Plant in the 200 East Area, Hanford Site (1960s photo).

to the ground was a planned practice for which the semiarid climate, unconsolidated surface soils, and slow groundwater movement beneath the Hanford Site were well suited. As the radionuclides and chemicals penetrated the ground, their concentrations were reduced by ion exchange on soil, dispersion, radioactive decay, and other physicochemical phenomena (Brown and Isaacson 1977; Robertson et al. 1983). Radionuclides that reached groundwater were diluted and further decayed as they moved slowly towards the Columbia River. As a legacy from years of disposal of soil in the 200 Areas, diluted vestiges of the most mobile radionuclides (particularly tritium) and chemicals such as nitrates now enter the Columbia River via the groundwater.

Underground tanks with 1-million-gallon capacity were built in the 200 Areas to store high-level radioactive wastes (Figure 3.4). Some of the older single-wall tanks eventually developed leaks. Radioactive materials were confined to the upper layers of soil near the leaks and did not reach

Fig. 3.4. Underground tanks (shown under construction) are used to store high-level radioactive wastes from reprocessing of irradiated reactor fuel in the 200 Areas.

groundwater. Single-wall tanks were replaced by double-wall tanks in the 1970s, and the contents of the single-wall tanks have been either transferred to the double-wall tanks or solidified by removing excess water through evaporation. Automated monitoring systems provide immediate warning of any new or recurring leak (Graham 1981). Today, an extensive program is underway to identify, classify, and permanently isolate all high-level wastes disposed of on the Hanford Site in earlier years (DOE 1987a,b, 1988; WHC 1988).

300 Area

Wastes from preparing reactor fuels in the 300 Area (Figure 3.5) were typical of wastes encountered in other metal processing industries. The 300 Area wastes consisted mostly of cooling water, pickling rinses, and dilute caustics. Up through 1971, wastes from fabrication of fuels used in

Fig. 3.5. Fuel preparation and research facilities in the 300 Area on the Hanford Site, Circa 1980.

the single-purpose reactors contained low concentrations of uranium and were discharged to two settling ponds a few hundred feet from the Columbia River (Foster et al. 1961). These ponds also received some chemical solutions and low-level radioactive liquids from nearby research laboratories. Some solid wastes also were buried near the 300 Area. Today, small amounts of these wastes are transported to the Columbia River when intergravel water under the 300 Area is forced to rise by high or fluctuating river flows.

400 Area

A commercial power plant, Washington Public Power Supply System Nuclear Project No 2 (WNP-2) is sited in the 400 Area near the Columbia River (Figure 3.6). WNP-2 began operation in 1984 and was designed to minimize environmental impacts. The plant uses Columbia River water for cooling, but has a closed-cycle system with six mechanical draft cooling towers and cooling ponds (Supply System 1977). The cooling

Fig. 3.6. The nuclear power plant (WNP-2) operated by Washington Public Power Supply System in the 400 Area on the Hanford Site.

water intakes are perforated pipes specially designed to prevent impingement and entrainment of fish. All liquid effluents released to the river, including cooling tower blowdown, are controlled to comply with regulatory standards for protection of water quality.

The Fast Flux Test Facility (FFTF), constructed to test fuels and materials for advanced nuclear technology for the U.S. Department of Energy (DOE), is farther inland in the 400 Area (Figure 3.7). It began operation in 1979 and was dedicated in 1982. The FFTF is cooled by recirculating liquid sodium (a metal) rather than water. Waste heat is exchanged with the air in cooling towers.

Water quality considerations

Release of effluent from the Hanford Site to the Columbia River has always been a cause of concern, but not cause for alarm. Good use was

Fig. 3.7. The Fast Flux Test Facility, sited well inland from the Columbia River in the 400 Area.

made of the river's dilution capacity. Because of its sizable volume, the Columbia River can dilute and disperse relatively large amounts of liquid effluent containing low-level contaminants from the Hanford Site. The assimilation of many wastes and their reduction to harmless levels, a feature of all river ecosystems, is reflected today in National Pollution Discharge Elimination System (NPDES) permits. These permits are issued for controlled, point-source releases by the U.S. Environmental Protection Agency under the National Environmental Protection Act.

Environmental monitoring has been conducted onsite or in the adjacent countryside since the establishment of the Hanford Site. Monitoring provides the data to 1) keep controlled releases of materials within reasonable limits or in compliance with existing, applicable standards; 2) prevent adverse effects on human health and well-being; and 3) avoid significant changes in the characteristics of air, water, and soils. Long before the 1980s, radiological monitoring at Hanford had evolved into a comprehensive program involving collection of data from the air, groundwater, surface water, foodstuffs, wildlife, soil, and vegetation. Data also were obtained on penetrating radiation and its potential effects on native plants, native animals, and people as a result of passage through the food web. At the same time, nonradiological monitoring provided data on the overlying air, underlying groundwater, and the adjacent Columbia River (Price 1986; PNL 1987).

Baseline water quality data were first obtained by E. I. du Pont de Nemours and Company, Inc., in 1943 for plant operation purposes. From the spring of 1951 to the spring of 1953, water quality in the Hanford Reach and adjacent areas was extensively surveyed by the U.S. Public Health Service (PHS). This survey also identified effects from radioactivity on physical, chemical, and biological characteristics of the river (Robeck et al. 1954). Much work was done between Priest Rapids Dam and the town of Paterson, a section that included what was called the "Hanford Works of the Atomic Energy Commission" and Lake Wallula. The PHS detected no effect on any water quality characteristic from the single-purpose reactor discharges. But they recommended further reduction in the amount of radioactivity entering the river from atomic energy installations. This recommendation preceded today's "As Low as Reasonably Achievable" principle for releases of artificially produced radioactivity to the environment.

Today, groundwater beneath the Hanford Site that reaches the Columbia River contains limited amounts of radionuclides and chemicals from disposal of low-level and intermediate-level liquid wastes in the 200 East and 200 West Areas. In some places, the groundwater also contains radioactive materials and chemicals from active and inactive waste disposal sites in the 100 and 300 Areas (PNL 1987). The movement and composition of water in the unconfined aquifer at Hanford are closely monitored by an extensive network of wells (Graham 1981; Law and Allen 1984; Cline et al. 1985; McGhan et al. 1985). Contaminated groundwater emerges as springs and seeps at some places along the Columbia River shoreline, particularly when river flows are low (McCormack and Carlile 1984). These uncontrolled sources may contain certain contaminants. Yet concentrations where the water is used downstream remain well within federal and state standards for public drinking water. The presence of such contaminants is difficult or impossible to detect downriver after dilution in river water.

References

Brown, D.J., and R.E. Isaacson. 1977. *The Hanford Environment as Related to Radioactive Waste Burial Grounds and Transuranium Waste Storage Facilities.* ARH-ST-155, Atlantic Richfield Hanford Company, Richland, Washington.

Cline, C.S., J.T. Rieger, J.R. Raymond, and P.A. Eddy. 1985. *Ground-Water Monitoring at the Hanford Site, January-December 1984.* PNL-5408, Pacific Northwest Laboratory, Richland, Washington.

DOE. See U.S. Department of Energy.

ERDA. See U.S. Energy Research and Development Administration.

Foster, R.F. 1972. "The History of Hanford and Its Contribution of Radionuclides to the Columbia River". In: *The Columbia River Estuary and Adjacent Ocean Waters*, eds. A. T. Pruter and D. L. Alverson, pp. 1–18. University of Washington Press, Seattle.

Foster, R.F., R.L. Junkins, and C.E. Linderoth. 1961. "Waste Control at the Hanford Plutonium Production Plant." *J. Water Pollut. Control Fed.* 35:511–529.

Graham, M.J. 1981. "The Radionuclide Ground-Water Monitoring Program for the Separations Area, Hanford Site, Washington State." *Ground Water Monit. Rev.* 1:51–56.

Groueff, S. 1967. *Manhattan Project. The Untold Story of the Making of the Atomic Bomb.* Little, Brown, and Company, Boston.

Jones, V.C. 1985. Manhattan: *The Army and the Atomic Bomb. Special Studies, United States Army in World War II.* Center of Military History, U.S. Army, Washington, D.C.

Law, A.G., and R.M. Allen. 1984. *Results of the Separations Area Ground-Water Monitoring Network for 1983.* RHO-RE-SR-81–24 P, Rockwell Hanford Operations, Richland, Washington.

McGhan, V.L., P.J. Mitchell, and R.S. Argo. 1985. *Hanford Wells.* PNL-5397, Pacific Northwest Laboratory, Richland, Washington.

McCormack, W.D., and J.M.V. Carlile. 1984. *Investigation of Ground-water Seepage from the Hanford Shoreline of the Columbia River.* PNL-5289, Pacific Northwest Laboratory, Richland, Washington.

Pacific Northwest Laboratory. 1987. *Environmental Monitoring at Hanford for 1986.* PNL-6120, Pacific Northwest Laboratory, Richland, Washington.

Price, K.R. 1986. *Environmental Monitoring at Hanford for 1985.* PNL-5817, Pacific Northwest Laboratory, Richland, Washington.

Robeck, G.G., C. Henderson, and R.C. Palange. 1954. *Water Quality Studies on the Columbia River.* Report of U.S. Public Health Service, Robert A. Taft Sanitary Engineering Center, Cincinnati, Ohio.

Robertson, D.E., A.P. Toste, K H. Abel, C.E. Cowan, E.A. Jenne, and C.W. Thomas. 1983. "Speciation and Transport of Radionuclides in Groundwater." In: *NRC Nuclear Waste Geochemistry '83*, eds. D. H. Alexander and G. F. Birchard, pp. 297–325. Nuclear Regulatory Commission, Office of Nuclear Regulatory Research, Washington, D.C.

Rostenbach, R.E. 1956. "Radioactive Waste Disposal at Hanford." *Sewage Indust. Wastes* 28:280–286.

U.S. Department of Energy. 1987a. *Environmental Survey Preliminary Report, Hanford Site, Richland, Washington.* DOE/EH/OEV-05-P, U.S. Department of Energy, Office of Environmental Audit, Washington, D.C.

U.S. Department of Energy. 1987b. *Disposal of Hanford Defense High-Level Transuranic and Tank Wastes, Final Environmental Impact Statement, Vols. 1–5.* DOE/EIS/0113, U.S. Department of Energy, Washington, D.C.

U.S. Department of Energy. 1988. *Hazardous Waste Management Plan, Defense Waste Management.* DOE/RL-88–01, U.S. Department of Energy, Richland Operation Office, Richland, Washington.

U.S. Energy Research and Development Administration (ERDA). 1975. *Final Environmental Statement, Waste Management Operations, Hanford Reservation, Richland, Washington.* ERDA-1538 (2 vol.), National Technical Information Service, Springfield, Virginia.

Washington Public Power Supply System (Supply System). 1977. *WPPSS Nuclear Project No. 2, Environmental Report - Operating License Stage, Docket No. 50–397.* Washington Public Power Supply System, Richland, Washington.

Westinghouse Hanford Company (WHC). 1988. *Hanford Environmental Management Program Implementation Plan.* WHC-EP-0180, Westinghouse Hanford Company, Richland, Washington.

Chapter 4

OPERATION OF THE SINGLE-PURPOSE REACTORS, 1943 TO 1971

Hanford's first plutonium-production "pile" (B Reactor) started up in September 1944, only 18 months after construction began. Low-power testing and eliminating startup problems continued through December until the reactor reached full power. Two additional prototype "piles" (D and F Reactors) started up in February 1945, and full-scale production was reached in March (Jones 1985). Five additional reactors were built along the high south bank of the Columbia River in the years following World War II. These eight reactors were designed, with appropriate safety considerations, to produce Pu-239 and other special nuclear materials by neutron bombardment of uranium fuel.

The single-purpose reactors functioned safely through their early years, and confidence in their safe operation grew with experience. No accidents involving radioactive materials occurred that led to serious injury or loss of human life. Further, the surrounding environment was not degraded from accidental release of radioactivity or chemicals. In fact, the first three prototype reactors (B, D, and F) provided more than 20 years of reliable operation (Figure 4.1).

In early 1964, federal officials decided to gradually phase out and shut down the older reactors. The last single-purpose reactor (KE Reactor) was shut down in January 1971. The N Reactor, a dual-purpose facility, started up in December 1963 and remained operational until its shutdown for safety improvements in January 1987. N Reactor was placed on cold standby in March 1988 when the estimated amount of plutonium needed in defense programs changed under political scrutiny.

Observations on the effects of radioactive materials in the Columbia River, which received the cooling effluent, were probably the most extensive ever made in rivers and estuaries of the United States (Eisenbud 1973). Today, the maximum radiation dose to humans near the nation's nuclear energy installations is limited to a few millirems per year, much less than the dose received from background radiation.

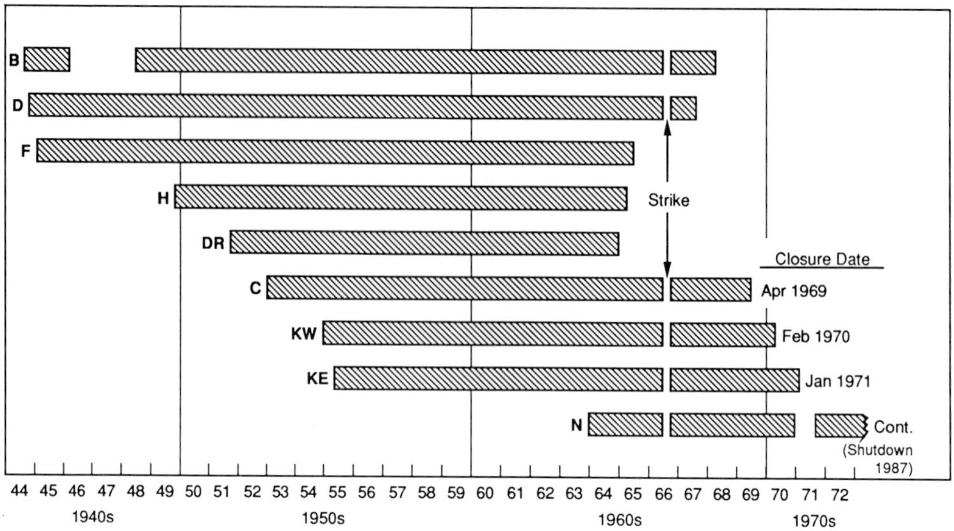

Fig. 4.1. Operational history of the single-purpose, plutonium-production reactors at Hanford. Capital letters represent the designation of each reactor (e.g., B Reactor).

Radiation doses required to induce perceptible injury to lower life forms (aquatic organisms) are higher by many orders of magnitude (Auerbach et al. 1971; Templeton et al. 1971).

Operational features

The eight single-purpose reactors at Hanford were graphite-moderated piles fueled with uranium and cooled with "light water" (natural water). The fissioning of uranium atoms under neutron bombardment in the metal-encased (or metal-cladded) fuel (aluminum or aluminum-zirconium alloy) released large amounts of energy in the form of heat. To avoid damaging the cladding encasing the reactor fuel, excess heat was removed by large amounts of water passing through spaces between the fuel elements and the tube wall. The metal cladding around the fuel elements prevented contact between uranium and the water. The outer wall of the tubes consisted of two layers through which water also passed. Loss of containment by rupture of a fuel element led to reactor shutdown and retention of water in special areas. Hanford's single-purpose reactors used treated river water in "once-through" cooling.

Proper functioning of the water intake and water treatment facilities was critical to each reactor's operation. Because the passages for coolant

Fig. 4.2. Paired intakes of the KE and KW Reactors where water was withdrawn from the Hanford Reach for once-through cooling.

in the reactor cores were small, water of low quality could rapidly plug the cooling system, and the reactor would shut down to avoid escalating temperatures. Also, rapid corrosion of the fuel cladding and reactor process tubes caused by impure water would cause early failures, leading to increased tube replacement costs (Young 1956).

The cooling water intakes for each reactor were built along the Columbia River shoreline (Figure 4.2). Barred racks across the intake prevented large pieces of trash from entering with the river water. Traveling screens behind the racks removed smaller debris. The raw river water was pumped upward from seal wells behind the screens and passed to filter plants located back from the riverbank.

Conventional methods of municipal water treatment were used to remove impurities from the raw river water before it entered the reactor cores (Figure 4.3). Treatment included alum flocculation when turbidity was high (spring and early summer). After the flocculus deposited in settling basins, the water was filtered through layers of sand, anthracite, and gravel. The pH was adjusted toward neutrality with sulfuric acid, and sodium dichromate was added to inhibit corrosion (Conley 1954;

Fig. 4.3. Water treatment facilities at the KW and KE Reactors in the 100-K Area. For KE Reactor:
A = main reactor building with adjacent stack
B = main pumphouse supplying raw river water for cooling
C = raw water filtration plants
D = retention basins to allow decay of radioactivity and heat.

Young 1956; Foster et al. 1961). Treated water was circulated through the reactor core by high-pressure, high-volume pumps.

After use, the effluent was routed to large basins where the water remained until the short-lived radionuclides decayed. In addition, all effluent entering the river was monitored to ensure that the radioactivity was within safe limits (Groves 1962).

Retention of cooling effluent in concrete or steel basins reduced gross radioactivity in the cooling water effluent by up to 50%. Retention times varied with operating power levels. From 1951 to 1960, the single-purpose reactors released a maximum of nearly 24,000 megawatts of heat (the equivalent of 24 typical power-producing reactors) and several thousand curies of radionuclides to the Columbia River each day (Rickard and Watson 1985).

The cooling effluent from each single-purpose reactor returned to the Columbia River through large pipes on the river bed. This effluent was soon mixed in the much larger flow (average of 120,000 cubic feet per second) of the Hanford Reach.

Effluent surfaced near midriver and moved downstream as relatively narrow plumes. Lateral dispersion was initially limited. The plumes remained near midriver, where they could not re-enter any cooling water intake, until they passed the reactor areas farthest downstream. Below this point, lateral dispersion was aided by curves in the river's course and by several islands. Effluent was well mixed in river water before reaching the cities of Richland, Pasco, and Kennewick some 50 kilometers (31 miles) downriver. Minimum travel times to Richland were about 11 hours at high river flows and about 22 hours at low river flows (Honstead et al. 1960).

With respect to radiation, the artificial radionuclides in the cooling water were reduced to environmentally insignificant levels before reaching the municipal intakes of Richland, Kennewick, and Pasco. Since closure of KE Reactor, cooling effluent from other energy-production facilities at Hanford (i.e., the effluent from N Reactor, HGP, and WNP-2) has released only small amounts of radioactive materials.

Background radiation

Naturally occurring radioactive elements (background radiation) have always contributed to the total radioactivity of the Columbia River. Natural radioactivity of water is derived primarily from radioactive rocks and minerals, although tritium originates in the atmosphere. Water in the Columbia River contains small amounts of uranium, thorium, radium, tritium, and K-40 that leach from soil deposits and rock formations. From 1948 to 1950, the initial years of Hanford operations, background radioactivity in the Columbia River averaged 3×10^{-9} microcurie per milliliter (μCi/mL) (Davis et al. 1956).

Uranium from natural sources occurs in Columbia River water at very low levels – about 1 microgram per liter (μg/L). But this amount was significant to the operation of the single-purpose reactors. Natural uranium led to the formation of Np-239, which decayed to Pu-239, as well as to activation and fission products when the cooling water was bombarded by neutrons in the reactor cores. Background concentrations of tritium and Sr-90 were about 10^{-3} and 1 picocurie per liter (μCi/L), respectively (Foster and Soldat 1966). Further, natural uranium occurs in groundwater beneath the Hanford Site at 2 to 30 $\mu g/L$ (Soldat 1961).

Before the first nuclear devices were detonated in atmospheric tests, resulting in worldwide fallout, total alpha activity (primarily from uranium) in Columbia River water was less than 5×10^{-9} $\mu Ci/mL$, and total beta activity was less than 5×10^{-8} $\mu Ci/L$, (Foster and Rostenbach 1954). Background levels in the Snake River were about the same. These values were below average when compared with other major rivers in the United States. Average background levels were equal to or less than 1×10^{-6} $\mu Ci/mL$ in aquatic organisms and 1×10^{-8} $\mu Ci/mL$ in river water (Robeck et al. 1954).

Today, the situation has changed. Relatively high levels of radioactivity from worldwide atmospheric fallout, primarily from nuclear tests by the United States and the Soviet Union in the 1950s, were added to relatively low background levels in the Columbia River. In fact, by the end of 1958, 250 known nuclear explosions, with a cumulative yield of nearly 75 megatons, had taken place in the atmosphere (Carter and Moghissi 1977). Since then, when a moratorium on nuclear tests in the atmosphere began, the Soviet Union and other nations have exploded nuclear devices above ground, while the United States has emphasized underground tests (Kathren 1984).

Over the last three decades, radioactivity from weapons testing has been scattered over the entire world, including the Columbia River watershed. A small portion has been carried into the river and downstream via erosion and sediment transport. Some radionuclides that may affect human health form strong associations with particulate matter in freshwater ecosystems.

In the early 1980s, most of the cobalt-60 activity in Columbia River sediments from Hanford to the river's estuary could be attributed to earlier releases from the single-purpose reactors. However, most activity from Pu-239/240 and Cs-137 and all activity from Am-241 in the sediment downstream of Hanford were probably derived from atmospheric fallout (Beasley and Jennings 1984). In 1977, an estimated 20% to 25% of the total plutonium inventory in the sediment of Lake Wallula below Hanford could be ascribed to past reactor operation. Most of the small amounts of plutonium reaching the Columbia River from the Hanford reactors were a result of the decay of Np-239 to Pu-239 in the reactor effluent (Beasley et al. 1981). On the other hand, most of the plutonium in Columbia River sediments today comes from atmospheric fallout.

Areas of concern

The impact of cooling water effluent from the single-purpose reactors at Hanford drew immediate attention during the startup and expansion years because they contained

- radioactive materials created by neutron bombardment in the reactor core (the nuclear reaction)
- heat from the energy released by neutron bombardment
- chemicals used in cooling water treatment.

Radioactivity from Reactor Effluent

No nuclear-fueled, water-cooled reactor existed before startup of B Reactor at Hanford in September 1944. However, the presence of a complex mixture of radionuclides in the cooling effluent was predicted on the basis of reactor design. This prompted the first studies in 1943 at the University of Washington on radiation effects on fish.

The radionuclides that appeared in the cooling effluent included both activation and fission products. Virtually all materials inside a reactor core were bombarded by neutrons and became, to a degree, radioactive.

Activation products, the major source of radionuclides in the cooling effluent, resulted from neutron activation of elements dissolved in river water. An additional smaller source of activation products was materials corroded from the reactor piping and fuel elements (Davis 1958; Honstead et al. 1960; Foster et al. 1961; Foster 1972).

Intricate adjustments of water chemistry were necessary to minimize the radionuclide content of effluents that entered the Columbia River (Conley 1954). Because the purpose of the once-through cooling was to remove heat, scale or film buildup on the hot aluminum surfaces of the fuel elements and the tubes had to be prevented. Thus, sodium dichromate was added to restrict corrosion of piping metals. Furthermore, impurities that would be activated into troublesome radionuclides by neutron bombardment had to be removed.

Different water treatment techniques were investigated as long as the single-purpose reactors remained in operation. Research on treatment techniques to prolong the effectiveness of fuel cladding and minimize activation products was begun before operations at Hanford actually started and continued until all the single-purpose reactors were shut down.

Amounts of fission products entering the Hanford Reach were low relative to amounts of activation products. The dominant and continuing source of fission products was "tramp" uranium in Columbia River water. The other source was fission products that escaped from fuel assemblies when protective cladding failed. Special instruments were installed to detect fuel cladding failures. When a failure was detected, the reactor was shut down and the offending fuel element replaced. Release of long-lived fission products to the Columbia River from fuel cladding failures were intermittent and infrequent, but short-lived activation products were released continuously as long as each reactor operated (Honstead et al. 1960; Foster et al. 1961; Foster 1972).

Radioactive isotopes, once generated, return to their stable chemical form only through physical decay. Different isotopes decay at different rates. Most radioactive materials entering the Hanford Reach in cooling effluent were beta- and gamma-emitting activation products with relatively short half-lives (Robeck et al. 1954; Parker et al. 1965). But concentrations of alpha emitters (including radium and uranium) remained similar in river water above and below the discharges (Foster and Rostenbach 1954). No significant amounts of alpha emitters (including uranium and plutonium) escaped from the reactors under normal operating conditions (Foster and Davis 1956).

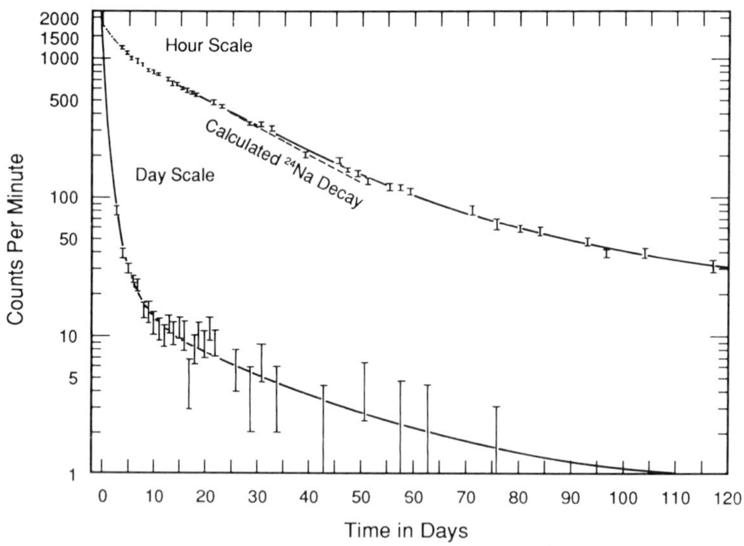

Fig. 4.4. Decay of radioactivity in a sample of Columbia River water from the Hanford Reach. The level of radioactivity is rapidly reduced due to decay of short-lived radionuclides. The decay rate of Na-24 (half-life of 15.1 hour) allows comparison with decay of total radioactivity (from Olson and Foster 1952).

Table 4.1.
Relative Abundance of More than 60 Radionuclides in the Cooling Effluent of the Single-Purpose Reactors at Hanford after 4-hour Retention in 1964 and 1968 (modified from Woolridge 1969; Watson and Templeton 1973).

Radio-nuclide	Physical Half-life	Radiation Emitted		Physical Half-life	Radiation Emitted
Major Occurrence (90%)			Trace Occurrence (1%)		
Na-24	15.0 h	B,G	Nb-95	35.0 d	B,G
Si-31	2.6 h	B,G	Mo-94	67.0 h	B,G
Cr-51	27.8 d	G	Ru-103	40.0 d	B,G
Mn-56	2.6 h	B,G	Ru-106	1.0 y	B,G
Cu-64	12.9 h	B,G	Sb-124	60.2 d	B,G
			I-131	8.1 d	B,G
Minor Occurrence (9%)			I-133	21.0 h	B,G
			I-135	6.7 h	B,G
P-32	14.3 d	B,G	Cs-136	13.0 d	B,G
Sc-46	83.8 d	G	Cs-137	30.0 y	B,G
Zn-69m	14.0 h	B,G	Ba-140	12.8 d	B,G
Ga-72	14.1 h	B,G	Ce-141	32.5 d	B,G
As-76	26.5 h	B,G	Ce-143	33.0 h	B,G
Sr-92	2.7 h	B,G	Ce-144	285.0 d	B,G
Sb-122	2.8 d	B,G	Pr-142	19.2 h	B,G
I-132	2.3 h	B,G	Pr-143	13.7 d	B
La-140	40.2 h	B,G	Nd-147	11.1 d	B,G
Eu-152m	9.2 h	B,G	Pm-147	2.6 y	B,G
Sm-153	47.0 h	B,G	Pm-149	53.0 h	B,G
Dy-165	2.35 h	B,G	Pm-151	28.0 h	B,G
Np-239	2.35 d	B,G	Eu-152	12.4 y	B,G
			Eu-156	15.2 d	B,G
Trace Occurrence (1%)			Gd-153	204.0 d	B,G
			Gd-159	18.0 h	B,G
H-3	12.3 y	B	Tb-160	72.0 d	B,G
C-14	5730.0 y	B	Tb-161	6.9 d	B,G
S-35	86.7 d	B	Ho-166	1.2×10^{-3} y	B,G
Ca-45	163.0 d	B	Er-169	9.4 d	B,G
Mn-54	312.0 d	G	Er-171	7.5 h	B,G
Fe-59	45.0 d	B,G			
Co-60	5.2 y	B,G			
Ni-65	2.6 h	B,G			
Zn-65	243.0 d	B,G			
Sr-87m	2.8 h	G			
Sr-89	50.6 d	B			
Sr-90	28.8 y	B			
Sr-91	9.7 h	B,G			
Y-90	64.2 h	B,G			
Y-91	59.9 d	B,G			
Y-93	10.1 h	B,G			

Key:
y = years
d = days
h = hours
B = beta
G = gamma

Table 4.2.
Biologically Significant Radionuclides Accumulating in Columbia River Biota from the Cooling Effluent of the Single-Purpose Reactors at Hanford, 1941–1971 (from Davis et al. 1958).

Radionuclide	Physical Half-life
As-76	26.8 h
Ba-140	12.8 d
Ce-141	33.1 d
Cu-64	12.8 h
Co-60	5.3 y
Cr-51	27.8 d
Fe-59	45.1 d
La-140	40.0 h
Mn-54	310.0 d
Mn-56	2.6 h
Na-24	15.1 h
Np-239	2.3 d
P-32	14.2 d
Ru-103	39.8 d
Sc-46	85.0 d
Sr-90	28.0 y
Zr-95/Nb-95	65.6 d
Zn-65	245.0 d

Key:
h = hours
d = days
y = years

A diverse spectrum of radionuclides was produced by Hanford's single-purpose reactors. In fact, more than 60 radionuclides were identified in the cooling effluents (Table 4.1). Many had short half-lives and could not be detected in the effluent discharged. Other radionuclides, initially scarce in the effluent, could not be detected after dilution in river water. Subsequently, the total amount of radioactive materials in the Columbia River below the reactors was rapidly reduced by decay of other short-lived radionuclides (Figure 4.4). Thus, radioactive decay greatly influenced the relative abundance of different radionuclides in river water downstream from Hanford. Only a few radionuclides accumulated in aquatic organisms and thus had biological significance (Table 4.2). While Zn-65 and P-32 were the most important of the "biologically active" radionuclides because they entered the food chain, Cr-51 was the most abundant (Watson et al. 1970). Activity levels of these radionuclides in

Table 4.3.
Estimated Activity of Three Biologically Active Radionuclides in Columbia River Water Below Hanford, 1951–1967 (from Beasley and Jennings 1984).

Radioactivity, pCi/L			
Year	Cr-51	P-32	Zn-65
1959	4198	154	206
1960	5404	210	298
1961	5774	264	341
1962	4285	176	224
1963	6736	194	308
1964 [a]	5722	86	144
1965	4043	87	161
1966	2671	93	135
1967	2375	85	140

[a] 1964 was the last year that all single-purpose reactors were operated at Hanford (Figure 4.1).

the Columbia River downstream of Hanford varied from year to year (Table 4.3).

The actual amount of radioactivity entering the Hanford Reach from the single-purpose reactors varied substantially from 1944 to 1971. Changes resulted from startup and shutdown of reactors at different

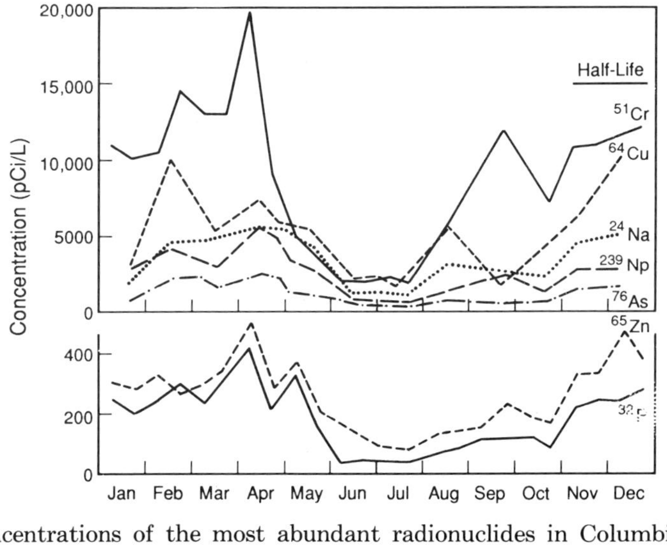

Fig. 4.5. Concentrations of the most abundant radionuclides in Columbia River water during 1964 at Richland – a point about 50 kilometers (31 miles) from the most downstream single-purpose reactor (from Foster 1972).

times (see Figure 4.1), operation of individual reactors at different power levels, closures for maintenance and refueling, water treatment modifications, frequency and severity of fuel cladding failures, and other operational features. The amounts of radionuclides released declined after deactivation of the reactors began in early 1965. Radioactivity in Columbia River water downstream from Hanford also varied seasonally (Figure 4.5).

At Richland, about 50 RKm below the reactors, the cooling effluent was well diluted, and activity levels of short-lived radionuclides were reduced by radioactive decay during transit (12 to 36 hours). The most abundant radionuclides remaining at Richland were As-76, Cm-64, Cr-51, Cu-64, I-131, Na-24, Np-239, P-32, and Zn-65 (Foster 1972). Of these, Cr-51 accounted for nearly half of the radioactivity; Zn-65 about 2%; P-32 about 1%; I-131 for less than 1%; and Cm-64, Np-239, As-76, and Na-24 for most of the rest (Foster and Soldat 1966).

Other factors, including adsorption on suspended or settled sediment and absorption or ingestion by microscopic organisms, also reduced radionuclide concentrations as they passed downstream. All factors were in operation all the way to the Columbia River outlet and into the Pacific Ocean. Yet all gamma emitters and the beta-emitter P-32 entering the Pacific Ocean in the mid-1960s, the period of maximum reactor operation, totaled 300,000 curies per year (Haushild et al. 1971). This was approximately 600 times more than the amount entering the Pacific Ocean the first year after all single-purpose reactors were shut down (Robertson et al. 1973).

Thermal Increments from Reactor Effluent

Once-through cooling meant that heat generated in the single-purpose reactors was transferred to the Columbia River. The temperature of the cooling effluent was many degrees greater than ambient river temperatures when discharged. A large amount of heat added to any body of water is a "thermal increment" that, if sufficiently high, could be detrimental to fish and other cold-blooded aquatic biota.

Studies on thermal effects gained importance in the 1960s, the peak period of reactor operations, but bioassays involving fish and reactor effluent had already been under way for 15 years. Cold-blooded aquatic organisms normally respond to moderately warmed water by an increase in growth rate.

Information on the radioactivity and heat content of each reactor's effluent was classified as secret (Nakatani 1969) and remains largely unavailable. Thermal increments up to 15°C occurred in swirls of water within 25 meters (82 feet) of each outfall, but they lasted only a few seconds (Jaske et al. 1970; Foster et al. 1972). Whenever flows were low during summer and all reactors were operating, the ambient river temperature was increased about 2.5°C over 90 kilometers (56 miles) downstream (Jaske and Synoground 1970). An agreement with dam operators allowed higher releases of cool water from Grand Coulee Dam at critical times when river temperatures peaked.

Effluent from each reactor merged into flows ranging from 0.8 to 3.0 meters per second (2.4 to 9.8 feet per second) at points where the Hanford Reach was 250 to 350 meters (820 to 1150 feet) wide. The ports of the discharge pipes faced upward so that downstream currents and turbulence promoted vertical mixing. On the other hand, the effluent was not completely mixed horizontally until carried well downstream (Honstead et al. 1960; Foster et al. 1972).

Because the effluent plumes passed downstream in narrow bands, large areas of unaffected water remained along the shoreline for the upstream migration of adult salmonids. For the most part, mixing was complete by the time the flow reached Ringold, a collection of small farms several kilometers below the most downstream reactor. Thus, Ringold represented the location where temperatures across the entire river were highest as a result of the reactor discharges (Jaske and Synoground 1970).

Chemicals in Reactor Effluent

Sodium dichromate, a corrosion inhibitor, was the only chemical found in reactor effluent in sufficient amounts to measurably influence water quality in the Hanford Reach. It was routinely added to the cooling water during pretreatment after the raw river water was filtered to prevent pitting of the aluminum in the reactor core piping (Conley 1954). Sodium dichromate had two adverse side effects. First, neutron activation of stable chromium already in the water resulted in appearance of Cr-51 in the effluent (Junkins 1969; Hall et al. 1970). Second, as the effluent emerged from the discharge ports, it contained hexavalent chromium at levels toxic to fish (Foster et al. 1961). When onsite bioassays indicated that hexavalent chromium inhibited fish growth, amounts were limited to 0.02 mg/L of chromium in the reactor effluent. This limit was considerably below that permitted in drinking water.

Other chemicals (e.g., sodium sulfate, sodium phosphate) used in cooling water treatment had little or no effect on water quality in the Columbia River (Conley 1954; Hall et al. 1970). Some nitrates entered the

Hanford Reach via groundwater from the 200 Areas, but their influence was masked by the large amounts of nitrate from other sources (Hall et al. 1970).

Some compounds used for pretreatment of raw river water also were returned to the Hanford Reach. For example, flocculating agents were used at each water treatment plant. The accumulated flocculus was discharged to the river whenever the settling basins were backflushed. This flocculus was highly turbid and produced a visible trail for several miles downriver. Dispersal of flocculus in river water in early years revealed the path of the effluent plume and allowed estimates of how radionuclides spread laterally when passing downriver (Soldat 1962).

Dilution capacity of the Columbia River

Throughout recorded history, the Columbia River has always had one major flow and temperature cycle each year. Downstream flows peak in early spring after snowmelt and rain in headwater tributaries. Temperatures peak in summer and early fall in association with insolation and low flows.

Today, annual extremes in the flow and temperature cycles in the mainstem Columbia River have been moderated by impoundments behind dams throughout the drainage system (PNWRC 1979). Grand Coulee Dam, a large storage reservoir below the Canadian border, effectively regulates all flows downstream. Lake Roosevelt, the large reservoir behind Grand Coulee Dam, has 5.2 million acre feet of active storage, 9.6 million acre feet of total storage, and a theoretical "flushing rate" of about 45 days (Ebel et al. 1988).

Grand Coulee Dam was completed in 1941. Studies two decades later showed that, on the average, water temperatures had not changed significantly after river-run reservoirs were filled on the middle and upper Columbia River. However, storage and release of water from Lake Roosevelt had delayed the timing of peak seasonal temperatures below Grand Coulee Dam. With respect to the Hanford Reach, the river was slightly cooler in summer and slightly warmer in winter than before the dams were built (Jaske and Goebel 1967; Jaske and Synoground 1970). Temperatures in the Hanford Reach today range from about 2°C in late January or February to about 20°C in August (Whelan and Newbill 1983).

The capacity of river water in the Hanford Reach to dilute the cooling effluent was essential to operation of the single-purpose reactors from

1944 to 1971. Annual flows through the Hanford Reach averaged 120,000 cubic feet per second (Supply System 1978). Dilution enabled the radionuclides, heat, and chemicals in the effluent to be reduced to relatively harmless levels as they passed downstream (Rostenbach 1956).

The effectiveness of dilution was a function of seasonally changing flows. Under regulations existing through the 1950s, no permits were required to release cooling effluent from the reactors. Today, point-source discharges entering the Columbia River from the Hanford Site are regulated under National Pollutant Discharge Elimination System permits (Price 1986). Dilution capacity provides assurance that any contaminants in onsite effluents do not breach water quality standards set by the state of Washington for the Columbia River.

The Federal Energy Regulatory Commission has established a minimum administrative flow for the Hanford Reach of 1020 cubic meters per second (36,000 cubic feet per second). Combined release and spill at Priest Rapids Dam above Hanford should not be less under ordinary conditions. Today, average daily discharges may reach 4530 cubic meters per second (160,000 cubic feet per second). In the past, average daily discharges

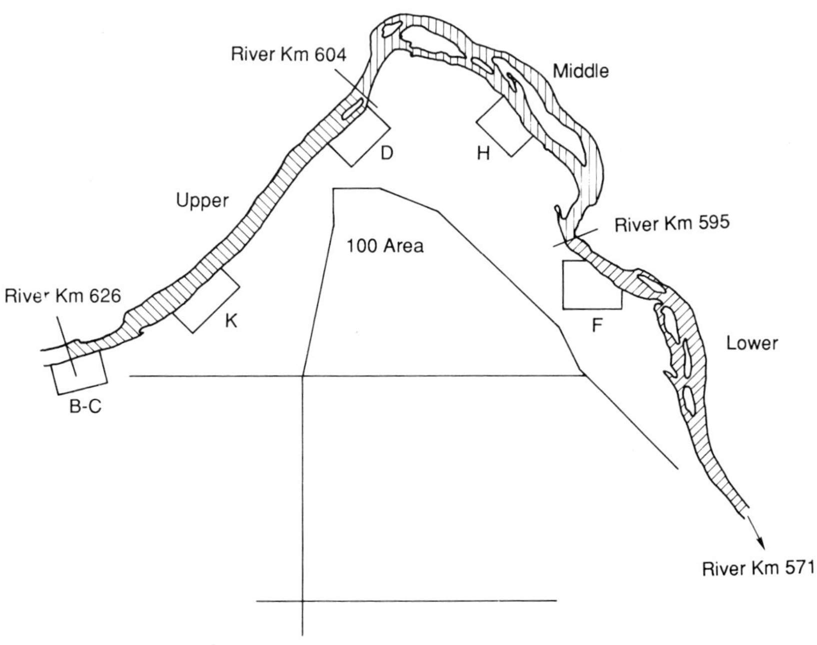

Fig. 4.6. The single-purpose reactors were distributed along the south shoreline of the Hanford Reach above the extent of potential floods. When discharged, cooling effluent was diluted in large quantities of river water.

during spring runoff were as high as 18,418 cubic meters per second (650,000 cubic feet per second). The 55-year average daily flow rate for the Hanford Reach is 3422 cubic meters per second (120,800 cubic feet per second). Minimum flows today are approximately 2266 cubic meters per second (80,000 cubic feet per second) or less during late summer and early fall (ERDA 1975).

Placement of the single-purpose reactors along the northern margin of the Hanford Site was ideal for dispersing cooling effluent in the Hanford Reach (Figure 4.6). All reactors were built between RKm 616 and RKm 587. From RKm 616 to 605, the river remained essentially straight and constant in width. From RKm 605 to 587, the river bent 90 degrees and flowed around a number of islands. Below the reactor farthest downstream (at RKm 587), the river first turned eastward and then flowed around a number of small islands (Sonnichsen et al. 1970). The uppermost straight section allowed effluents to be released in midstream. The lowermost braided sections aided effluent mixing.

References

Auerbach, S.I., D.J. Nelson, S.V. Kaye, D.E. Reichle, and C.C. Coutant. 1971. "Ecological Considerations in Power Plant Siting." In: *Proceedings, Environmental Aspects of Nuclear Power Siting, 1971.* International Atomic Energy Agency, Vienna, Austria.

Beasley, T.M., L.A. Ball, J.E. Andrews III, and J.E. Halverson. 1981. 'Hanford-Derived Plutonium in Columbia River Sediments." *Science* 214:913–915.

Beasley, T.M., and C.D. Jennings. 1984. "Inventories of 239,240 Pu, 241 Am, 137 Cs, and 60 Co in Columbia River Sediments from Hanford to the Columbia River Estuary." *Environ. Sci. Tech.* 18:201–212.

Carter, M.W., and A.A. Moghissi. 1977. "Three Decades of Nuclear Testing." *Health Phys.* 33:51–71.

Conley, W.R. 1954. "Hanford Atomic Energy Plant – Water Supply." *J. Am. Water Works Assoc.* 46:621–633.

Davis, J.J. 1958. "Dispersion of Radioactive Materials by Streams." *J. Am. Water Works Assoc.* 50:1501–1515.

Davis, J.J., R.W. Perkins, R.F. Palmer, W.C. Hanson, and J.F. Cline. 1958. "Radioactive Materials in Aquatic and Terrestrial Organisms Exposed to Reactor Effluent Water." In: *Proceedings of the Second United Nations International Conference on the Peaceful Uses of Atomic Energy*, Vol. 18, pp. 4211–428. United Nations, Geneva, Switzerland.

Davis, J.J., D.G. Watson, and C.C. Palmiter. 1956. *Radiobiological Studies of Columbia River Through December 1955.* HW-36075, Hanford Works, Richland, Washington.

Ebel, W.J., C.D. Becker, J.W. Mullan, and H.L. Raymond. 1988. "The Columbia River – Towards a Holistic Understanding." In: *Proceedings of the International Large River Symposium*, ed. D. P. Dodge, pp. 205–219. Can. Spec. Publ. Fish. Aquat. Sci. 106, Toronto, Canada.

Eisenbud, M. 1973. *Environmental Radioactivity. 2nd Ed.* Academic Press, New York.

ERDA. See U.S. Energy Research and Development Administration.

Foster, R.F. 1972. "The History of Hanford and Its Contribution of Radionuclides to the Columbia River." In: *The Columbia River Estuary and Adjacent Ocean Waters*, eds. A. T. Pruter and D. L. Alverson, pp. 1–18. University of Washington Press, Seattle.

Foster, R.F., and J.J. Davis. 1956. "The Accumulation of Radioactive Substances in Aquatic Forms." In: *Proceedings of the First International Conference on Peaceful Uses of Atomic Energy*, Vol. 13, pp. 364–367. International Atomic Energy Agency, Geneva, Switzerland.

Foster, R.F., and R.E. Rostenbach. 1954. "Distribution of Radioisotopes in Columbia River." *J. Am. Water Works Assoc.* 46:631–640.

Foster, R.F., and J.K. Soldat. 1966. "Evaluation of the Exposure that Results from the Disposal of Radioactive Wastes into the Columbia River." In: *Disposal of Radioactive Wastes into Seas, Oceans and Surface Waters*, ed. A. Guillon, pp. 696–696. International Atomic Energy Agency, Vienna, Austria.

Foster, R.F., R.T. Jaske, and W.L. Templeton. 1972. "The Biological Cost of Discharging Heat to Rivers." In: *Peaceful Uses of Atomic Energy*, Volume II, pp. 631–640. International Atomic Energy Agency, Vienna, Austria.

Foster, R.F., R.L. Junkins, and C.E. Linderoth. 1961. "Waste Control at the Hanford Plutonium Production Plant." *J. Water Pollut. Control Fed.* 35:51–529.

Hall, R.B., J.P. Corley, J.K. Soldat, and R.T. Jaske. 1970. "Environmental Effects of An Extended Plant Shutdown, Appendix E." In: *Effect of Hanford Plant Operations on the Temperature of the Columbia River 1964 to Present*, eds. R. T. Jaske and M. O. Synoground. BNWL-1345, Battelle, Pacific Northwest Laboratories, Richland, Washington.

Haushild, W.L., H.H. Stevens, Jr., J.L. Nelson, and G.R. Dempster, Jr. 1971. *Radionuclides in Transport in the Columbia River from Pasco to Vancouver, Washington.* Open file report, U.S. Geological Survey, Portland, Oregon.

Honstead, J.F., R.F. Foster, and W.H. Bierschenk. 1960. "Movement of Radioactive Effluents in Natural Waters at Hanford." In: *Disposal of Radioactive Wastes II.* Proceedings of the Scientific Conference on the Disposal of Radioactive Wastes, 11–21 November 1959, Monaco, pp. 381–399. International Atomic Energy Agency, Vienna, Austria.

Jaske, R.T., and J.B. Goebel. 1967. "Effects of Dam Construction on Temperatures of the Columbia River." *J. Am. Water Works Assoc.* 59:931–942.

Jaske, R.T., and M.O. Synoground. 1970. *Effect of Hanford Plant Operations on the Temperature of the Columbia River 1964 to Present.* BNWL-1345, Battelle, Pacific Northwest Laboratories, Richland, Washington.

Jaske, R.T., W.L. Templeton, and C.C. Coutant. 1970. "Methods for Evaluating Effects of Transient Conditions in Heavily Loaded and Extensively Regulated Streams." *Chem. Eng. Prog. Symp. Ser.* 67:31–39.

Jones, V.C. 1985. *Manhattan: The Army and the Atomic Bomb. Special Studies, United States Army in World War II.* Center on Military History, U.S. Army, Washington, D.C.

Junkins, R.L. 1969. "Reactor Releases of Radionuclides." In: *Biological Implications of the Nuclear Age*, B. Shore and F. Hatch, Chairmen, pp. 133–143. CONF-690303, U.S. Atomic Energy Commission, Washington, D.C.

Kathren, R.L. 1984. *Radioactivity in the Environment: Sources, Distribution, and Surveillance*. Harwood Academic Publishers, New York.

Nakatani, R .E. 1969. "Effects of Heated Discharges on Anadromous Fishes." In: *Biological Aspects of Thermal Pollution*, eds. P. A. Krenkel and F. L. Parker, pp. 291–317. Vanderbilt University Press, Nashville, Tennessee.

Olson, P.A., Jr., and R.F. Foster. 1952. *Accumulation of Radioactivity in Columbia River Fish in the Vicinity of the Hanford Works*. HW-23093, Hanford Engineering Works, Richland, Washington.

Pacific Northwest Regional Commission (PNWRC). 1979. *Water Today and Tomorrow, Volume II, The Region*. PNWRC, Vancouver, Washington.

Parker, H.M., R.F. Foster, I.L. Ophel, F.L. Parker, and W.C. Reinig. 1965. "North American Experience in the Release of Low-Level Wastes to Rivers and Lakes." In: *Proceedings of the Third International Conference on the Peaceful Uses of Atomic Energy*, Vol. 14, Environmental Aspects of Atomic Energy and Waste Management, pp. 61–71. United Nations, New York.

Price, K.R. 1986. *Environmental Monitoring at Hanford for 1985*. PNL-5817, Pacific Northwest Laboratory, Richland, Washington.

Rickard, W.H., and D.G. Watson. 1985. "Four Decades of Environmental Change and Their Influence upon Native Wildlife and Fish on the Mid-Columbia River, Washington, USA." *Environ. Conserv*. 12:241–248.

Robeck, G.G., C. Henderson, and R.C. Palange. 1954. *Water Quality Studies in the Columbia River*. Special Report, U.S. Department of Health, Education and Welfare, Washington, D.C.

Robertson, D.E., W.B. Silker, J.C. Langford, M.R. Petersen, and R. W. Perkins. 1973. "Transport and Depletion of Radionuclides in the Columbia River." In: *Radioactive Contamination of the Marine Environment, Proceedings of a Symposium*, pp. 141–158. International Atomic Energy Agency, Vienna, Austria

Rostenbach, R.E. 1956. "Radioactive Waste Disposal at Hanford." *Sewage Ind. Wastes* 28:280–286.

Soldat, J.K. 1961. *Some Radioactive Materials Measured in Various Waters in the United States – A Literature Search*. HW-70706, Hanford Atomic Products Operation, Richland, Washington.

Soldat, J.K. 1962. *A Compilation of Basic Data Relating to the Columbia River*. Section 8, Dispersion of Reactor Effluent in the Columbia River. HW-69369, Hanford Atomic Products Operation, Richland, Washington.

Sonnichsen, J.C., Jr., D.A. Kottwitz, and R.T. Jaske. 1970. *Dispersion Characteristics of the Columbia River Between River Miles 383 and 355*. BNWL-1477, Battelle, Pacific Northwest Laboratories, Richland, Washington.

Supply System. See Washington Public Power Supply System.

Templeton, W.L., R.E. Nakatani, and E.E. Held. 1971. "Radiation Effects." Chapter 9 in *Radioactivity in the Marine Environment*, ed. A. H. Seymour. National Academy of Sciences-National Research Council, Washington, D.C.

U.S. Energy and Research Development Administration (ERDA). 1975. *Final Environmental Statement, Waste Management Operations, Hanford Reservation, Richland, Washington*. ERDA-1538 (2 vols), National Technical Information Service, Springfield, Virginia.

Washington Public Power Supply System (Supply System). 1978. *Supplemental Information on the Hanford Generating Project in Support of a 316(a) Demonstration*. Washington Public Power Supply System, Richland, Washington.

Watson, D.G., and W.L. Templeton. 1973. "Thermoluminescent Dosimetry of Aquatic Organisms." In: *Radionuclides in Ecosystems, Proceedings of the 3rd National Symposium on Radioecology*, pp. 1125–1129. CONF-710501-P2, Atomic Energy Commission, Washington, D.C.

Watson, D.G., C.E. Cushing, C.C. Coutant, and W.L. Templeton. 1970. *Battelle, Radioecological Studies on the Columbia River, Part I.* BNWL-1377, Battelle, Pacific Northwest Laboratories, Richland, Washington.

Whelan, G., and C.A. Newbill. 1983. *Update of Columbia River Flow and Temperature Data Measured at Priest Rapids Dam and Vernita Bridge*. PNL-4868, prepared for UNC Nuclear Industries, Inc. by Pacific Northwest Laboratory, Richland, Washington.

Woolridge, G.B., ed. 1969. *Evaluation of Radiological Conditions in the Vicinity of Hanford for 1967.* BNWL-983, Battelle, Pacific Northwest Laboratories, Richland, Washington.

Young, J. R. 1956. *Operational Problems of the Original Hanford Production Reactors.* HW-56230, Hanford Engineering Works, Richland, Washington.

Chapter 5

UNIVERSITY OF WASHINGTON STUDIES, 1943 TO 1960

The Manhattan District, U.S. Army Corps of Engineers (Corps) was responsible for all research and development contracts relating to the development of the atomic bomb. To disguise its real purpose, the program was first called the Development of Substitute Materials (DSM) Project. It later became known as the Manhattan Project (Groves 1962; Groueff 1967; Williams and Cantelon 1984).

Work on the first single-purpose plutonium-production reactor at Hanford began on June 7, 1943. During the planning process, officers and engineers of the Corps discussed possible adverse impacts on the Columbia River from using its water for cooling. The volume of water required to cool each reactor was so great that such factors as temperature and chemical toxicity of the effluent might be critical to aquatic life.

Of greater importance, because so little was known at the time about the effects of ionizing radiation, was the expected appearance of diverse radionuclides in the cooling effluent. Even if no environmental effects were found, it was still essential to collect scientific data to assure those involved in the DSM Project that the reactors were operating safely. The public would require similar assurance after World War II ended, when the need for secrecy had passed.

The secret beginnings

The nature of activities at Hanford, particularly as they related to Columbia River water, required that initial investigations on radiation effects be conducted elsewhere to avoid any apparent link with the secret DSM Project. This effort required neither engineers nor nuclear physicists, but people trained and specializing in aquatic biology. And it had to begin immediately.

On May 20, 1943, more than 2 weeks before work began on the first Hanford reactor, possible contamination of the Columbia River was discussed in a meeting at the University of Chicago (Hines 1962). In attendance were some 20 top-level persons representing various sectors of the atomic energy program. At this meeting, the need to conduct aquatic studies related to planned releases of radionuclides at Hanford was brought up. A logical organization to conduct such studies already existed at the University of Washington's College of Fisheries in Seattle, some 320 kilometers (200 miles) northwest of Hanford across the Cascade Range. A logical candidate to head the aquatic effort was Dr. Lauren R. Donaldson, then an Assistant Professor of Fisheries. Donaldson was contacted by Dr. Stafford L. Warren, Chief of the Medical Section for the Manhattan Project.

On August 21, 1943, a proposal was tendered by the National Defense Committee. Donaldson was asked to undertake special studies dealing with the effects of radioactivity on aquatic organisms, especially fish. The basis for these studies could not be stated or described, but they were related to the Columbia River, involved relatively large amounts of radioactivity, and were important to the United States Government. Further, these studies could not be linked with the Columbia River, and they had to be conducted in a normal research setting on the University of Washington campus (Hines 1962; Groueff 1967). The tentative contract fee was $65,000, and the terms were backdated to be effective August 15, 1943. The program was identified as an "Investigation of the Use of X-Rays in the Treatment of Fungoidal Infections in Salmonid Fishes."

The University of Washington group was organized as the Applied Fisheries Laboratory (AFL) to disguise the true nature of the group's work. The initial team included Donaldson, Kelshaw Bonham, Richard F. Foster, and Arthur D. Welander (Figures 5.1 and 5.2). Subsequently, Foster transferred to Hanford and Allyn H. Seymour joined the group. Elsewhere in the United States, other groups were organized to study the effects on mammals of neutron, alpha, beta, and gamma radiation; the ingestion and inhalation of fission products; and related radiobiological phenomena.

The AFL worked only with X-radiation, or X-rays. At that time no artificial radionuclides were available in sufficient quantity for experimental work. Furthermore, the specific isotopes that might appear in the cooling water from the unique "piles" at Hanford were not known. Research at the AFL initially focused on potential damage from radioactivity introduced into a river, and staff examined the elusive effects of

Fig. 5.1. Individuals helping to establish the Applied Fisheries Laboratory (AFL) at the University of Washington, Seattle; photo was taken April 21, 1944. (Photo contributed by the Laboratory of Radiation Ecology, University of Washington.)
Left to Right, Rear: Dr. William F. Thompson, Director, College of Fisheries; Mr. Gerald B. Talbot, representing the U.S. Bureau of Sport Fisheries and Wildlife; Mr. Hanford Thayer, U.S. Army Corps of Engineers and Liaison Officer to AFL; Dr. Richard F. Foster, AFL.
Left to Right, Front: Dr. Stafford L. Warren, Chief, Medical Section of U.S. Army Corps of Engineers; Mr. Francis W. Bishop, University of Rochester radiation technician who supervised installation of X-ray equipment at AFL; Mrs. Marie K. Beach, Secretary for the AFL; Dr. Kelshaw Bonham, AFL; Dr. Lauren R. Donaldson, Director, AFL.

low-level radiation (Hines 1962). Most important, the research centered on fish and other cold-blooded aquatic forms.

Those involved soon realized that observations in the Columbia River at the location of cooling water discharges were also important. As a result, du Pont de Nemours and Company, Inc. (du Pont) opened a small, onsite facility at Hanford. Foster transferred to du Pont to direct the effort, while staff at the AFL served as advisors. As a result, an Aquatic Biological Laboratory was opened during June 1945 in the 100-F Area on the Hanford Site.

Fig. 5.2. Scientists at the Applied Fisheries Laboratory who conducted studies on radiation effects related to cooling water discharges at Hanford in the 1940s. Clockwise, upper left: Dr. Lauren R. Donaldson, Director; Dr. Kelshaw Bonham; Dr. Allyn H. Seymour; Dr. Arthur D. Welander. (Photos contributed by the Laboratory of Radiation Ecology, University of Washington.)

Two years lapsed between the time the AFL was established and the end of World War II in August 1945. Few studies were completed in that brief span. Results from the first investigations were not published until 1947 and 1948. Studies then under way at the AFL included the effects of X-rays on adult, eyed embryos, and fingerling chinook salmon; adult, embryos, and fingerling rainbow trout; and snails, algae, and crustacea.

After World War II, the AFL's connection with the Manhattan Project and production of plutonium at Hanford was acknowledged. The University of Washington's AFL group was renamed the Laboratory of Radiation Biology to accurately reflect its main line of work. The Laboratory continued contract research related to problems of radiobiology and expanded its sphere of investigations. The Manhattan District of the Corps was phased out by the end of 1946, and the new U.S. Atomic Energy Commission, created by the Atomic Energy Act of 1946, assumed responsibility for contracts, facilities, and management of activities at Hanford.

Studies with X-radiation

The AFL first examined radiation effects on chinook salmon (Figure 5.3) in October 1943, almost a year before the first reactor at Hanford began operation. Years of observations would be required, however, corresponding to the extended life cycle of an anadromous fish, before the work would be complete. Adults from the first experimental groups would

Fig. 5.3. Adult, sexually mature, female chinook salmon from the Green River Hatchery, Auburn, Washington. Fertilized eggs from this hatchery were used in initial studies at the Applied Fisheries Laboratory.

not return from the sea until 1945, 1946, or 1947. In retrospect, the initial choice was unfortunate. Radiobiological data applicable to Columbia River fish were urgently needed, and salmonoids with a shorter, more direct life cycle were available (e.g., rainbow trout). The AFL soon initiated other studies to help define the effects of radioactivity in aquatic environments.

Initial studies used a Picker-Waite radiation therapy machine, with a capacity of 25 mega-amps at 200 kilovolts, to provide experimental exposures (Donaldson 1945). The use of X-rays to provide ionizing radiation was taken for granted, not only because exposures could be measured and controlled, but because other types of ionizing radiation were unavailable. An X-ray is a form of electromagnetic energy with considerable penetrating power. X-rays are usually formed in an electrical device, while gamma rays are emitted by unstable or radioactive isotopes (Hall 1984).

The Laboratory of Radiation Biology (1951–1966) continued investigations related to releases of radioactivity at Hanford after the end of World War II. However, its major research emphasis shifted to other topical areas (Anonymous 1963). Today, the University of Washington group functions as the Laboratory of Radiation Ecology. Major projects in which the group was involved through the 1950s were

- Pacific Studies – Investigation and documentation of the residual radiobiological developments at nuclear bomb test sites, and at Rongelap atoll in the Marshall Islands.
- Project Chariot Studies – Completion of marine biological inventories and radiobiological analyses in the Chukchi Sea off the northwest coast of Alaska.
- Columbia River Studies – Investigations in the Columbia River estuary and along the neighboring coastlines of the deposition of radionuclides transported downriver from Hanford to the Pacific Ocean.
- Fern Lake Studies – Investigation, with the use of short-lived radionuclides, of the biological patterns and nutritional cycles in a Washington lake and its watershed.
- Low-Level Irradiation Studies – Examination of genetic effects in successive generations of salmon after exposing eggs during embryogenesis to chronic, low-level radiation.

"Columbia River Studies" at the AFL in early years involved the effects of X-rays on various development phases of salmonids. Phases exposed included gametes (Foster et al. 1949), eyed eggs (Welander et al. 1948), various phases of embryogenesis (Welander 1954), fingerlings (Bonham et al. 1948), and adults (Welander et al. 1949). Later work included exposures to radiation from a cobalt source (Donaldson and Bonham 1964, 1970; Bonham and Donaldson 1966).

The initial decision to investigate radiation effects in developing phases of salmonids was, in many ways, fortuitous. Scientists concluded three decades later that the eggs and young of some species of teleost fish were the most sensitive of aquatic organisms to ionizing radiation (IAEA 1976).

Effects of X-Rays on Fingerling Chinook Salmon

The first study conducted at the AFL to appear in scientific literature was on the acute effects on young fish exposed to relatively high doses of X-rays. It established a foundation for further studies with chronic effects involving lower, more subtle exposures.

Eight groups of fingerling chinook salmon were exposed to 0 (control), 100, 250, 500, 750, 1000, 1250, 2500, and 5000 roentgens (R) (Bonham et al. 1948). Twelve weeks after exposure, the doses above which significant effects occurred were 250 R for mortality, 500 R for weight, and 1000 R for length. The number of blood cells in circulation significantly declined two and three weeks after irradiation in groups of fish exposed to 750 and 1250 R. Concentration of hematopoietic cells in the kidney declined in a similar manner in the group exposed to 750 R 1 and 2 weeks after irradiation. [a]

Effects of X-Rays on Embryos and Alevins of Chinook Salmon

Another early study examined the effects of acute radiation on fertilized eggs taken from a salmonid. Data were obtained on mortality and appearance of abnormalities as the eggs hatched and young developed.

Eyed eggs of chinook salmon were exposed to X-rays at a rate of 37.2 roentgens per minute (R/min), accumulating in single doses of 0 (control), 250, 500, 1000, 2500, 5000, and 10,000 R. Control and exposed groups were observed for 125 days. Analysis of effects was based on gross pathology,

---

[a] Radiation units in early studies with ionizing radiation were sometimes measured with less precision than is available today. Radiation doses in early studies were often measured in terms of R; however, for most penetrating photons, 1 R nearly equals 1 rad and 1 rem, which are units used most often in more recent studies (Kathren 1984). The methods used to expose aquatic organisms and to measure radiation are described by the researcher when reporting results.

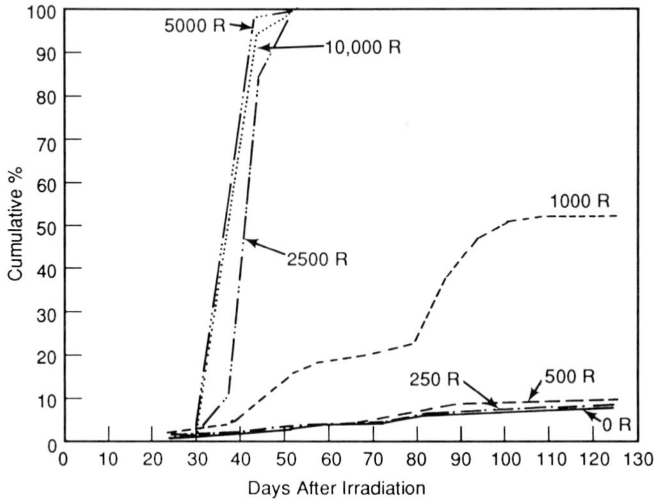

Fig. 5.4. Cumulative mortality of young chinook salmon exposed during embryogenesis to X-irradiation at total doses ranging from 0 (control) to 10,000 roentgens (R) (from Welander et al. 1948).

mortality, growth, development of various body structures, and histopathology (Welander et al. 1948).

Doses of 2500 R and above were 100% lethal. However, the onset of mortality was delayed four weeks until after the eggs hatched, and degenerative symptoms and death did not occur until 30 to 51 days after exposure (Figure 5.4). Doses of 1000 R retarded the development of cutaneous pigment, vascular system, fins, eyes, and body length and weight, and eventually caused 51% mortality. Doses of 500 R slightly retarded growth and pigment formation. Cumulative losses among groups exposed to 250 and 500 R, the low-dose groups, were slightly higher than among controls.

Tissue studies (histopathology) showed that the gonads were most sensitive to X-rays, followed by the hematopoietic tissues (kidney and spleen). Development of cells in hematopoietic tissues was temporarily retarded after exposures of 250, 500 and 1000 R, roughly in proportion to dose. Blood-forming cells were destroyed at 2500 R and above.

Effects X-Rays on Embryos and Young from Adult Rainbow Trout

Rainbow trout, 20 months old and containing partially mature eggs and sperm, were exposed to X-rays. Rainbow trout were used because they mature faster than anadromous salmon and can be held in fresh water their entire life.

Subsequently, the survival, teratology, and growth of fertilized eggs and hatched alevins were monitored. Adult fish were exposed at 8.25 R/min to give cumulative, whole-body doses of 0 (control), 50, 100, 500, 750, 1000, 1500, and 2500 R. After exposure, the fish were reared until sexually mature, when their eggs were stripped and fertilized. The combined eggs and sperm were from fish receiving identical exposures to radiation (Foster et al. 1949).

Some adult fish exposed to 2500 R had internal damage and died before they spawned. As the amount of radiation received by the parents increased, the degree of development attained by the embryo decreased. Mean egg mortalities were significantly greater when parents had been exposed to 500 R or more. Most eggs lost during incubation contained abnormal embryos.

No characteristic abnormalities were associated with radiation. The types of abnormal embryos among progeny from control fish were almost

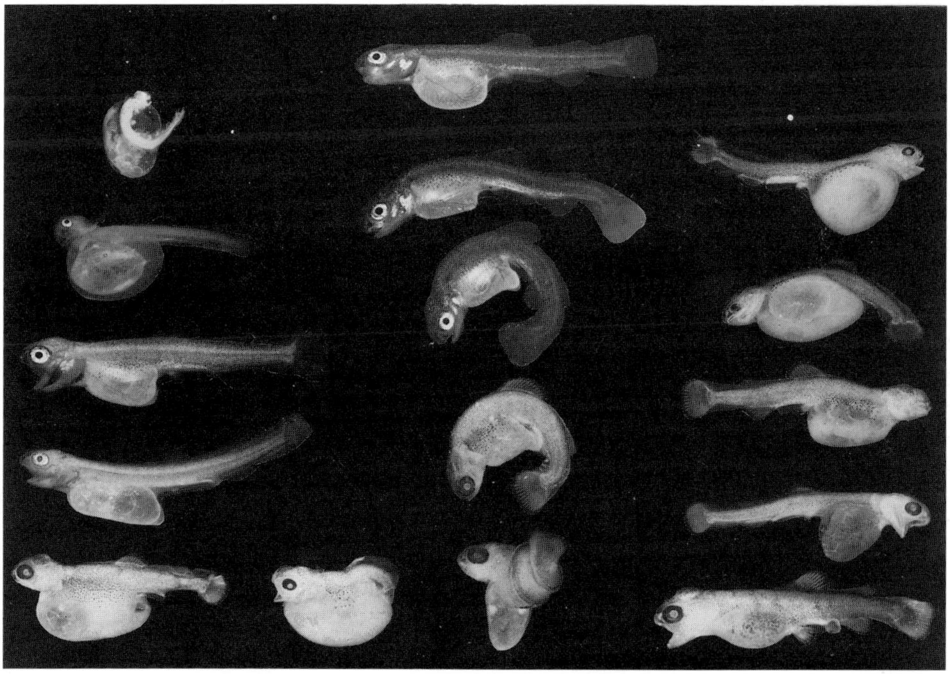

Fig. 5.5. Abnormal embryos of rainbow trout resulting from exposing sexually mature fish to excessive (500 or more roentgens) X-irradiation. (Photo contributed by the Laboratory of Radiation Ecology, University of Washington.)

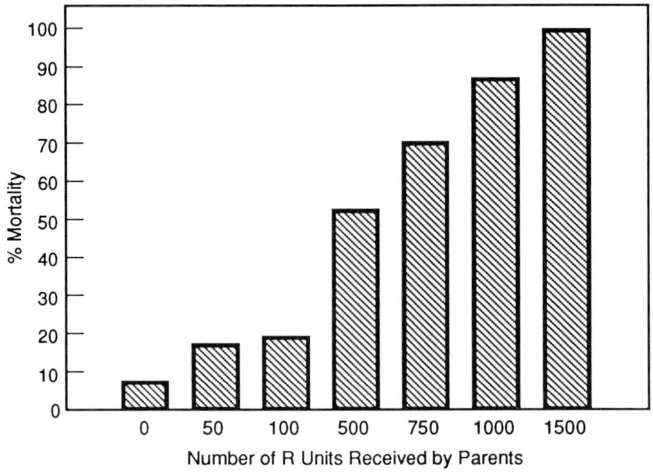

Fig. 5.6. Cumulative mortality of young rainbow trout reared from parents exposed to X-irradiation at total doses ranging from 0 (control) to 1500 roentgens (R) (from Foster et al. 1949).

identical to those from parents exposed to either low or high radiation. However, the number of abnormal embryos increased, and the stage of development attained decreased, as radiation dose increased (Figure 5.5). Almost all embryos from parents exposed to 1500 and 2500 R were so abnormal they died before the egg blastopore closed.

During the alevin stage, mortality in every group from exposed parents, even those receiving only 50 R, were significantly greater than in controls (Figure 5.6). Increased mortality rates persisted as long as 6 months after the eggs hatched in groups where exposure to adults was 500 R or more.

Growth of young trout during their first year was affected by the irradiation received by their parents. Adults exposed to as low as 100 R produced offspring with slightly retarded growth, while adults exposed to 500 R produced offspring with appreciably slower growth.

Effects of X-Rays on Adult Rainbow Trout

A study where adult rainbow trout were exposed to acute doses of X-rays filled another gap. Nearly mature rainbow trout (about 21 months old) were continuously exposed at 8.25 R/min to provide total, whole-body doses of 0 (control), 50, 100, 500, 1000, 1500, and 2500 R. Postexposure monitoring examined mortality, growth, and gross pathology (Welander et al. 1949).

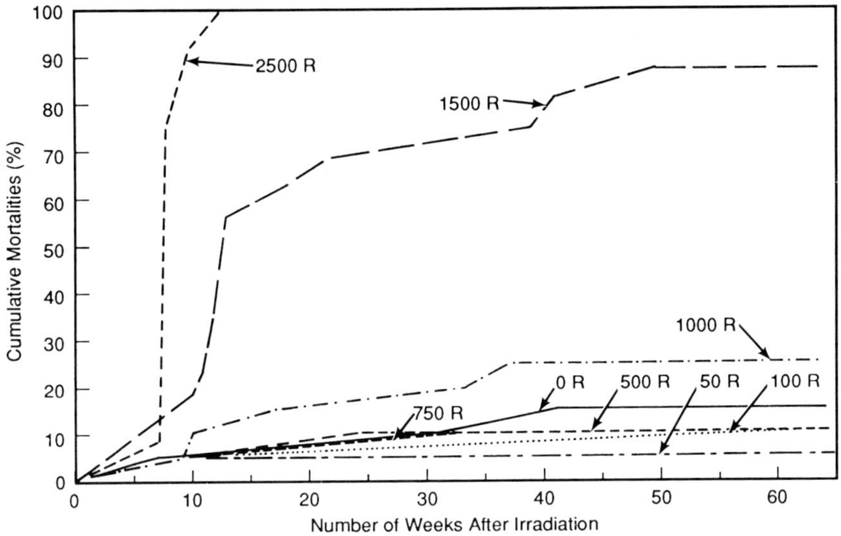

Fig. 5.7. Cumulative mortality of adult rainbow trout exposed to direct whole-body X-irradiation at total doses ranging from 0 (control) to 2500 roentgens (R) (from Welander et al. 1949).

Mortality was significantly higher among fish exposed to 1000 R or more. Doses of 2500 R killed all fish in 13 weeks, with most dying during the eighth week. Doses of 1500 R killed 56% of the fish in 13 weeks and 87% in 64 weeks (Figure 5.7). Mortality among the controls was 15%. All autopsied fish exposed to 500 R or more suffered radiation injuries, which were generally in proportion to radiation dose. Gross pathology included hemorrhage, petechiae, necrosis, appearance of fungi, and intestinal damage. Growth was retarded in all exposed fish, even at exposures as low as 50 R, in comparison with the controls.

Effects of X-Rays on Snails, Crustacea, and Algae

The relative sensitivity of selected snails, crustacea, and algae to acute doses of ionizing radiation was important to evaluating effects in aquatic systems. Furthermore, the factors contributing to the lethal action of radiation among these organisms needed evaluation.

Test organisms were obtained from localities not necessarily related to the Columbia River at Hanford. They included freshwater snails (*Lymnaea* or *Radix* sp.), marine snails, saltwater crustacea (*Artemia, Calliopius*, and *Allorchestes*), and freshwater algae (*Chlorella, Ankistrodesmus, Chroococcus*, and *Synechococcus*). The common goal was to determine

Table 5.1.
Relative Sensitivity of Algae, Snails, and Crustacea to Ionizing Radiation (in roentgens) (after Donaldson and Foster 1957).

Group	50% Kill,R	100% Kill,R	Latent period
Algae	8,000–100,000	25,000– 600,000	45days
Protozoa [(a)]	10,000–300,000	18,000–1,250,000	45 min to 40 days
Molluscs	5,000– 20,000	10,000– 50,000	45 min to 40 days
Crustacea	500– 90,000	10,000– 80,000	5 to 80 days
Fish [(a)]	600– 3,000	370– 20,000	14 to 460 days

[(a)] Based on the cited reference, data are included in the table to expand coverage of taxonomic groups. Circumstances under which experiments were conducted, particularly in relation to irradiation source, and variability in the response of different individuals may well account for the wide mortality ranges.

levels where 50% [Lethal $Dose_{50}$ (LD_{50})] of the organisms died (Bonham and Palumbo 1951), but actual exposures to X-rays (both hard and soft types) varied with test species and conditions.

Algae were the most resistant group (Table 5.1). Algae and invertebrates were generally more tolerant of ionizing radiation than fish. The mortality decreased as exposure time shortened. This study demonstrated that similar groups of aquatic organisms differ widely in their susceptibility to radiation. Further, doses causing harm to one species will not necessarily harm a different species.

Effects of X-Rays on Trout During Embryogenesis

The relative sensitivity of six embryonic stages of rainbow trout to X-rays was determined. Because ionizing radiation penetrates all organs and tissues, some structures appearing during early development of fish might be more sensitive than others. The period or periods during embryogenesis when radiation was most damaging might be crucial.

Developing trout embryos were separated into six arbitrary prehatch stages (#1 to #6) ranging from one cell to hatching. Each stage was exposed to graded doses of radiation, ranging from 0 (control) to 2570 R, at an intensity near 133 R/min. Effects were evaluated at the time the embryos hatched (Welander 1954).

On the basis of cumulative mortality, the embryos in the one-cell stage (#1) were most sensitive to radiation, with half being killed by exposure to 58 R (LD_{50}) at the time the yolk sac was absorbed. Mortalities generally decreased as the embryos developed (LD_{50} from 300 to 900 R).

Growth was retarded during the early eyed stage (#5) at doses as low as 38 R, but other stages seemed unaffected at doses less than 200 R. The number of parr marks appearing in hatched fry was usually reduced in all six prehatch stages after doses of 300 R or more. The number of fin rays was reduced by doses from 75 to 100 R in the 32-cell (#2), late germ-ring (#4), and early eyed (#5) stages.

Effects of X-Rays on Embryonic Snails

The effects of X-rays on early cleavage stages of the freshwater snail were explored. Previous studies at the AFL had indicated that adult snails were more resistant to ionizing radiation than embryo snails (Bonham and Palumbo 1951).

Embryonic snails develop more rapidly, in about 10 to 11 days, than do salmonid eggs. Thus, the time that early cleavage stages are selected for exposure is more critical. Recently deposited egg masses were irradiated and then photographed periodically to quantify development (Bonham 1955).

Sensitivity of snail embryos depended on the particular mitotic stage being irradiated. On the basis of mortality to hatching, eggs undergoing mitosis were more sensitive (half killed at 100 R) than eggs in the resting stage (half killed at 300 to 400 R). Later embryonic stages (trochophore through early shelled embryo) were even less sensitive (half killed at doses between 500 and 1000 R).

Studies with cobalt-60 gamma rays

In the 1960s, the Laboratory of Radiation Biology (formerly the AFL) at the University of Washington diversified its research dealing with freshwater ecosystems. Initially, the Laboratory had examined the radiosensitivity of salmonid eggs and juveniles by exposing them to relatively strong doses of X-rays (acute effects). But little information had been obtained on effects of low-level exposures (chronic effects) produced in a mildly radioactive river, such as the mainstem Columbia downstream from Hanford in the 1960s.

The appearance of worldwide fallout from atmospheric tests with atomic devices added to the need for continued study. Gamma irradiation from a Co-60 source, rather than X-rays, was used in exposures. The tests were designed to cover the entire life cycle of an anadromous fish – from the time eggs and juveniles were exposed in fresh water, through their sojourn in the sea, to their return as maturing adults.

Effects of Chronic Irradiation on Embryogenesis of Salmon

In late 1960, fertilized eggs of chinook and coho salmon were exposed continuously during embryogenesis to low-level gamma irradiation from radioactive cobalt at 0.5 R daily. Fingerling fish produced by eggs of irradiated and control groups were released to migrate to the sea. Adult fish were examined when they returned to spawn in 1962 and subsequent years.

Irradiation doses totaled 33 to 37 R for chinook salmon and 40 R for coho salmon. Exposure was planned from in situ measurements (Foster and Nelson 1961) that gave an estimate of about 10 picocuries per milliliter (pCi/mL) from the combined activity of the five common radionuclides in Columbia River at the city of Pasco. At this level of exposure, total dose to young fish by the time they migrated to the sea as smolts from the Hanford Reach was probably less than 1 R. Compared with a normal background radiation of 0.02 milliroentgen (mR) per hour (before the onset of weapons testing), an experimental dose of 20 mR per hour (0.5 R per day) was 10,000 times greater.

Effects on juvenile salmon were assessed by comparing survival and growth rates, number of vertebrae, vertebrae abnormalities, opercular effects, and sex ratios. The first report gave results up to the time the young chinook and coho salmon were released (Donaldson and Bonham 1964).

No significant radiation effect was demonstrated. Control fish proved superior in some values measured, exposed fish superior in others. Opercular defects were slightly higher among exposed fish. With both chinook and coho salmon, mortality was greater at times among control fish than among exposed fish, yet accumulative losses were similar in both groups at liberation.

In later years, adult chinook and coho salmon returning to ponds at the Laboratory were assessed to determine the effects of irradiation during embryogenesis. Twenty-five features were compared in control and exposed groups, with the number of fish and their fecundity given greatest importance. The second report gave data from returns in 1962, 1963, 1964, and 1965 (Bonham and Donaldson 1966). Additionally, some eggs from the returning adults were exposed during embryogenesis to an even greater dose of cobalt gamma irradiation, totaling 95 R.

Exposure to gamma irradiation during embryogenesis at 0.5 R daily did not impair the reproductive capacity of adults over one generation. Although abnormalities were slightly higher among young fish from the exposed group (first report), adult survival was not impaired, nor did abnormalities appear among returning adults (second report). On the contrary, irradiated fish returned in greater numbers and produced more viable eggs than the controls.

When fertilized eggs from some adult returnees were irradiated at a 2.5-time greater dose, there was no increase in mortality or abnormalities among the smolts, nor was their growth retarded.

Follow-Up Studies with Chronic Irradiation and Young Salmonids

Because the Columbia River was so important for producing large numbers of Pacific salmon and steelhead, studies involving radiation and early development of salmonids continued into the 1970s. An estimated dose of less than 1 R in young fish from the Columbia River at Pasco remained the standard for experimental exposures. Irradiation was provided by a Co-60 gamma source.

During successive years, large numbers of chinook salmon eggs and alevins were irradiated at different levels each day through embryogenesis, from the time of fertilization until the yolk sac was absorbed. Over this span, exposures of successive broods increased from 0.5 roentgen per day (R/day) in 1960 to 1.3 R/day in 1965, 2.8 R/day in 1966, 5.0 R/day in 1967, and 10.0 R/day in 1968. Some eggs and alevins from adults irradiated during embryogenesis in preceding tests were irradiated again. Results were compared between experimental and control groups when adults returned from the ocean years later.

The damage noted previously in studies of acute irradiation exposures was not observed among juvenile salmon when chronic exposures nearly doubled from 1.3 to 10.0 R/day (total dose of 820 R). In fact, gamma-irradiated eggs at 0.5 R/day produced adults that returned in greater numbers and mature females that produced more viable eggs (Donaldson and Bonham 1970).

In 1969 and 1970, eggs were exposed during embryogenesis to Co-60 irradiation at doses up to 50 R/day. Results from examining the gonads of premigratory smolts in all eight experiments were reported separately (Bonham and Donaldson 1972). Sex ratios up to and including the 1967 brood that received a dose of 5 R/day were not effected by irradiation. However, gonad development was abnormal in most of the 1968 brood exposed at 10 R/day, and in the 1969 and 1970 broods exposed to 50 R/day. No radiation effect was noted at doses up to 10 R/day.

Returns of adult salmon through 1977 did not change significantly until exposures reached 10 or more R/day (Hershberger et al. 1978). At these exposure levels (10, 20, and 17 to 50 R/day), small fish grew slower and had higher mortality before migrating seaward as smolts than did controls. Further, numbers of adults returning to spawn were lower than the controls, their ages were greater, and some adult males appeared to be sterile. The dose-response relationship was not linear. There seemed to be a threshold with an exponential response with higher doses of gamma irradiation.

The Laboratory's studies revealed no effect from radiation on young salmon at levels considerably above those actually present in the Columbia river below Hanford during the 1960s (Templeton et al. 1971). In fact, significant effects did not appear until exposures were about 800 times the calculated maximum dose that young salmon would receive before migrating seaward (Hershberger et al. 1978).

Significance of the university effort

Studies at the AFL and its successors contributed greatly to clarifying the effects of penetrating radiation on aquatic organisms, particularly on salmonid fish. Exposures resulting in adverse effects were considerably higher than those in the Hanford Reach from radionuclides in cooling water discharges of the production reactors. But findings at the AFL went far beyond determination of lethal limits from radiation exposure (Donaldson and Foster 1957). Other aspects of radiation biology were explored that proved equally important to assessing radiation effects in river ecosystems. Significant generalizations emerged.

Relative Sensitivity of Taxonomic Groups

Lower or more primitive forms of aquatic organisms were usually more resistant to ionizing radiation than higher, more complex forms such as vertebrates (Table 5.1). The algae and protozoa were most resistant, with LD_{50} values at many thousand roentgens. Mollusks and crustacea were somewhat more sensitive, with LD_{50} values at a few thousand R. Fish were the most sensitive, with LD_{50} values at about 1000 R.

Relative Sensitivity of Development Stages

Sensitivity of salmonids to ionizing radiation decreased as they grew older. Approximate values for 50% mortality were 50 to 100 R for gametes (eggs and sperm) within mature rainbow trout (Foster et al. 1949), 1000 R for fertilized and developing eyed eggs of chinook salmon (Welander et al. 1948), 1250 to 2500 R for fingerling chinook salmon (Bonham et al. 1948), and 1500 R for adult rainbow trout (Welander et al. 1949). Furthermore, eggs of rainbow trout were more sensitive to irradiation during the initial, one-cell stage than during later phases of embryogenesis.

Aquatic snails also differed in radiosensitivity with age. Eggs of freshwater snails were more susceptible than adults (Bonham and Palumbo 1951). Eggs of a marine snail undergoing mitosis were two to four times more sensitive to radiation than eggs in the one- and two-cell stage, and later embryonic stages were even less sensitive (Bonham 1955).

Retardation of Development by Irradiation

Low-level exposures showed the potential to reduce growth rates of aquatic organisms significantly. For example, groups of aquatic snails (Bonham and Palumbo 1951) and young salmonids (Welander et al. 1949) displayed slower growth at certain levels of irradiation than did control groups. Retarded growth was a very sensitive measurement of the effects of X-rays on fish and often varied in proportion to exposure dose. Effects on growth were not confined to exposed fish, but also could be reflected in their progeny (Foster et al. 1949).

Growth of rainbow trout fingerlings was retarded when eggs were irradiated during early development (Welander 1954). Young fish from eggs irradiated during the 32-cell, late germ-ring, and early eyed stages often displayed abnormally large heads and eyes. Further, the number of parr marks was reduced by doses of 300 R or more, and the number of dorsal and anal fin rays was reduced by doses of 75 to 100 R.

Pathology of Radiation Damage

Acute exposures of rainbow trout to 1500 and 2500 R of whole-body irradiation produced typical radiation symptoms of mass hemorrhage, petechiae, and ecchymosis. Exposures to 1000 or more R produced muscular hemorrhage, exposures to 750 R produced hemorrhage in the peritoneum lining the body cavity, and exposures to 500 R produced hemorrhage in the gonads (Welander et al. 1949).

Gross abnormalities among young rainbow trout unexposed and exposed to X-rays during embryogenesis were similar. But abnormalities were usually more numerous among exposed fish, with the exception of anomalies in the dorsal and adipose fins produced by 200- and 400-R exposures of 32-cell embryos (Welander 1954). The number of cells was reduced and development was retarded, roughly in proportion to dose, in hematopoietic tissue of the anterior kidney of chinook salmon reared from eggs exposed to 250, 500, and 1000 R of X-rays. The number of primordial germ cells in the gonads of chinook salmon was greatly reduced after eggs were exposed to 250 R (Welander et al. 1948).

Generally, radiation effects in the tissues of fish were similar or identical to effects produced in tissues of nonaquatic animals. The fish tissues most sensitive to radiation damage were those undergoing rapid division and growth. Dividing gonadal and hematopoietic tissues were many times more sensitive than other tissues that divided less rapidly (Donaldson and Foster 1957). An explanation might be that different organ systems become more actively dividing and, consequently, more sensitive to ionizing radiation at different stages of development.

References

Anonymous. 1963. *The Twentieth Year*. Laboratory of Radiation Biology, University of Washington, Seattle.

Bonham, K. 1955. "Sensitivity to X-Rays of the Early Cleavage Stages of the Snail *Helisoma subcrenatum*." *Growth* 19:1–18.

Bonham, K., and L.R. Donaldson. 1966. "Low-Level Chronic Irradiation of Salmon Eggs and Alevins." In: *Disposal of Radioactive Wastes into Seas, Oceans and Surface Waters*, pp. 861–883. International Atomic Energy Agency, Vienna, Austria.

Bonham, K., and L.R. Donaldson. 1972. "Sex Ratios and Retardation of Gonadal Development in Chronically Gamma-Irradiated Chinook Salmon Smolts." *Trans. Am. Fish. Soc.* 101:421–434.

Bonham, K., L.R. Donaldson, R.F. Foster, A.D. Welander, and A.H. Seymour. 1948. "The Effect of X-Ray on Mortality, Weight, Length, and Counts of Erythrocytes and Hematopoietic Cells in Fingerling Chinook Salmon, *Oncorhynchus tshawytscha* Walbaum." *Growth* 12:101–121.

Bonham, K., and R. Palumbo. 1951. "Effects of X-Rays on Snails, Crustacea, and Algae." *Growth* 15:151–188.

Donaldson, L.R. 1945. *Equipment and Procedures Used in the Study of the Effects of Irradiation of Fish with X-Rays*. Report UWFL 1, Laboratory of Radiation Biology, University of Washington, Seattle.

Donaldson, L.R., and K. Bonham. 1964. "Effects of Low-Level Chronic Irradiation of Chinook and Coho Salmon Eggs and Alevins." *Trans. Am. Fish. Soc.* 93:331–341.

Donaldson, L.R., and K. Bonham. 1970. "Effects of Chronic Exposure of Chinook Salmon Eggs and Alevins to Gamma Radiation." *Trans. Am. Fish. Soc.* 99:111–119.

Donaldson, L.R., and R.F. Foster. 1957. "Effects of Radiation on Aquatic Organisms." In: *The Effects of Atomic Radiation on Oceanography and Fisheries*, Publ. No. 551, pp. 91–102. National Academy of Sciences, National Research Council, Washington, D.C.

Foster, R.F., and I.C. Nelson. 1961. *Evaluation of Radiological Conditions in the Vicinity of Hanford, April–June 1961.* HW-70552, U.S. Atomic Energy Commission Report, Office of Technical Services, U.S. Department of Commerce, Washington, D.C.

Foster, R.F., L.R. Donaldson, A.D. Welander, K. Bonham, and A.H. Seymour. 1949. "The Effect on Embryos and Young of Rainbow Trout from Exposing the Parent Fish to X-Rays." *Growth* 13:111–142.

Groueff, S. 1967. *Manhattan Project. The Untold Story of the Making of the Atomic Bomb.* Little, Brown, and Company, Boston.

Groves, L.R. 1962. *Now It Can Be Told. The Story of the Manhattan Project.* Da Capo Press, Inc., New York.

Hall, E.J. 1984. *Radiation and Life.* 2nd Ed. Pergamon Press, New York.

Hines, N.O. 1962. *Proving Ground. An Account of the Radiobiological Studies in the Pacific, 1941–1961.* University of Washington Press, Seattle.

Hershberger, W.K., K. Bonham, and L.R. Donaldson. 1978. "Chronic Exposure of Chinook Salmon Eggs and Alevins to Gamma Irradiation: Effects on Their Return to Freshwater as Adults." *Trans. Am. Fish. Soc.* 107:621–631.

International Atomic Energy Agency (IAEA). 1976. *Effects of Ionizing Radiation on Aquatic Organisms and Ecosystems.* Technical Report Series No. 172, International Atomic Energy Agency, Vienna, Austria.

Kathren, R.L. 1984. *Radioactivity in the Environment: Sources, Distribution, and Surveillance.* Harwood Academic Publishers, New York.

Templeton, W.L., R.E. Nakatani, and E.E. Held. 1971. "Radiation Effects." In: *Radioactivity in the Marine Environment*, pp. 221–239. National Academy of Science, Washington, D.C.

Welander, A.D. 1954. "Some Effects of X-Irradiation of Different Embryonic Stages of the Trout (*Salmo gairdneri*)." *Growth* 18:221–255.

Welander, A.D., L.R. Donaldson, R.F. Foster, K. Bonham, and A.H. Seymour. 1948. "The Effects of Roentgen Rays on the Embryos and Larvae of the Chinook Salmon." *Growth* 12:201–242.

Welander, A.D., L.R. Donaldson, R.F. Foster, K. Bonham, A.H. Seymour, and F. G. Lowman. 1949. *The Effects of Roentgen Rays on Adult Rainbow Trout.* UWFL-17, Applied Fisheries Laboratory, University of Washington, Seattle, Washington.

Williams, R.C. and P.L. Cantelon, eds. 1984. *The American Atom. A Documentary History of Nuclear Policies from the Discovery of Fission to the Present 1931–1984.* University of Pennsylvania Press, Philadelphia.

Chapter 6

SETTING FOR BIOENVIRONMENTAL STUDIES IN THE HANFORD REACH, 1945 TO 1971

The startup of the first single-purpose reactor at Hanford in 1944 presented a unique situation. Never before had a large variety of artificial radionuclides been created artificially and discharged to an aquatic ecosystem. Their presence in the cooling water effluent, distribution and fate in the Columbia River, and the eventual passage of the long-lived components to the Pacific Ocean all required long-term radioecological investigations.

Initial studies soon evolved to include not only radioactivity in fish but also in plankton, benthic invertebrates, river sediments, and pathways leading to humans. Scientists also investigated the effects of heat and chemicals in reactor effluent on fish. Eventually, several integrated studies were conducted in the laboratory and field, and ecological functions were explored in lower portions of the Columbia River and in coastal areas of Oregon and Washington.

Opportunities

Many different approaches were taken to examine bioenvironmental effects during the years the single-purpose reactors operated at Hanford. Research findings from 1945 to 1971 are reviewed under the following topics:

- reactor effluent monitoring (Chapter 7)
- field studies with radioactivity (Chapter 8)
- laboratory studies with radioactivity and aquatic organisms (Chapter 9)
- thermal effect studies in the Hanford Reach (Chapter 10).

Integration of field and laboratory research at Hanford was important. Frequently, field studies revealed ecological mechanisms that needed to be explored and quantified under controlled laboratory conditions, and vice versa. On the other hand, data obtained in the laboratory were not always valid for field conditions where the functional ecological system is more complex. Yet laboratory studies were essential to show that actual dose rates from radioactivity in the Columbia River and exposure to process chemicals and heat were not apt to cause adverse effects.

Many potential applications of artificial radionuclides in aquatic ecological studies were soon recognized. Because the chemical properties of radioactive and nonradioactive elements are similar, the artificial radionuclides could be used as "tracers" to obtain information that could be gathered in no other way. For example, in ecological studies, tracers are used to follow passage of an element through various components of an ecosystem. Therefore, studies that monitored levels of radioactivity in aquatic organisms also provided information on nutrient cycles and metabolic rates and on complex physical, chemical, and biological processes in the river itself.

The species of aquatic organisms used in experimental studies at Hanford varied. Generally, organisms present in the Hanford Reach or in the Columbia River downstream from Hanford were used, particularly endemic species of fish (see Glossary). And often, but not always, they were species most susceptible to environmental stressors such as radioactivity, heat, and chemicals (indicator organisms).

By 1949, onsite studies had shown that the fisheries resources of the Columbia River were not affected by the plutonium-production facilities and that there was no health hazard to people who used the river and its fish (Foster 1972). The limited potential for adverse effects was primarily due to rapid dilution of reactor effluents in the Columbia River, and rapid decay of short-lived radionuclides during downstream transport. But many questions remained that could be resolved only by continued studies.

The eight single-purpose reactors at Hanford reached maximum operation in the 1960s. At the same time, research on thermal effects in aquatic ecosystems was emphasized nationwide. In 1965, Battelle Memorial Institute (Battelle) took over research and development (R&D) responsibilities on the Hanford Site from General Electric Company.

Facilities

An aquatic biology group was organized at Hanford in 1945, with assistance from the Applied Fisheries Laboratory at the University of Washington. A plan for initial aquatic studies at Hanford was outlined on

June 9, 1945, at a meeting held at the University of California (Foster 1946). Participants agreed that the best way to determine the effect of effluent from the reactors was to set up a series of dilutions in which fish could be reared. This effort and corresponding field work began in 1946. Under the direction of Dr. Richard F. Foster, [a] this small pioneering group was the first to actually examine in situ the effects arising from an atomic energy installation on a river ecosystem.

Initial laboratory and field research was funded by the Manhattan Project, U.S. Army Corps of Engineers (Corps), which was overseeing all activities at the Hanford Works. After 1947, this effort continued with support from the U.S. Atomic Energy Commission (AEC) and its successors, the U.S. Energy Research and Development Administration (ERDA) and the U.S. Department of Energy (DOE).

Rearing facilities and "wet labs" were established in the 100-F Area. The first aquatic biology building was the 146-F Hut with 1280 square feet of floor space and 20 wooden troughs for holding fish (Figure 6.1). This structure was replaced by the 146-FR Building, a better and larger facility, in 1952 (Figures 6.2 and 6.3). The 146-FR Building was destroyed by fire on November 1964, but it was soon rebuilt (Figures 6.4 and 6.5). Facilities in the 100-F Area for aquatic studies were dismantled after 1971, when the staff moved to new buildings in the 300 Area just north of Richland.

---

[a] Richard F. Foster had a long and distinguished career at Hanford. He arrived at Hanford in 1945 as a fisheries graduate from the University of Washington to plan and direct the first aquatic studies on the effects of discharges from the prototype reactors, especially radioactivity (see Chapters 7, 8, and 9). In July 1979, Foster reminisced about those early years: "Nuclear energy and plutonium production were creating a whole new family of environmental concerns that had to be resolved. Everything was brand new. The first piles (reactors) had just started and, because of the war, red tape was nonexistent. You could map out a project, get the equipment, run the tests and issue a report all in a matter of weeks. It was an exciting time."

Foster's career accomplishments at Hanford were many and varied. In 1960, he transferred to the Radiation Protection Group under General Electric Company to set up an Environmental Evaluations Program. He was later asked to head the new Earth Sciences Section at PNL, and he directed initial environmental evaluations for nuclear power plants in relation to the National Environmental Policy Act. Foster, who advanced to the rank of Scientist V at PNL, also served on advisory and special task groups for the National Academy of Sciences, the U.S. Public Health Service, the State of Washington, and the International Atomic Energy Agency. He was also a consultant to the NRC's Advisory Committee on Reactor Safeguards. Foster became a member of the National Council on Radiation Protection and Measurements in 1969, and was elected to its board of directors in 1979.

Fig. 6.1. The first aquatic "wet laboratory," 146-F Hut, was used at Hanford from 1945 to 1952. The photo shows the wet laboratory on the right, a botany laboratory on the left, and rearing ponds for fish in the center foreground (contributed by R.F. Foster).

Briefly in 1964 and 1965, effluent monitoring was conducted in a small "wet lab" at the 1706-KE Building, 100-K Area. This laboratory was planned in the event the original reactors (including F Reactor) would be shut down, eliminating the source of effluent to the larger 146-FR lab. At the 1706-KE lab, heated effluent was diverted from the heat exchange pit on the KE Reactor discharge pipeline to a head tank in the laboratory. The effluent was then cooled or mixed with raw river water, as required. This temporary facility contained eight, 5-foot-long troughs for rearing fish and had a few small outdoor ponds.

Personnel

In the 1950s, the group directed by Foster was formally named the Aquatic Biology Section. Staff scientists included Calvin L. Cooper,

Fig. 6.2. The 146-FR Building, 100-F Area, was built in 1952 at Hanford to replace the 146-F Hut as a facility for studies with aquatic organisms related to discharges from the single-purpose reactors.

Raymond W. Coopey, Jared J. Davis, Philip A. Olson, Jr., Roy E. Nakatani, Clair C. Palmiter, Robert C. Pendleton, Robert H. Schiffman, Donald G. Watson (Figure 6.6), and Robert H. Whittaker. Karl E. Herde, a member of another Hanford group, often collaborated with Foster. The number of supporting research technicians varied, but included Robert G. Genoway, Clarence O'Malley, Albert C. Schroeder, and Carl H. Hemphill. In addition, scientists on temporary assignment from academic institutions and other organizations contributed to various aquatic studies.

The initial aquatic staff (Foster, Olson, and Watson) was part of E.I. du Pont de Nemours, Inc.'s "P-Department," which operated the first reactors for the Corps. Responsibility for managing all activities at Hanford was transferred from the Corps to General Electric Company in 1946. At that time, the aquatic group was reassigned to the Hanford Laboratories Department directed by H. M. Parker. In 1948, the aquatic group was placed under a new Biology Section, which later became part

Fig. 6.3. Interior of the 146-FR Building, also called the Aquatic Ecology Laboratory. Scientist Phillip A. Olson (left) and technician Carl H. Hemphill are transferring experimental fish.

of the Hanford Biological Laboratories Division, and was directed by Harry A. Kornberg for more than 20 years.

In the 1960s, the original group conducting studies in or related to the Columbia River at Hanford underwent many changes. The scientific staff then included C. Dale Becker, Charles C. Coutant, Colbert E. Cushing, John M. Dean, M. Paul Fujihara, David H. W. Liu, Philip A. Olson, Jr., Roy E. Nakatani, Robert H. Schiffman, Mark J. Schneider, John A. Strand, William L. Templeton, and Donald G. Watson. Technicians included Robert G. Genoway, Carl H. Hemphill, Jerry D. Maulsby, Clarence O'Malley, Albert C. Schroeder, and Alan J. Scott. Foster left the Aquatic Biology Section in 1960 to establish a section responsible for determining and reporting the radiation dose received by people living near the Hanford Site.

Battelle assumed responsibility for managing aquatic research programs at Hanford in January 1965. Research and development programs

Fig. 6.4. The 146-FR Building, which contained the aquatic laboratory, was destroyed by fire on November 4, 1964.

on the Hanford Site became more diversified and, environmentally, greater emphasis was placed on basic ecology studies. The ERDA replaced AEC in January 1975. The ERDA, in turn, was replaced by DOE in 1977.

Artificial radioactivity

The most emotional issue related to the Hanford Site since its beginning has always been radioactivity. Even today, radioactivity and its effects remain a predominant and little-understood issue among non-nuclear scientists, public officials, and concerned citizens.

The radioactivity causing concern at Hanford was not natural radioactivity, to which people have always been exposed (Hall 1984; Kathren 1984), but artificial radioactivity arising from the development and use of

Fig. 6.5. Interior of the 146-FR Building after its restoration in 1966. This facility was used until 1971, when all laboratory studies with aquatic organisms were moved from the 100-F Area, 30 miles north of Richland, to the new Life Sciences Laboratory I Building in the 300 Area near Richland.

nuclear power. Radioactivity does not impair water quality or threaten human health in the same way as do organic and inorganic compounds when released to an aquatic ecosystem such as the Columbia River. Adverse effects arise primarily from ionizing radiation in sensitive organs and tissues. The effects of ionization on cells and tissues differ greatly from the effects of toxic compounds.

For the purpose of environmental monitoring, detection and measurement of radioactivity are relatively simple processes compared to detection and measurement of toxic compounds. This is due, in large part, to the early development of instruments that accurately measured radionuclides and the vast amount of R&D that has taken place over 40 years to understand radiation effects. Instruments to detect toxic chemicals in the environment were extremely crude until just recently, and even now

Fig. 6.6. Scientists with the Aquatic Biology Group who conducted bioenvironmental studies associated with cooling water discharges from the single-purpose reactors at Hanford in the 1960s. Clockwise, from upper left: Richard F. Foster, director; Jared J. Davis, and Donald G. Watson.

they remain less sensitive than the instruments used to detect ionizing radiation.

Ionizing radiation imposes the most severe health effects of all radiation. Ionization disrupts the chemical bonds (orbital electrons) of the molecules that make up the cells and tissues of living organisms and damages DNA. The most penetrating types of ionizing radiation artificially produced are X-rays (from electrical devices) and gamma rays (from radionuclides). Both X-rays and gamma rays represent energy packets (photons) transmitted as a wave without any movement of material – just as heat and light from the sun cross space to reach the earth. Furthermore, X-rays and gamma rays are penetrating and can pass through thick barriers. Medical practitioners use X-rays to examine bones or teeth, and X-rays and gamma rays to treat cancer.

Beta and alpha rays are other forms of ionizing radiation. Beta rays can pass through a human hand but, unless of very high energy, they can be stopped by a modest barrier. Alpha rays are less penetrating, and can be stopped by a thin barrier, even a piece of paper. However, significant cellular damage can occur if a beta emitter, such as strontium, or an alpha emitter, such as plutonium, enters the body by inhalation or ingestion. Further, biological effects are enhanced by a radionuclide's chemical affinity for a particular body organ or tissue, such as the affinity of strontium for calcium in bone.

The artificial radionuclides in the cooling effluent of the single-purpose reactors at Hanford were a mix of alpha, beta, and gamma emitters. Appearance of artificial radioactivity in the atomic program served as a catalyst for in-depth studies of the effects of ionizing radiation lasting to this day. Modern scientists know more about the cancer-producing potential of ionizing radiation than about any other environmental carcinogen (Hall 1984). Much of this information began with R&D activities at Hanford and other national laboratories since the start of the "nuclear age."

Measurement of radioactivity in early years at Hanford, even though representing the state of the art, was primitive by today's standards. Many early studies in the Columbia River provided data only on "total alpha," "total beta," or "total activity," the measurements most practical at that time. Now that individual radionuclides can be measured with precision, scientists realize that data from Hanford's formative years were limited in terms of assessing public safety, environmental transfer, or uptake by river organisms. By the same token, early data on radioactivity at Hanford should be refined, reinterpreted, and reassessed under more exacting criteria. Only part of this effort has been possible to date.

Under international agreement, the units used to measure radioactivity for four decades at Hanford were recently replaced by the International System of Units, or SI. The International Commission on Radiological Units and Measurements decreed that SI units for radiology would be introduced in 1980 and used side by side with the old units until 1984, when the old units would be phased out (Hall 1984). The transition caused much confusion, and integration of the new units remains incomplete.

For these reasons, and to avoid errors in transcribing original data, the units used to quantify radioactivity in this document are those given in reference sources. They were not altered.

Role of advisory groups

After World War II ended in 1945, an effort was made to inform other government agencies, interested scientists, and the public of studies conducted at Hanford to determine the effect of reactor discharges on the Columbia River. The AEC, then responsible for research activities at Hanford, removed restrictions on release of data from bioenvironmental studies, except for data that might disclose the capacity for plutonium production or certain technical details of the plutonium production process (Foster 1972).

Late in 1945 and during 1946, representatives of the U.S. Fish and Wildlife Service and the Oregon and Washington Department of Fisheries and Game were invited to Hanford to observe and review studies involving the Columbia River. This move initiated a process of information transfer. Thereafter, until closure of the last single-purpose reactor in 1971, various ad hoc groups were established to plan and coordinate specific studies related to the Columbia River. The objectives of each group varied. In all cases, however, aquatic scientists at Hanford held key roles.

Columbia River Advisory Group (CRAG)

The AEC set up the Columbia River Advisory Group (CRAG) in 1949 to review its Columbia River research program and provide advice on program direction and waste disposal practices. Members of CRAG were senior officers of the State of Washington Pollution Control Commission, the State of Washington Department of Health, the State of Oregon Sanitary Authority, and the Portland office of the U.S. Public Health Service. These persons were given security clearance so that no pertinent information was withheld. CRAG met with Hanford representatives periodically for about 15 years (Foster 1972).

Amounts of radioactivity in the Columbia River downstream from Richland were reported quarterly to CRAG in a series of unclassified letters. Subsequent reports included not only data but evaluations on the effects of radioactivity, heat, chemicals, and sewage effluent from the Hanford Works (Clukey 1957). Periodic meetings were held to provide CRAG with the information needed to evaluate ongoing programs (i.e., Singlevich 1950).

Working Committee for Columbia River Studies

The AEC organized the Working Committee for Columbia River Studies (WCCRS) in 1962. At that time, the number of organizations examining radioactivity in the Columbia River had increased to nine. Additionally, other federal and state agencies were funding or conducting similar investigations. Some coordinating body was needed to avoid unnecessary duplication and to ensure that the various studies provided desired information (Wildman 1966).

Specifically, the WCCRS reviewed the scope and results of research related to radioactivity released in the Hanford Reach and other investigations dealing with radioactivity in the northeast Pacific Ocean. The WCCRS included 18 individuals representing the AEC, U.S. Public Health Service, U.S. Geological Survey, U.S. Bureau of Commercial Fisheries, U.S. Army Corps of Engineers, State of Washington Department of Health, State of Washington Pollution Control Commission, State of Oregon Board of Health, Oregon State University, and University of Washington. The Pacific Northwest Laboratory (PNL) was included after Battelle took over management of Hanford activities in 1965. The WCCRS lost momentum in the late 1960s, however, as financial support diminished and the number of operating reactors declined.

Columbia River Thermal Effects Study

The Columbia River Thermal Effects Study (CRTES) was initiated in July of 1968 in response to inconsistent temperature standards adopted for the Columbia River by the states of Washington and Oregon. To resolve the differences, state and federal agencies dealing with water pollution control needed additional information on the temperature requirements of Pacific salmon and improved techniques to predict thermal regimes in the Columbia River (EPA 1971). The Federal Water Quality Act of 1965, which required establishment of water quality standards, added force to this effort.

Researchers from the main resource management agencies in the Pacific Northwest participated in the CRTES. At the forefront were the U.S. Environmental Protection Agency, the AEC (represented by PNL), and the National Marine Fisheries Service. A Technical Advisory Committee, consisting of 16 representatives from federal, state, and power and water management agencies, was formed to review and coordinate research effort. The CRTES effort lasted 2 years.

References

Clukey, H.V. 1957. *The Hanford Atomic Project and Columbia River Pollution*. HW-54243-Rev, Hanford Works. Available from Public Reading Room, Hanford Science Center, Federal Building, Richland, Washington.

Foster, R.F. 1946. *Some Effects of Pile Area Effluent Water on Young Chinook Salmon and Steelhead Trout.* U.S. Atomic Energy Commission Report, HW 7–4759, Hanford Works. Available from Technical Library, Pacific Northwest Laboratory, Richland, Washington.

Foster, R.F. 1972. "The History of Hanford and its Contribution of Radionuclides to the Columbia River." In: *The Columbia River Estuary and Adjacent Waters*, eds. A. T. Pruter and D. L. Anderson, pp. 1–18. University of Washington Press, Seattle.

Hall, E.J. 1984. *Radiation and Life*. 2nd Ed. Pergamon Press, New York.

Kathren, R.K. 1984. *Radioactivity in the Environment: Sources, Distribution, and Surveillance*. Harwood Academic Publishers, New York.

Singlevich, W. 1950. "Radiochemical Study of the Columbia River." In: *Meeting of the Columbia River Advisory Group March 1–7, 1950*, pp. 21–32. HW-17595, Hanford Works, Richland, Washington.

U.S. Environmental Protection Agency (EPA). 1971. *Columbia River Thermal Effects Study*, Vol. I, Biological Effects Study. U.S. Environmental Protection Agency, Washington, D.C.

Wildman, R.D. 1966. "The United States Atomic Energy Commission's Columbia River Program." In: *Disposal of Radioactive Wastes into Seas, Oceans and Surface Waters*, ed. A. Guillon, pp. 671–682. International Atomic Energy Agency, Vienna, Austria.

Chapter 7

REACTOR EFFLUENT MONITORING, 1945 TO 1971

The simplest, most direct way to examine the biological effect of cooling water effluent from the prototype reactors was in laboratory bioassays that exposed fish to various dilutions of reactor effluent.

From as early as 1945, juvenile salmonids (salmon and trout) were reared in full-strength and diluted reactor effluent. The main objectives of this effort were to quantify the conditions when adverse effects first appeared, and to determine if such effects were due to radioactivity, chemical toxicity, temperature, or some combination of these factors (Foster 1952). Information was then extrapolated to actual dilutions in the Hanford Reach below the discharge points.

Three features of the cooling water effluent were soon identified as potentially harmful to aquatic organisms in the Hanford Reach. The first was chemicals added during pretreatment of raw river water. The second was heat extracted while cooling the reactor core. The third was radioactivity produced by neutron bombardment in the reactor.

Monitoring reactor effluent with fish

Bioassays with reactor effluent began when the single-purpose reactors started operation in 1945 and continued for more than 20 years. These bioassays usually involved the rearing of eggs, fry, and fingerling fish in laboratory troughs through which flowed either river water (control), various dilutions of reactor effluent, or undiluted effluent. Effects were evaluated primarily by comparing mortality and growth rates from control and exposed groups of fish. Eventually, effects of chemicals, elevated temperatures, and radioactivity were examined independently.

Rearing Chinook Salmon and Steelhead Trout

In the initial bioassay series, from July 1945 to July 1946, eggs and young of chinook salmon and steelhead trout were exposed to concentra-

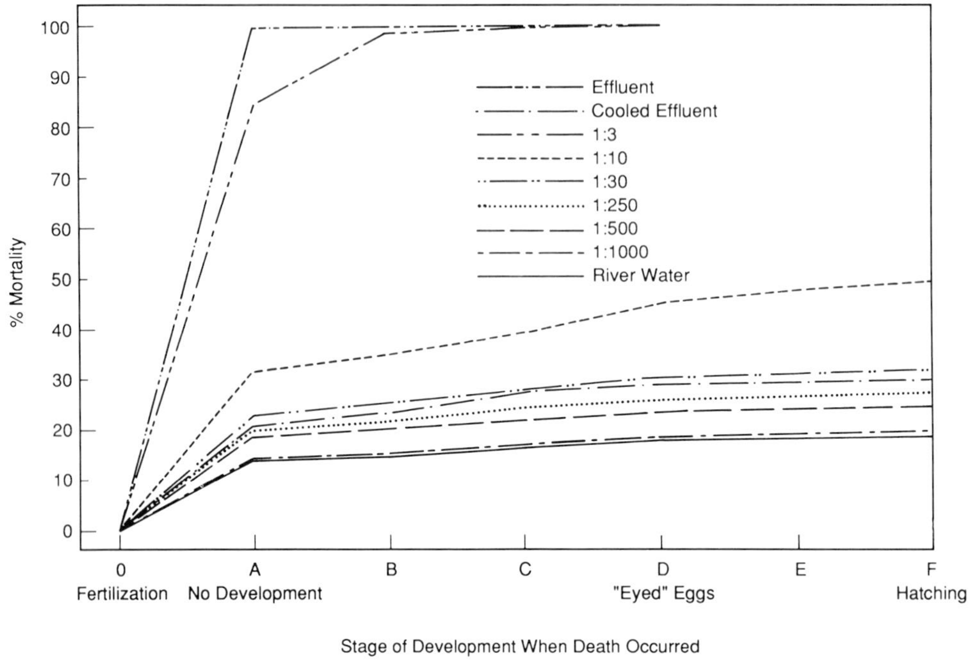

Fig. 7.1. Mortality of chinook salmon eggs incubated in various concentrations of reactor effluent during initial bioassays at Hanford (from Foster 1946).

tions of uncooled and cooled reactor effluent. Because facilities for rearing fish in the first "wet laboratory" at Hanford were new, problems with the water supply occurred once rearings were under way. Nevertheless, the effort produced the first information on the toxicity of reactor effluent and provided a foundation for future studies.

Results from rearing both chinook salmon and steelhead trout were generally similar. Essentially, fry and fingerlings were killed by exposure to 1) undiluted reactor effluent, with its accompanying higher temperatures; 2) cooled, undiluted reactor effluent; and 3) some 1:3 (33%) and 1:10 (10%) dilutions (Figures 7.1 and 7.2). However, dilutions of 1:50 (2%) or greater did not adversely effect mortality or growth (Foster 1946). Trout actually grew faster at low dilutions because of the slightly warmer water. Adverse effects on fish were detected primarily from exposure to elevated temperatures.

Proprietary chemicals used when fuel elements were replaced in a reactor core produced "off-standard" effluent. While off-standard ef-

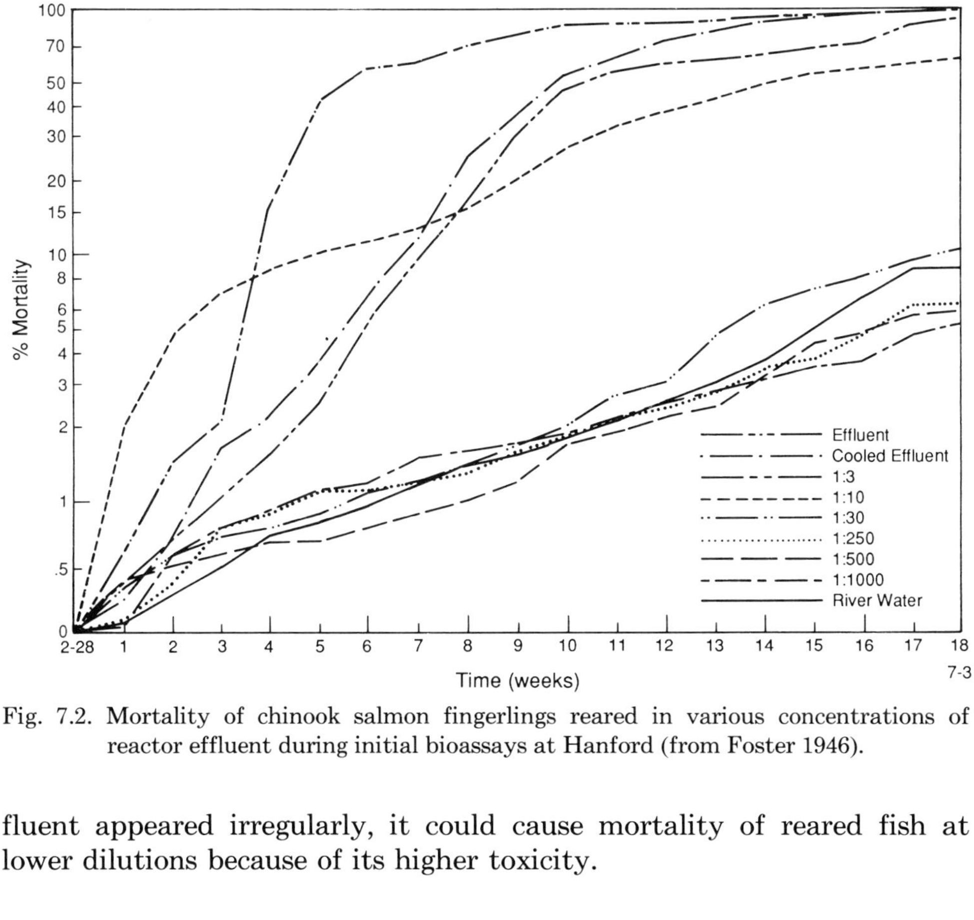

Fig. 7.2. Mortality of chinook salmon fingerlings reared in various concentrations of reactor effluent during initial bioassays at Hanford (from Foster 1946).

fluent appeared irregularly, it could cause mortality of reared fish at lower dilutions because of its higher toxicity.

Rearing Coho Salmon

In the second bioassay series, eggs, fry, and fingerlings of coho salmon were exposed to dilutions of uncooled and cooled reactor effluent. Fish were reared from December 1946 to October 1947. Eggs survived in cooled effluent. However, eggs held in partially cooled effluent and in warmed 1 : 5 (20%) and 1 : 10 (10%) dilutions had mortalities significantly higher than did controls held in river water or eggs held in cooled effluent (Olson 1948). Temperature was a major factor.

Fry (posthatch stage) were more susceptible to effluent water than eggs or fingerlings. Mortalities were statistically significant in 1 : 50 (2%) dilutions of effluent and at all higher concentrations. Cooled effluent alone caused significant mortalities, compared to controls held in river water. Mortality of fingerling coho salmon did not increase further when effluent was diluted to 1:25 (4%) or more. Growth was not retarded in dilutions of 1:25 (4%) or more.

Mortality patterns suggested that both heat and chemicals caused chronic losses of eggs, fry, and fingerling salmonids at intermediate concentrations of effluent. Mortalities were acute at high concentrations (low dilution in river water) and negligible at low concentrations (high dilution). The dilutions used in the bioassays represented considerably higher concentrations than those resulting in the Hanford Reach after the effluent was mixed. However, evidence was obtained that heat and process chemicals (e.g., sodium dichromate) in the effluent could impact aquatic life if discharges were increased by perhaps an order of magnitude (Foster 1972).

Extended Rearing of Rainbow Trout

In subsequent efforts, rainbow trout were reared in 5% reactor effluent over a complete generation covering 3 years, from 1949 to 1952. To examine the effects of radioactivity, one group of trout was fed algae contaminated with radionuclides from the effluent. Exposed trout underwent slightly higher mortalities than control fish during the first year. Growth was slower during the first few months, but the ultimate size of the fish was not affected. Adult trout reared in reactor effluent spawned less successfully than did controls reared in river water. Adverse effects were probably due to elevated temperatures and process chemicals, because feeding of radioactive algae appeared to cause no harm (Olson and Foster 1953).

Long-Term Monitoring of Effluent with Salmonids

From 1945 to 1965, eggs and young of chinook salmon, coho salmon, rainbow trout, and some other salmonids were reared in dilutions of reactor effluent as part of the long-term monitoring program at Hanford. The objectives and methods of the different bioassays changed somewhat from year to year. When adverse effects did occur, they were related to unfavorably warmed temperatures and chemical toxicity, and not to low levels of radioactivity.

In the Hanford Reach, the cooling effluent mixed with river water to concentrations much lower than those causing adverse effects during monitoring bioassays. Therefore, the discharges of reactor cooling effluent were considered unlikely to harm fish in the Columbia River. Results of different monitoring studies were reported each year in annual reports of the Biology Department (Olson and Foster 1952a; Foster and Olson 1953; Olson and Foster 1953; Olson and Foster 1955a,b).

In general, effluent concentrations greater than 4% caused excessive mortality of fish during overextended periods of exposure. Fish growth accelerated at all concentrations up to 6% because the water was warmed. When concentrations exceeded about 7%, growth was depressed by the presence of hexavalent chromium. Radioactivity caused no demonstrable damage at these dilutions (Nakatani 1969).

In 1958, as Priest Rapids Dam neared completion, the possible effects of the changing river flow from power generation on the toxicity of reactor effluent was examined in bioassays. Young chinook salmon and mountain whitefish were exposed to 2.5-fold changes in effluent concentrations by simulating daily fluctuations in dilution capacity of the Hanford Reach. Results indicated that discharges at Priest Rapids Dam could change dilution levels in the Hanford Reach sufficiently to cause mortality among river fish (Olson 1959).

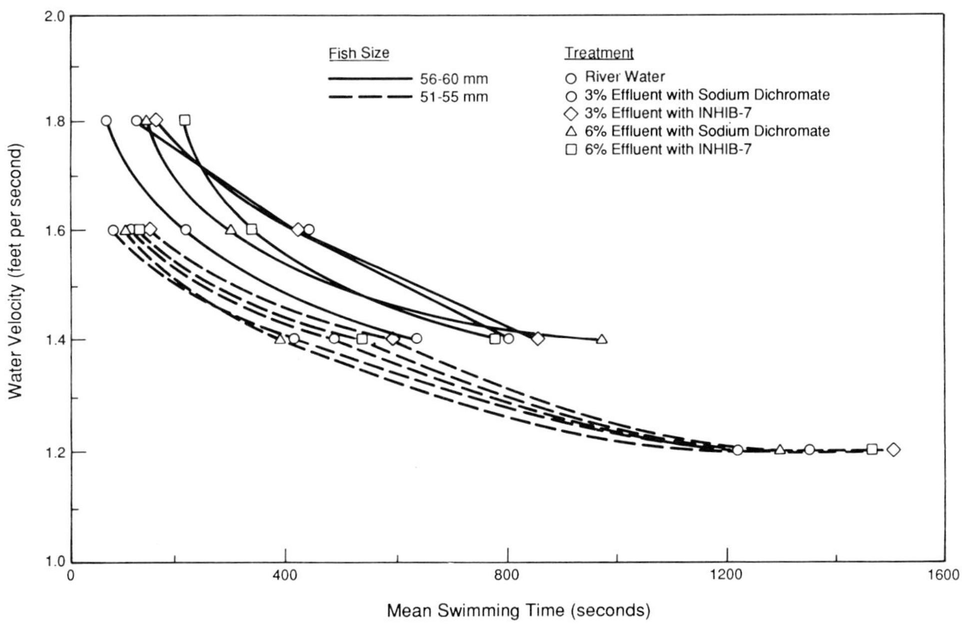

Fig. 7.3. Swimming performance of juvenile chinook salmon reared in diluted reactor effluent containing additives (from Nakatani 1964b).

Swimming Performance of Rainbow Trout

Fish capture food, escape predators, and perform other essential functions by swimming. In 1962 and 1963, possible effects of reactor effluent on the ability of salmonids to swim were examined. The test called for rearing groups of juvenile chinook salmon in 0 (control), 3%, and 5% or 6% reactor effluent for several months. Their swimming performance was then quantified in a specially constructed hydraulic test flume. Reared fish were placed in the upper end of the flume through which water flowed at a known velocity. Performance was measured as the time a fish could swim before being overcome by fatigue.

The swimming performance of control and exposed fish similar in size did not differ at three flow velocities (Nakatani 1963, 1964a,b). Size had an influence. Regardless of treatment, the larger 56- to 60-millimeter (2.2- to 2.4-inch) fish swam better than the smaller 51- to 55-millimeter (2.0- to 2.2-inch) fish. Further study indicated that swimming performance was impaired very little when corrosion inhibitors, such as sodium dichromate or INHIB-7, were included in the effluent-diluted water used for rearing (Figure 7.3).

Chemical effects during monitoring

Chemicals added to the cooling water during pretreatment were the most important factor influencing the survival of fish exposed to reactor effluent. However, removing impurities from raw river water and reducing corrosion and biofouling with chemicals were necessary steps for the efficient, safe operation of the reactors.

Toxicity of Sodium Dichromate

Sodium dichromate, an additive used to inhibit corrosion, was the primary chemical of concern in reactor effluent. In addition to its toxicity to aquatic life, neutron bombardment of sodium dichromate in the reactor core produced Cr-51, a radionuclide.

Initial, long-term, chronic bioassays were conducted in 1954. Young chinook salmon and rainbow trout were reared from eggs in low concentrations of sodium dichromate. Eggs hatched in the highest concentration of 0.18 part per million (ppm) hexavalent chromium (Figure 7.4). However, survival of young chinook salmon and trout was adversely

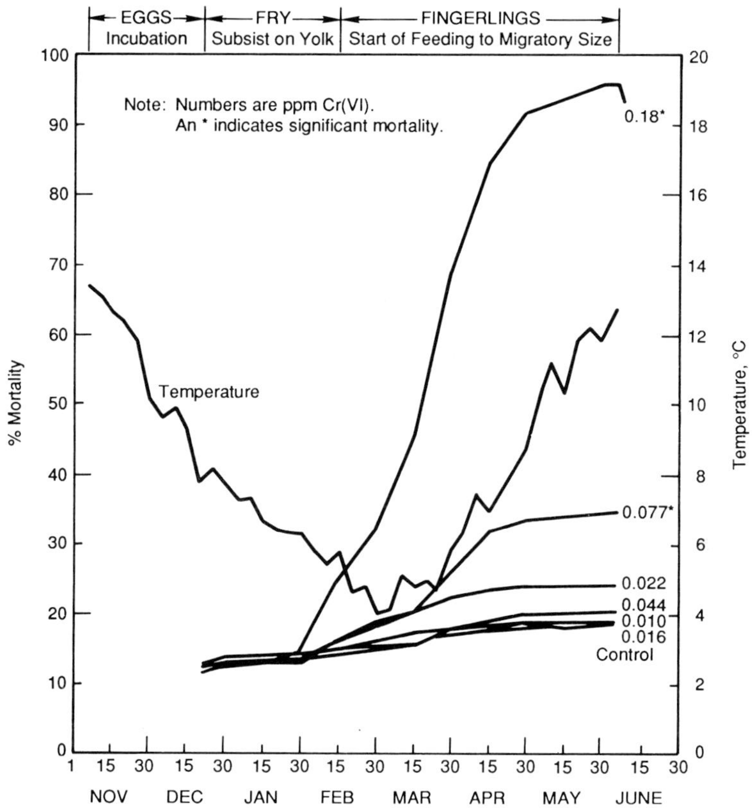

Fig. 7.4. Chronic mortality of juvenile chinook salmon exposed to hexavalent chromium solutions (from Olson and Foster 1956).

affected by 0.08 ppm chromium. Growth appeared to be retarded at the lowest concentration of 0.013 part per million chromium (Olson and Foster 1956). Growth, a sublethal effect, was a more sensitive indicator than mortality.

Subsequent studies with sodium dichromate (as hexavalent chromium) confirmed that 1) chinook salmon eggs were not affected by concentrations that were lethal to young fish after hatching, 2) effects on young salmon were somewhat less from intermittent exposure than from constant exposure, and 3) young salmon were less tolerant of chromium at 5°C (41°F) than at 10°C (50°F) (Olson and Foster 1957). The presence of 0.09 ppm hexavalent chromium markedly retarded growth at each temperature. At 0.02 ppm, hexavalent chromium was toxic, but trivalent chromium was not (Olson 1958a).

The chronic bioassays conducted at Hanford with chromium and sensitive life stages of fish led to a locally recommended limit of 0.02

milligram per liter hexavalent chromium in the Columbia River. This limit was below the level permitted for human consumption, and was perhaps more than 100 times more stringent than a limit derived from short-term, acute bioassays (Foster et al. 1961).

Uptake and Metabolism of Chromium in Trout

Biochemical methods were later applied to identify the mechanisms of toxicity in fish exposed to hexavalent chromium. Accumulation and metabolism were compared in adult rainbow trout reared in hatchery water, water containing 2.5 ppm hexavalent chromium, and Columbia River water taken downstream of the reactors. When Cr-51 was used as a tracer, it was discovered that fish accumulated hexavalent chromium from the water through their gills.

Chromium rapidly entered fish through gill tissues. Concentrations in trout exposed to 2.5 ppm hexavalent chromium reached equilibrium with that of the water in 2 to 4 days. Tissues and fluids involved in adsorption (gills), transport (plasma), and excretion (kidney, gall bladder, bile) acquired the greatest concentrations. But the metal also appeared in the brain, caeca, small intestine, and spleen of exposed fish. Substantial amounts of chromium were excreted. Trout exposed to levels as low as 0.18 ppm also accumulated chromium (Buhler et al. 1969).

The Use of Chlorine

Chlorine was added continuously to the incoming river water to prevent growth of algae and other microorganisms in filters and other equipment used in the water treatment process. At the outset, it was noted that water entering the reactors was more toxic than the effluent leaving them because of the chlorine. While chlorine was never a problem in the effluent discharge, its effect on monitoring bioassays needed to be evaluated.

A special pipeline was installed between the pumphouse and the aquatic laboratory to provide prepassage (influent) cooling water. Water that passed through the core was cooled to match the temperature of incoming water. Chlorine was removed from the influent by passage through a charcoal filter. In the bioassays, chinook salmon eggs were reared in dilutions of dechlorinated river water (control), chlorinated river water (influent), and dilutions of cooling effluent for 7 months. Young rainbow trout also were exposed.

Egg survival was slightly impaired in 2.5% effluent, probably because of unfavorable warming. If residual chlorine was present, as in the influent water, mortalities were significant. Few mortalities occurred in dechlorinated effluent until chromate was present. Juvenile rainbow trout showed a similar response, a slight retardation of growth, but mortality did not increase in 2.5% reactor effluent or in 5% influent water (Olson and Foster 1954).

Results clearly showed that residual chlorine added to the incoming river water would kill fish at relatively low concentrations. However, the discharged effluent had little residual chlorine. As a dissolved gas, chlorine was driven off by the high temperatures in the reactor core followed by exposure of effluent to the atmosphere in the shoreline retention basins before its discharge. As a result, chlorine treatment had no effect on the Hanford Reach.

Other Industrial Additives

Chemicals other than sodium dichromate were considered for use as corrosion inhibitors in the cooling water. Before their use, they were evaluated for potential toxicity to fish. Some were chromium compounds, some were used to condition the water, and others were used to purge radioactive materials from pipes. Because many additives were proprietary compounds, their chemical makeup was unknown.

Bioassays were useful in establishing "median tolerance limits," based on concentrations that experimentally killed 50% of the exposed fish in a few days. Tolerance limits varied with each type of compound. Concentrations unlikely to harm river fish for each compound were estimated, and a safety factor was applied to provide a conservative limit (Nakatani 1964a; Liu and Nakatani 1964).

Temperature effects during monitoring

Temperature was second in importance of the three factors affecting the survival of fish exposed to reactor cooling effluent. In several monitoring bioassays, mortality of all eggs, embryos, and young was caused by elevated temperatures (in designed thermal increments). A cold-blooded aquatic organism is limited in its range of thermal tolerance, and any thermal increment is likely to affect growth and survival. Excessively high temperatures cause mortality.

Columbia River water (at ambient temperature) warmed during the spring and summer, then cooled during the fall and winter in a seasonal cycle. Heat in a reactor core raised temperatures in the effluent to well above ambient. A temperature increase tolerated by fish when the river water was cold might be detrimental when the river water was warm.

Rearing Chinook Salmon at Elevated Temperatures

The maximum temperatures tolerated by eggs and young of chinook salmon during rearing were first examined. Exposures were based on thermal increments that paralleled seasonal changes in the river water. Eggs were taken in October 1953 from adult salmon in the Hanford

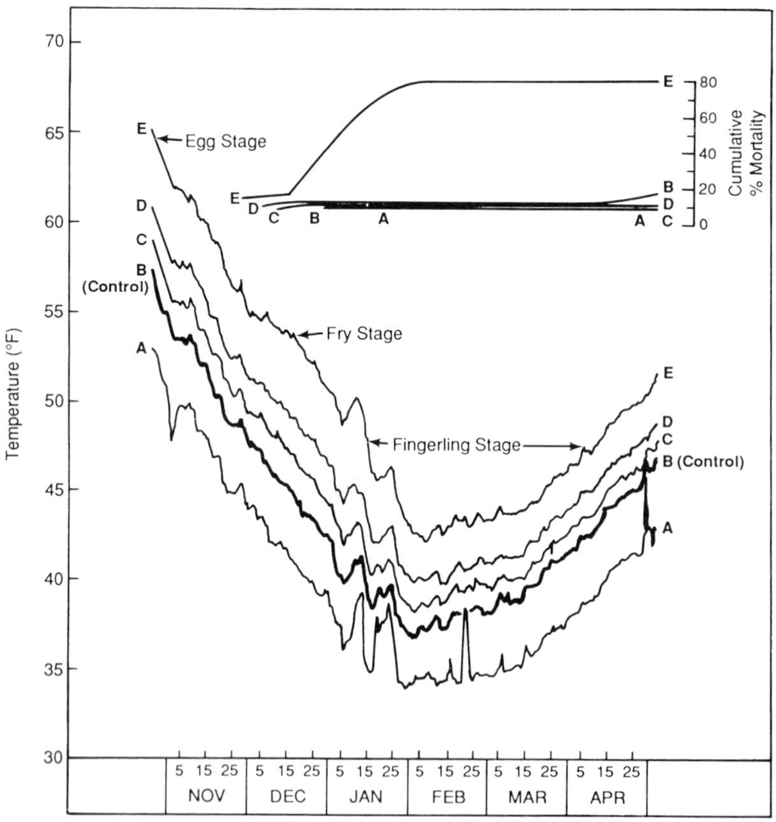

Fig. 7.5. Mortality of young chinook salmon reared under thermal increments that followed seasonal temperature changes in the Hanford Reach (from Olson and Foster 1955c).

Reach and reared until May 1954. Groups of eggs were held at ambient river temperature (control), 2°C below ambient, and at three incremental temperatures.

Mortalities significantly higher than those in the controls group occurred only in the group reared at the highest incremental temperature, 4.4°C (8°F) above ambient. About 90% of the eggs in this group hatched successfully. However, fry and alevins underwent heavy mortalities, even though they were exposed to temperatures averaging less than 10°C (50°F) during late winter (Figure 7.5). The delayed mortality was a probably a stress factor resulting from earlier exposure of the eggs to unseasonally high temperatures (Olson and Foster 1955c).

Rearing Whitefish at Elevated Temperatures

This experiment examined the effect of increased temperatures on survival of eggs and young of mountain whitefish, a sportfish spawning in the Hanford Reach. Eggs were stripped for rearing from whitefish taken during December 1956. Experimentally, thermal increments were designed to parallel seasonal changes in the Hanford Reach, and ambient temperatures in control groups ranged from 2.7° to 18.4°C (36.7° to 65.1°F) during rearing. Test groups were held at temperatures 2°C (35.6°F) and 3°C (37.4°F) above ambient.

The temperature level was crucial for survival of whitefish as they passed through embryogenesis to the fingerling stage. Cumulative mortality was high among controls (39%). However, mortality increased to 65% and 76% in the two groups reared at slightly elevated temperatures (Olson 1958b).

Thermal Resistance of Two Chinook Salmon Races

Fall chinook salmon from the Hanford Reach were not available for most earlier effluent monitoring studies. Instead, eggs and fingerlings from Puget Sound stocks were used. In 1964, chinook salmon fingerlings were obtained from a spawning channel at Priest Rapids Dam (above Hanford) and reared in diluted reactor effluent to examine the possibility that racial differences might exist.

Mortalities were low (3% to 6%) in fingerling salmon from Priest Rapids held in 0% (control), 2%, 4%, and 6% effluent that had been cooled. However, Priest Rapids fish in uncooled effluent underwent higher mortalities in 4% and 6% effluent (11% and 20%, respectively). It appeared that local stocks of fall chinook salmon might actually be more sensitive to elevated temperatures than Puget Sound stocks (Olson and Nakatani 1965).

For confirmation, fall chinook salmon from both Priest Rapids and Puget Sound were reared simultaneously in diluted reactor effluent in 1966 under conditions identical to those in 1964. This time, no marked racial differences between the two stocks were found in terms of mortality, growth, or accumulation of radionuclides. While growth and uptake of radionuclides increased at warmer temperatures, exposure to 6% effluent was tolerated (Olson 1967). Because of effective dilution in the Hanford Reach, concentrations of effluent below the reactors were well below 6%.

Radioactivity effects during monitoring

Radioactivity (initially an unknown quantity) in the reactor effluent was the least important of the three factors influencing growth and survival of reared fish. In fact, the radioactive content of the effluent proved to be an advantage to scientific inquiry. For example, mechanisms of uptake and elimination of radionuclides could be examined in reared fish under controlled conditions. Biological results derived from monitor-

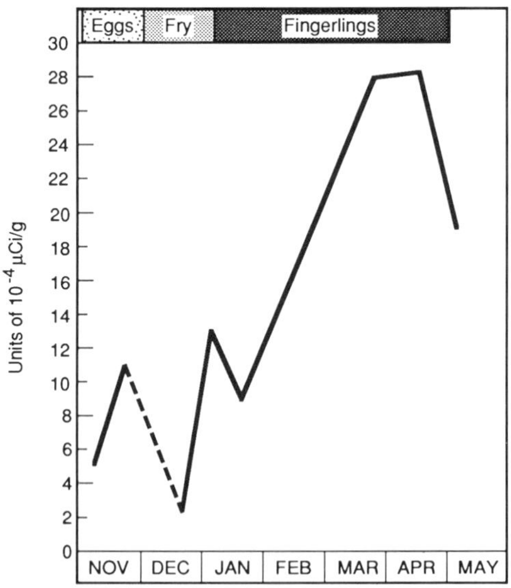

Fig. 7.6. Gross beta activity of young chinook salmon reared from eggs in 10% reactor effluent (from Olson and Foster 1955a).

ing bioassays and field studies did lead to reductions in the amount of P-32 (a biologically active radionuclide) and sodium dichromate discharged to the Hanford Reach in reactor effluent (Nakatani 1969).

About 80% of the radioactivity accumulated by salmon reared in 5% effluent came from Na-24, about 10% from Cu-64, and lesser amounts from As-76, rare earths, and radionuclides in the iron, zinc-cobalt, calcium, and phosphorus groups. Gross beta activity of young salmon reared in 10% reactor effluent generally increased with exposure time as the fish grew (Figure 7.6). When salmon eggs hatched, the shells with their adsorbed radioactivity were lost, and the total activity of the developing eggs dropped (Olson and Foster 1955a).

Algae grown in undiluted reactor effluent soon became radioactive. While most of this activity originated from Na-24 and the copper-arsenic group, about 40% consisted of the longer-lived P-32, and the zinc-cobalt, calcium, iron, and rare earth groups. Algae simply immersed for a few days in the effluent acquired 85% of the short-lived radionuclides, probably by adsorption rather than by assimilation (Olson and Foster 1953).

Accumulation of Radioactivity by River Fish

Large numbers of fish from the Hanford Reach were collected and analyzed for radioactivity between April 1948 and June 1950. This effort showed how much radioactivity was acquired by different species and sizes (ages) of fish. It also allowed uptake from exposure to river water to be separated from ingestion of food organisms. In contrast, the juvenile salmonids exposed in monitoring studies with reactor effluent took up radionuclides, for the most part, directly from the water.

Most radioactivity bioaccumulating in different fish was from P-32 in scales, bone, and certain visceral organs. Activity was influenced by the size of the fish, feeding habits, and metabolism, in addition to the amount of radioactivity in the water. Generally, activity densities increased from year to year as more reactors were built and operated at Hanford. Gross radioactivity was higher in food remains from the stomachs of fish than in their tissues, pointing to the aquatic food web as the most important way for fish to accumulate radionuclides. Levels of radioactivity appearing in fish were not high enough to be hazardous to the fish themselves or to people eating them (Olson and Foster 1952b).

Temperature and Uptake of Radioactivity

Through the 1960s, the uptake of radioactivity by young salmonids reared in diluted reactor effluent was routinely determined each year.

Table 7.1.
Uptake of Radionuclides by Juvenile Chinook Salmon from Priest Rapids Stocks in Three Dilutions of Reactor Effluent, 1965 (from Olson 1966)

Percent effluent [a]	Water temperature	Radionuclide uptake (pCi/g wet weight)		
		Na-24	Cr-51	Zn-65
0	river	12	11	3
	river + ΔT [b]	11	14	4
2	river	854	64	16
	river + ΔT	945	77	18
	river + $2\Delta T$			
4	river	1665	94	26
	river + ΔT	1840	96	28
	river + $2\Delta T$	1720	72	14

[a] Temperatures were highest at 4% effluent dilution.
[b] ΔT = delta T

The radionuclides commonly measured were Na-24, Cr-51, and Zn-65. Control fish held in raw river water acquired some radioactivity because some reactor outfalls were upstream of the water intake for the aquatic laboratory. By this time, study after study had established that reared salmonids took up radionuclides primarily through their gills. The artifi-

Table 7.2.
Uptake of Radionuclides by Juvenile Chinook Salmon from Priest Rapids and Puget Sound Stocks in Four Dilutions of Reactor Effluent, 1966 (from Olson 1967)

Stock	Percent effluent [a]	Mean length (mm)	Radionuclide uptake (pCi/g wet weight)		
			Na-24	Cr-51	Zn-65
Priest	0	39	77	19	6.8
Rapids	2	41	750	38	20
(5-month	4	45	1390	53	36
rearing)	6	46	2210	65	45
Puget	0	61	73	26	7.7
Sound	2	66	720	43	15
(3-month	4	69	1370	48	19
rearing)	6	68	1950	77	33

[a] Temperatures were highest at the 6% effluent dilution.

cial diet contained no radioactivity, and freshwater fish, in general, swallow little water except with their food.

In 1965, juvenile chinook salmon were reared for 6 months in 0% (control), 2%, and 4% dilutions of reactor effluent under different temperature increments that varied seasonally with river temperature. Concentrations of the three main gamma emitters, Na-24, Cr-51, and Zn-65, varied directly with effluent concentration (Table 7.1). Temperatures above ambient, which were related to the amount of effluent added, had no apparent effect on the uptake of radionuclides (Olson 1966).

In 1966, one group of juvenile chinook salmon was reared for 5 months and another group for 3 months in 0% (control), 2%, 4%, and 6% reactor effluent under four temperature increments. The control temperature and the experimental increments were allowed to change over the season. Again, uptake of Na-24, Cr-51, and Zn-65 varied with amount of effluent and temperature elevation (Table 7.2). Fish held in 6% effluent acquired the highest levels of radioactivity (Olson 1967). Calculations indicated that, as in preceding studies, concentrations of effluent in the Hanford Reach were considerably less than those used experimentally because of dilution in the river ecosystem.

Food Web Transfer of Radioactivity to Trout

Fish in laboratory troughs acquired radionuclides directly from the water through their gills. However, fish in the Hanford Reach acquired radionuclides primarily from the food they ate. Therefore, the food web was the main factor involved in uptake of radionuclides by fish in the Hanford Reach. The specific mechanisms needed further study.

One early study examined the transfer of radionuclides from reactor cooling water to fish food organisms, their assimilation by rainbow trout, and their deposition in various tissues (Olson 1952). Snails, crayfish, and young carp were reared in partially cooled reactor effluent, allowed to accumulate radionuclides, and fed each day to yearling trout for several months. The most common radionuclide concentrated by food organisms was P-32. Most of the radionuclide accumulated in the calcareous tissues of trout, the least in the fat and muscle.

While the half-life of P-32 was relatively short, compared to most radionuclides of concern, it was taken up readily by fish. Further, the half-life of P-32 (2 weeks) was sufficient that it might be transferred to people eating fish from the Hanford Reach. Hence, the environmental fate of P-32 drew considerable attention in other studies.

Significance of effluent environmental monitoring studies

Effluent monitoring was initiated at Hanford almost as soon as the single-purpose reactors started operating in 1944. It was a new field of endeavor. Never before had nuclear reactors been operated to produce an artificial element, plutonium, and never before had river water been used to cool these reactors. Initial concerns included the effects of radioactivity, the activation and fission byproducts that were released in the cooling water effluent.

Monitoring soon demonstrated that the primary effect of the cooling water effluent on aquatic life was from chemicals used to treat the raw river water, rather than from the addition of radioactivity or heat. Furthermore, bioassays with reactor effluent were designed to monitor separately, or in combination, the effects of chemicals, heat, and radioactivity in the cooling effluent. The bioassays also allowed investigation of effects related to seasonal conditions in the Hanford Reach. Bioassays of fish were the basic technique used in monitoring the reactor effluent at Hanford. Each bioassay provided quantitative data on the level of chemicals, heat, and radioactivity that might impair ecological functions in the river below the effluent discharges. Calculations that compared the laboratory-derived effect levels with concentrations expected in the Hanford Reach, taking into consideration seasonal variations in river flow and temperature, indicated a conservative safety margin before any significant effect would occur. The long-term conduct of bioassays with reactor effluent provided assurance to Hanford Site management and the government that the effluents were diluted to relatively safe levels.

Bioassays with reactor effluent by the Aquatic Biology Group corresponded with related investigations in other departments at Hanford. For example, the Biomedical Group and Health Physics Group conducted extensive research on the effects of radioactive river water on plants and warm-blooded animals, and they examined potential routes of transmission to people living near the Hanford Site. In one study, crops were irrigated with reactor effluent in experimental plots. No significant effects were observed in plants grown in soil that contained the equivalent of 25 years worth of accumulated effluent material.

Today, bioassays are widely employed as a basic technique to detect toxicological, behavioral, and physicological effects from effluent and other waste streams at industrial plants and sanitary treatment facilities. Lethal and sublethal effects detected by bioassays provide a basis for interpreting chemical analyses that routinely accompany them. The results of bioassays are extremely relevant to the growing problem of protecting aquatic life and human health.

Philip A. Olson, Jr. played a major role in conducting bioassays with reactor effluent and fish at Hanford beginning in 1946. He died in 1971. A memorial plaque in his honor at the Life Sciences I Building in the 300 Area reads: "He was in the vanguard of biologists who initiated research studies on the effects of nuclear reactors on aquatic life. His field of focus concerned the toxicity and temperature effects of effluents on all life stages of Columbia River salmon, and his scientific contributions in this area were significant."

References

Buhler, D.R., S.R. Caldwell, and R.M. Stokes. 1969. "Tissue Accumulation and Metabolic Effects of Hexavalent Chromium in Trout." In: *Pacific Northwest Laboratory Annual Report for 1968 to the USAEC Division of Biology and Medicine, Vol. I Life Sciences*, Part 2 Ecological Sciences, pp. 2.7–2.13. BNWL-1050 PT2, Battelle, Pacific Northwest Laboratories, Richland, Washington.

Foster, R.F. 1946. *Some Effects of Pile Area Effluent Water on Young Chinook Salmon and Steelhead Trout.* U.S. Atomic Energy Commission Report, HW 7–4759, available from Technical Library, Pacific Northwest Laboratory, Richland, Washington.

Foster, R.F. 1952. "Biological Problems Associated with the Discharge of Pile Effluent into the Columbia River." In: *Biology Research – Annual Report 1951*, pp. 11–13. HW-25021, Hanford Works, Richland, Washington.

Foster, R.F., R.L. Junkins, and C.E. Linderoth. 1961. "Waste Control at the Hanford Plutonium Production Plant." *J. Water Poll. Control Fed.* 35:511–529.

Foster, R.F. 1972. "The History of Hanford and its Contribution of Radionuclides to the Columbia River." In: *The Columbia River Estuary and Adjacent Waters*, eds. A. T. Pruter and D. L. Anderson, pp. 1–18. University of Washington Press, Seattle.

Foster, R.F. and P.A. Olson, Jr. 1953. "Effect of Reactor Effluent Water on Young Silver Salmon." In: *Biology Research – Annual Report 1952*, pp. 31–38. HW-28636, Hanford Atomic Products Operation, Richland, Washington.

Liu, D.H.W., and R.E. Nakatani. 1964. "Toxicity of Industrial Chemicals to Fish." In: *Hanford Biology Research Annual Report for 1963*, pp. 201–211. HW-80500, Hanford Atomic Products Operation, Richland, Washington.

Nakatani, R.E. 1963. "Swimming Performance of Chinook Salmon Reared in Reactor Effluent." In: *Hanford Biology Research Annual Report for 1962*, pp. 211–222. HW-76000, Hanford Atomic Products Operation, Richland, Washington.

Nakatani, R.E. 1964a. "Reactor Effluent Monitoring and Bioassay of Industrial Chemicals with Fish." In: *Report to the Working Committee for Columbia River Studies on Progress Since September 1962 for Projects Carried Out by General Electric Company at Hanford*, ed. R.F. Foster, pp. 1–18. HW-80645, Hanford Atomic Products Operation, Richland, Washington.

Nakatani, R.E. 1964b. "Swimming Performance of Chinook Salmon Reared in Reactor Effluent – II." In: *Hanford Biology Research Annual Report for 1963*. pp. 201–208. Hanford Atomic Products Operation, Richland, Washington.

Nakatani, R.E. 1969. "Effects of Heated Discharges on Anadromous Fishes." In: *Biological Aspects of Thermal Pollution*, eds. P.A. Krenkel and F. L. Parker, pp. 291–317. Vanderbilt University Press, Nashville, Tennessee.

Olson, P.A., Jr. 1948. *Some Effects of Pile Area Effluent Water on Young Silver Salmon*. HW-8944, Hanford Works Report. Available from Technical Library, Pacific Northwest Laboratory, Richland, Washington.

Olson, P.A., Jr. 1952. "Observations on the Transfer of Pile Effluent Radioactivity to Trout." In: *Biology Research – Annual Report 1951*, pp. 30–39. HW-25021, Hanford Works Report. Available from Technical Library, Pacific Northwest Laboratory, Richland, Washington.

Olson, P.A. 1958a. "Comparative Toxicity of Cr(VI) and Cr(III) in Salmon." In: *Hanford Biology Research Annual Report for 1957*, pp. 211–218. HW-53500, Hanford Atomic Products Operation, Richland, Washington.

Olson, P.A. 1958b. "Temperature Tolerance of Eggs and Young of Columbia River Fish." In: *Hanford Biology Research Annual Report for 1957*, pp. 211–214. HW-53500, Hanford Atomic Products Operation, Richland, Washington.

Olson, P.A. 1959. "Effects of Variable River Flow on the Toxicity of Reactor Effluent." In: *Hanford Biology Research Annual Report for 1958*, pp. 131–137. HW-59500, Hanford Atomic Products Operation, Richland, Washington.

Olson, P.A. 1966. "Reactor Effluent Monitoring – 1965." In: *Pacific Northwest Laboratory Annual Report for 1965 in the Biological Sciences*, pp. 131–133. BNWL-280, Pacific Northwest Laboratory, Richland, Washington.

Olson, P.A. 1967. "Reactor Effluent Monitoring: Comparison of Puget Sound and Priest Rapids Chinook Salmon." In: *Pacific Northwest Annual Report for 1966 to the USAEC Division of Biology and Medicine, Vol. I Biological Sciences*, pp. 180–181. BNWL-480, Pacific Northwest Laboratory, Richland, Washington.

Olson, P.A., and R.F. Foster. 1952a. "Effect of Pile Effluent Water on Fish." In: *Biology Research – Annual Report 1951*, pp. 41–52. HW-25021, Hanford Works, Richland, Washington.

Olson, P.A., and R.F. Foster. 1952b. *Accumulation of Radioactivity in Columbia River Fish in the Vicinity of Hanford Works*. HW-23093, Hanford Works Report. Available from Technical Library, Pacific Northwest Laboratory, Richland, Washington.

Olson, P.A., and R.F. Foster. 1953. "Extended Rearing of Rainbow Trout in Dilute Reactor Effluent." In: *Biology Research – Annual Report 1952*, pp. 20–29. HW-28636, Hanford Atomic Products Operation, Richland, Washington.

Olson, P.A., and R.F. Foster. 1954. "Reactor Effluent Monitoring with Young Chinook Salmon." In: *Biology Research – Annual Report 1953*, pp. 24–35. HW-30437, Hanford Atomic Products Operation, Richland, Washington.

Olson, P.A., and R.F. Foster. 1955a. "Effect of Reactor Area Effluent Water on Chinook Salmon Fingerlings." In: *Biology Research – Annual Report 1954*, pp. 19–23. HW-35917, Hanford Atomic Products Operation, Richland, Washington.

Olson, P.A., and R.F. Foster. 1955b. "Effect of Reactor Area Effluent Water on Migrant Juvenile Blueback Salmon." In: *Biology Research – Annual Report 1954*, pp. 21–27. HW-35917, Hanford Atomic Products Operation, Richland, Washington.

Olson, P.A., and R.F. Foster. 1955c. "Temperature Tolerance of Eggs and Young of Columbia River Chinook Salmon." *Trans. Am. Fish. Soc.* 85:201–207.

Olson, P.A., and R.F. Foster. 1956. "Effect of Chronic Exposure to Sodium Dichromate on Young Chinook Salmon and Rainbow Trout." In: *Biology Research – Annual Report*

1955, pp. 31–46. HW-415000, Hanford Atomic Products Operation, Richland, Washington.

Olson, P.A., and R.F. Foster. 1957. "Further Studies on the Effect of Sodium Dichromate on Juvenile Chinook Salmon." In: *Biology Research – Annual Report 1956*, pp. 211–224. HW-47500, Hanford Atomic Products Operation, Richland, Washington.

Olson, P.A., and R.E. Nakatani. 1965. "Reactor Effluent Monitoring – 1964." In: *Hanford Biology Research Annual Report for 1964*, pp. 191–201. BNWL-122, Pacific Northwest Laboratory, Richland, Washington.

Chapter 8

FIELD STUDIES WITH RADIOACTIVITY IN THE HANFORD REACH, 1945 TO 1971

From the beginning of the Hanford project, amounts of radioactivity in the Columbia River downstream near the Tri-Cities (Richland, Pasco, and Kennewick) were always less than those calculated from the amounts released in the cooling water effluent of the single-purpose reactors, allowing for the physical decay associated with downstream travel times. Clearly, the amount of radioactivity in the flowing water was rapidly reduced during downstream transport. Reduction was a result of the interaction of chemical, physical, and biological features of the river ecosystem, all of which were then unknown.

Ecological phenomena involving radioactivity in the Columbia River downstream from the single-purpose reactors were closely examined for more than two decades. Results of initial studies were usually detailed in internal documents listed as classified. Findings were reviewed periodically in the scientific literature after World War II ended.

This chapter reviews field studies of radioactivity in the Hanford Reach. Field studies are always influenced by uncontrolled ecological variables. Therefore, field studies required the support of controlled laboratory studies (discussed in Chapter 9) to precisely define ecological phenomena and interactions.

Radionuclide releases – early studies (1941–1962)

Sampling of Columbia River water at Hanford started in 1944, before the first reactor began producing plutonium, and routine checking of radioactivity in river organisms began in 1946 (Foster and Davis 1956). Researchers soon learned that the amounts and distribution of radioactivity in the Columbia River downstream from the reactors were controlled by complex ecological processes. Much later, raw data gathered from 1946 to 1962 on radioactivity in the Hanford Reach were first summarized in the open literature (Soldat 1962).

Early field studies with radioactivity included evaluations of aquatic communities upstream and downstream of the effluent outfalls, seasonal variations in these communities, interspecific relationships such as food chains, and possible exposure of local residents to radioactivity. Gross radioactivity was usually measured as total beta emissions.

Radioactivity also was measured in different Columbia River organisms. The measurements revealed which radionuclides in the cooling

Fig. 8.1. Drift plankton were collected from the Hanford Reach in the early 1950s (photo) for radioactivity analysis by filtering river water through a fine-meshed net by Raymond W. Coopey.

water discharges were concentrated in river biota, which organisms became most radioactive, the distribution of radioactivity in the river at various points below the discharges, the seasonal fluctuations in radioactivity, what river organisms or their progeny that received radiation (if any) were affected, and if humans using river fish for food were apt to be harmed (Foster 1952, 1959a).

Exploratory Surveys

Studies relating to radioactivity in Columbia River fish were initiated in 1946 (Healy 1946; Herde 1946, 1947). A preliminary field survey of radioactivity in other aquatic forms and additional studies on fish were carried out between November 1947 and April 1948 (Coopey 1948; Herde 1948). Data specific to benthic invertebrates and plankton were obtained next (Coopey 1951; Davis and Coopey 1951), followed by more extensive data on fish (Olson and Foster 1952) and crustacea (Coopey 1953a). Surveys also were made of radioactivity in aquatic organisms in the lower

Fig. 8.2. Bottom organisms were collected from the Hanford Reach in the early 1950s for radioactivity analysis by a heavy dredge. Left to right, Jared J. Davis, Calvin L. Cooper, and Clair C. Palmiter.

Columbia River (Foster et al. 1949). In 1951 and 1952, investigations were coordinated with a radiological survey of the Columbia River by the U.S. Public Health Service (Robeck et al. 1954) and, years later, with studies on the flux of radionuclides through the entire Columbia River and its estuary into the Pacific Ocean (Pruter and Alverson 1972).

For these studies, aquatic organisms were collected upstream (control sites) and downstream of the reactors, particularly in the paths of the discharge plumes where maximum exposure to radioactive materials occurred (Figures 8.1 and 8.2). Radioactivity was measured by the best analytical methods available at that time, usually within a few hours after samples were collected. Gross beta radioactivity was routinely measured by direct count. Information from initial studies allowed development of hypotheses, thus establishing the framework for more specific explorations.

Initial Studies on Radioactivity in Hanford Reach

Initially, surveys determined amounts of radioactivity in common plants and animals of the Hanford Reach and considered possible radiological hazards to humans. This effort led to increased knowledge about radionuclide transport and dispersion of radioactivity in the Columbia River ecosystem (Davis et al. 1952).

Short-lived activation products accounted for most radioactivity in the reactor cooling effluent. And most of these had decayed before release to the river. Generally, radionuclides appearing in aquatic organisms had longer half-lives than those in the effluent. As early as 1951, it was known that radioactivity levels varied among Columbia River organisms, and that the highest radioactivity was acquired by plankton drifting downstream through the effluent plumes (Figure 8.3).

Fish collected near the effluent discharges in 1951 contained primarily P-32, with small amounts of Na-24 and traces of long-lived radionuclides. Suckers, which fed on attached periphyton, contained the greatest amounts of radioactivity (Olson and Foster 1952). Aquatic insect larvae took up radioactivity composed primarily of P-32 (70% to 85%), Na-24 (15% to 30%), and traces ($<1\%$) of the long-lived radionuclides (Davis and Cooper 1951). Algae contained considerable radioactivity from the copper group (47%), P-32 (25%), rare earths (8%), Na-24 (7%), and the arsenic, iron, and zinc-cobalt groups (4%). Most radioactivity in plankton drifting downstream through the discharge plumes was from short-lived radionuclides (70% to 90%), probably Na-24 and Mn-56, and a complex

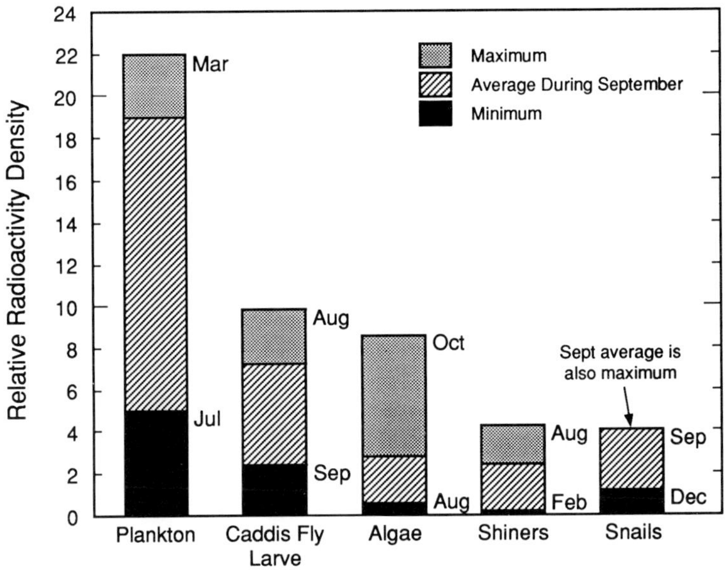

Fig. 8.3. Relative intensity of radioactivity in aquatic organisms from the Hanford Reach in 1951 (from Davis et al. 1952).

dominated by P-32 (5% to 30%). Naturally occurring K-40 (a gamma emitter) was present in all organisms (Coopey 1951).

Uptake of radioactivity by drift plankton was studied intensively from February 1949 through February 1950 (Coopey 1953b). With a few

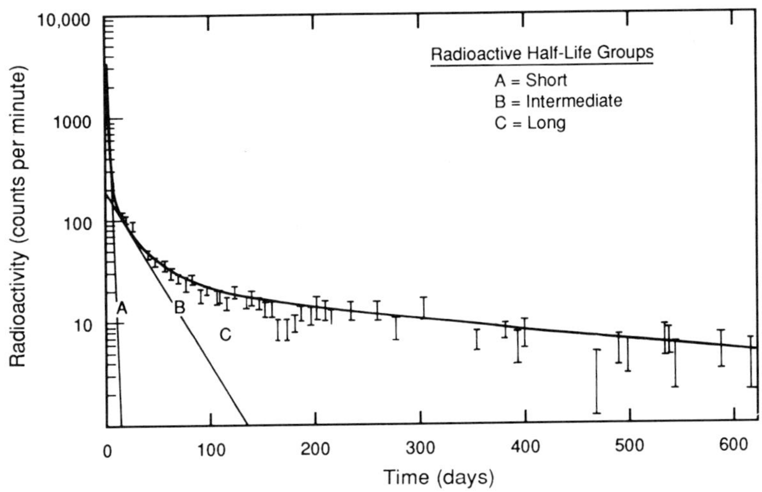

Fig. 8.4. Decay of radioactivity in a plankton sample taken from the Hanford Reach, January 1949 (from Coopey 1953b).

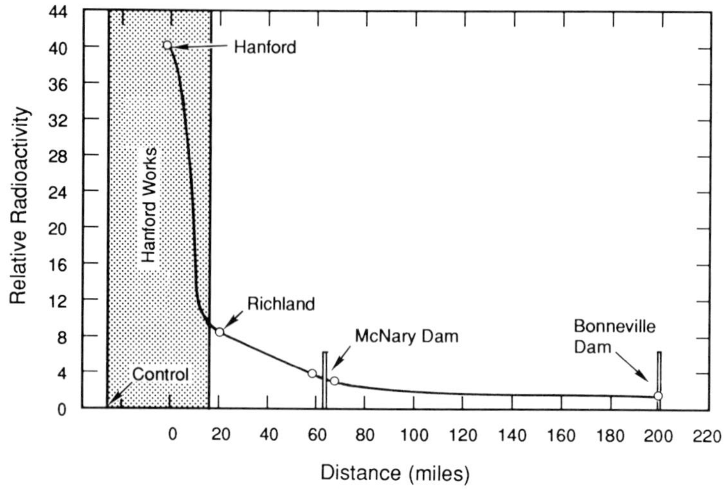

Fig. 8.5. Intensity of radioactivity in snails in relation to distance downstream from the Hanford single-purpose reactors, September 1951 (from Davis et al. 1952).

exceptions, drift plankton carried the greatest amounts of radioactivity of all river organisms, and uptake was rapid. Furthermore, radioactivity in plankton varied with changes in river flow that diluted the cooling water discharges. Decay of radioactivity in isolated plankton samples showed three isotope groups: a) one with a half-life of a day or less (70% to 95% of the radioactivity), b) one with half-lives of intermediate duration, and c) one with a half-life of more than 600 days (Figure 8.4).

Background levels in aquatic organisms above Hanford were less than 1×10^{-5} microcurie per gram (μCi/g) in 1951. In the Hanford Reach both upstream and downstream of the reactors, gross beta radioactivity in river water and aquatic organisms increased as effluent from each reactor joined the river flow. Maximum inshore levels occurred near the former Hanford townsite. Radioactivity in fish from the Hanford Reach peaked at 2.7×10^{-4} μCi/g, a level considered to be safe for human consumption (Davis et al. 1952). Radioactivity in biota then declined with distance downriver (Figure 8.5), a pattern that persisted until the last single-purpose reactor was shut down 20 years later.

Only trace amounts of radionuclides with less than 14-day half-lives appeared in fish (free movement) and benthic invertebrates (sessile) more than 32 kilometers (20 miles) downstream of the reactors. However, some short-lived radionuclides were detected in drift plankton 11 kilometers (7 miles) downstream from McNary Dam because river currents rapidly carried these organisms downstream (Davis et al. 1955).

Seasonal Variations in Radioactivity

Whenever the number of reactors operating remained constant, gross radioactivity in Columbia River water downstream from Hanford was related to the volume of river flow available to dilute the reactor discharges. Gross radioactivity was highest during the fall and winter, when river flows were low, and lowest during the spring freshet, when river flows were high (Foster and Davis 1956).

Gross beta activity in plankton (drifting) and periphyton (attached) corresponded to radioactivity in the river water because uptake involved direct absorption and adsorption of radionuclides. In contrast, radioactivity in larger aquatic organisms, such as fish, was usually related to their food intake and metabolic rate.

Radioactivity in plankton and periphyton decreased during late spring because high flows of the annual spring freshet diluted the radionuclides in effluent. Radioactivity in fish was lowest during winter, when the Columbia River was cold, and their consumption of contaminated food organisms (plankton, benthic invertebrates, and small fish) was at a minimum. In contrast, radioactivity in fish was greatest during late summer, when the river was warm and food consumption was high (Figure 8.6). Young, rapidly growing fish accumulated radionuclides faster than older, slowly growing fish.

Uptake relationships in fish generally extended to other cold-blooded aquatic organisms of the Hanford Reach, including insects, crustacea, and

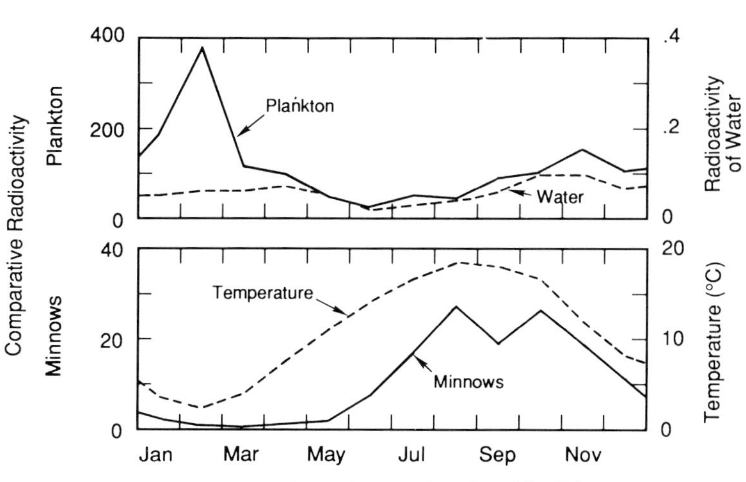

Fig. 8.6. Seasonal fluctuations in radioactivity of Columbia River water and organisms (from Foster and Davis 1956). The "minnows" were the redside shiner.

mollusks. However, the complex life cycles of individual species altered these relationships. For example, immature aquatic insects acquired less radioactivity during their resting phase of development than when actively feeding as larvae or nymphs (Foster and Davis 1956; Foster 1959a).

Effect of Time and Distance on Radioactivity Downstream from the Hanford Site

Initial field measurements of radioactivity in river water downstream from the Hanford reactors diminished with time and distance. Several interacting factors were involved. One factor was the physical decay of short-lived radionuclides. Much of the radioactivity in drift plankton was a result of adsorption of radionuclides with short half-lives. Therefore, radioactivity dropped rapidly as these microorganisms passed downstream with the river flow (Figure 8.7).

During the early years of reactor operation, beta radioactivity in drift plankton (primary producers) was about three times greater than in fish (consumers) in the lower portion of the Hanford Reach, based on picocuries per gram of wet weight. But fish were more radioactive 80 kilometers (50 miles) downstream and beyond (Davis 1958).

In larger aquatic organisms, containing primarily P-32 (14.3-day half life), the reduction of radioactivity with distance was more gradual. Large organisms took up and retained much P-32, reducing the amount of P-32 in river drift so that it passed downstream more slowly. Retention of P-32

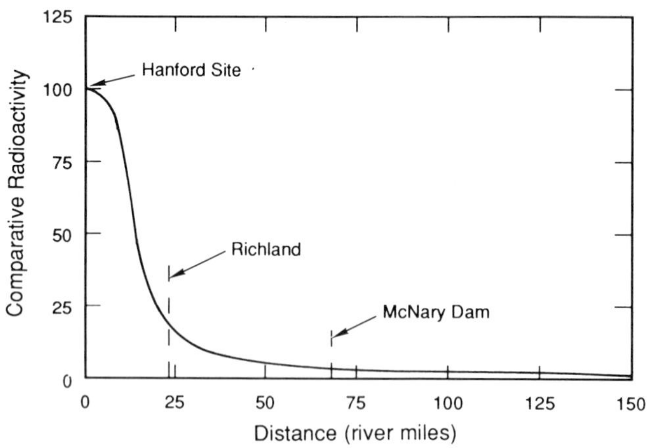

Fig. 8.7. Decline of radioactivity in drift plankton with distance downstream from the Hanford reactors (from Foster and Davis 1956).

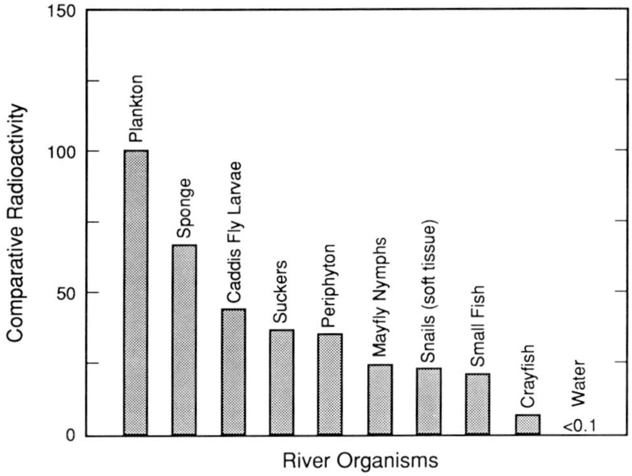

Fig. 8.8. Comparative radioactive density in Columbia River organisms downstream of the Hanford single-purpose reactors, 1955 (modified from Foster and Davis 1956).

in large organisms provided additional time for radioactive decay, but organisms living downstream became less radioactive than expected, based only on the time lapse for river flow (Foster and Davis 1956; Foster 1959a).

Small amounts of radioactivity were removed from the Hanford Reach when immature aquatic insects emerged as breeding adults, when mammals and birds fed on aquatic organisms, and when river water was withdrawn for irrigation and municipal use.

Some radioactivity was deposited in sediments on the bottom of the Columbia River. Deposition included the ions adsorbed on inanimate plankton particles, dead microorganisms that settled to the bottom, and excretia from river organisms (Foster and Davis 1956), as well as adsorption to particulate mineral material (Figure 8.8). Most of the long-lived radionuclides that remained in solution or were bound to suspended materials reached the Pacific Ocean.

Radioactivity in adult chinook salmon returning from the sea to spawn in the Hanford Reach was very low – less than 1×10^{-5} μCi/g. This was equal to the background level in Columbia River organisms upstream of Hanford at that time. Radioactivity in chinook salmon fry emerging after overwintering in the gravel was similar to the radioactivity in other young fish in the Hanford Reach (Watson 1952). Quantitative experiments showed that the amount of radioactivity tolerated by salmon fry was several times greater than the amount they encountered downstream from the Hanford reactors.

Uptake of Radioactivity by River Organisms

Amounts of radioactivity taken up by aquatic organisms varied widely (Figure 8.9). Uptake depended, in part, on the physical and chemical properties of different radionuclides and, in part, on the physiological requirements of aquatic organisms for the specific elements from which the radionuclides were derived. River temperatures and dilution of radionuclides by high spring flows were also factors affecting uptake.

Until 1955, gross radioactivity in drift plankton (mainly diatoms) was about 2000 times the radioactivity in water from the Hanford Reach (see Figure 8.8). Despite the large number of radionuclides in the cooling effluent, 30% to 50% of the radioactivity in plankton came from P-32, 25% to 50% from Cu-64, 5% to 15% from Na-24, and less than 10% from other combinations (Foster and Davis 1956).

Although P-32 accounted for less than 1% of radioactivity in river water, it contributed 70% to 95% of the radioactivity in most fish and benthic invertebrates. During the summer, when uptake was maximum, P-32 in small fish and caddis fly larvae was about 150,000 and 350,000 times greater, respectively, than in river water. The bioaccumulation of

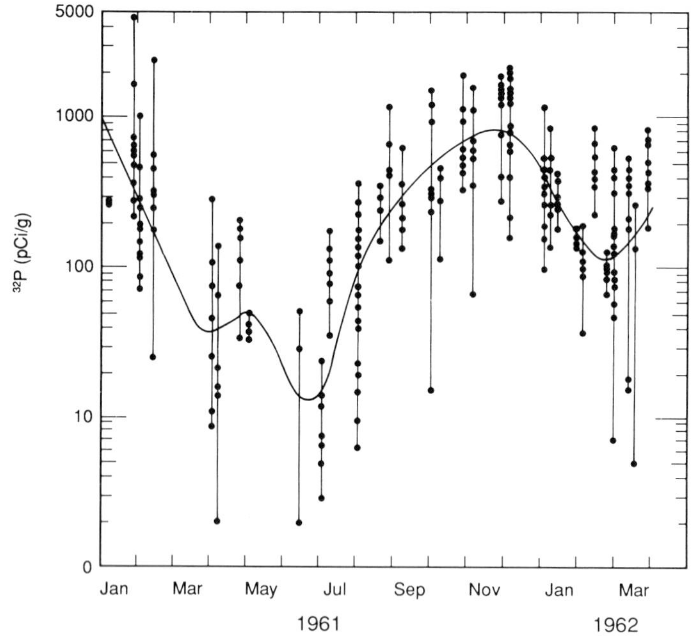

Fig. 8.9. Seasonal concentrations of P-32 in the tissues of mountain whitefish from the Hanford Reach, 1961 to 1962 (from Foster and McConnon 1965).

P-32 reflected the paucity of the phosphate ion in the Hanford Reach [about 0.03 part per million (ppm)] and the high metabolic requirement of river organisms for phosphate. Because radionuclides are deposited in tissues according to biological need, bone and scales acquired large amounts of P-32, while muscle and fat acquired little (Foster and Davis 1956).

The mechanisms involved in uptake of radioactivity from river water included adsorption, diffusion or absorption, and ingestion (Davis 1958; Davis and Foster 1958; Foster 1959a). Uptake also involved passage through the food web, when large river organisms ingested smaller ones; for example, when plankton were consumed by river vertebrates and invertebrates, which were in turn consumed by fish.

High amounts of radioactivity in Columbia River plankton and sponges were associated, in part, with their extensive surface areas, which facilitated adsorption. Surface texture, such as the gelatinous coverings and bacterial growths of some benthic invertebrates, modified adsorption patterns.

However, radionuclides that readily diffused through tissues were obtained directly from river water. In aquatic plants, the essential inorganic ions and some organic compounds needed for photosynthesis were taken up by diffusion. Animal membranes were more selective and absorbed only a few radionuclides in significant amounts. For example, fish immersed in cooling effluent absorbed about 130 times more Na-24 than any other radionuclide.

In supporting laboratory studies at Hanford, more than half of the P-32 added to water was taken up by plankton in the first hour. However, maximum levels were not reached until after 15 hours. Phosphorus-32 accumulated in sessile algae and bottom organisms much more slowly, and maximum amounts appeared in small fish in about 2 weeks. Drift plankton attained maximum radioactivity (from short-lived radionuclides) about 1 hour after entering an effluent zone (Foster and Davis 1956).

Aquatic insects downstream of the reactor discharges contained more radioactivity than Columbia River water at the same location. For example, larvae of the caddis fly within 5 kilometers (3 miles) of the reactors commonly acquired levels of gross beta radioactivity 1400 times greater than levels found in river water. Further, the radionuclides accumulated by aquatic insects were proportionately different in river water. In decreasing order, the five most abundant radionuclides in caddis fly larvae were P-32, Cu-64, Cr-64, Np-239, and Na-24 (Davis 1965).

Uptake of Radioactivity by Fish

Administrators at Hanford were always concerned with accumulation of radionuclides in aquatic organisms eaten by humans. Consumption of fish was one primary route through which radioactivity might be transferred. Feeding and migration habits greatly influenced uptake of radionuclides by fish (Davis and Foster 1958).

Amounts of radioactivity in different species of fish varied widely (Davis et al. 1958). However, the only radionuclides to accumulate significantly in edible muscle tissues of Columbia River fish were P-32 and Zn-65, both activation products (Foster and McConnon 1965; Foster and Soldat 1966).

Because the half-life of P-32 is only 14.3 days, concentrations in fish dropped rapidly whenever the fish slowed or ceased feeding. Thus, levels of P-32 in fish tissues varied seasonally (Figure 8.9). In contrast, Zn-65 had a half-life of nearly 9 months, and concentrations in fish tissues fluctuated less (Figure 8.10). Water temperature, which directly affected metabolism in fish, influenced the rate at which radionuclides accumulated. Uptake of P-32 was very slow at temperatures below 5°C, and peaked at 12 to 16°C (Foster and McConnon 1965).

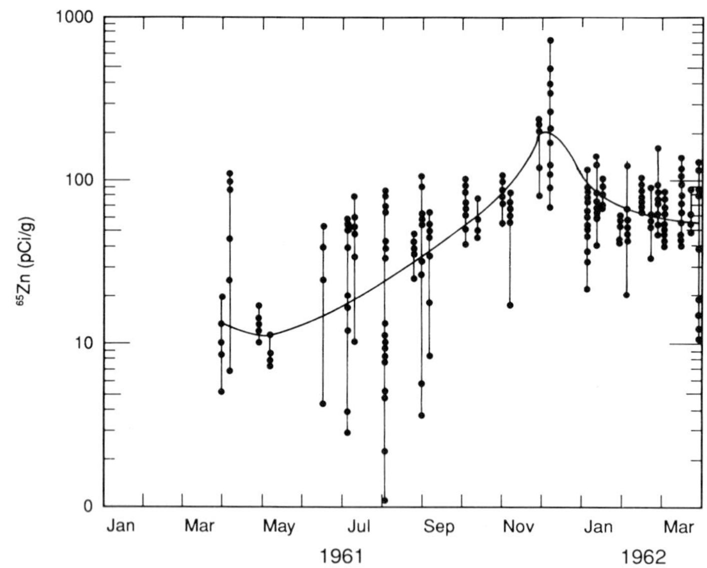

Fig. 8.10. Seasonal concentrations of Zn-65 in the tissue of mountain whitefish from the Hanford Reach, 1961 to 1962 (from Foster and McConnon 1965).

Radiobiological surveys at Hanford from 1951 through 1954 (Davis et al. 1952, 1953, 1954, 1955) revealed relatively high beta radioactivity during winter and spring in some whitefish that migrated each year from downstream to upstream of the reactor discharges. A maximum radioactivity of 5.3×10^{-4} μCi/g of muscle tissue was recorded from one whitefish in May, but only background or very low activities were recorded during the summer (Davis et al. 1954).

Radioactivity in whitefish near Hanford from June 1950 to December 1956 varied with collection area, season, age of fish, tissue, and dilution of reactor effluent. At sport-fishing areas available to the public, P-32 reached maximum concentrations in whitefish flesh of about 2×10^{-4} μCi/g. Maximum permissible concentrations of P-32 for humans at that time would be reached only if a person ate 2.7 pounds of whitefish each week (Watson and Davis 1957).

Beta radioactivity in smallmouth bass at Hanford peaked in the fall and dropped to background levels in April. Radionuclides were concentrated in all parts of bass, but were about 10 times higher in scales and bones than in skin and muscle. Phosphorus-32 accounted for more than 90% of the radioactivity, although it was less than 1% of the amount in river water. Most radioactivity in bass and other resident fish came from ingestion of other organisms, rather than by direct absorption. Radioactivity levels in bass were well below levels considered to be hazardous to people eating them. The limited half-life of P-32 precluded the buildup of radioactivity from year to year (Henderson and Foster 1957).

Adult salmon migrating up the mainstem Columbia River were exposed to radioactivity downstream from Hanford, but as they do not feed during upstream migration, they did not accumulate significant levels of beta radioactivity. On the other hand, juvenile salmon passing seaward through the Hanford Reach consumed aquatic invertebrates containing radionuclides and accumulated some radioactivity (Olson and Foster 1952). In the spring of 1957, outmigrating salmon fry averaged 4×10^{-4} μCi/g wet weight, compared to 6×10^{-4} μCi/g in redside shiners, a small resident fish (Davis 1958).

As long as the single-purpose reactors operated, suckers were the most radioactive species of large fish in the Columbia River (Olson and Foster 1952; Watson and Davis 1957; Davis et al. 1958). Suckers feed largely on sessile algae, so they took up radionuclides by a two-step process (water to algae, algae to fish). Uptake by whitefish, which feed primarily on aquatic insect larvae, was a three-step process (water to algae, algae to insect larvae, insect larvae to fish). Uptake by northern squawfish, which are piscivorous as adults, was a different three-step process (water to algae or insect larvae to prey fish to predator).

Measurement of Radioactivity in River Ecosystem

New techniques for analyzing radioactivity were developed by 1957, based on gamma-ray and coincidence gamma-ray spectrometry that permitted precise, sensitive, and rapid measurement of complex mixtures of radionuclides in biological samples without chemical separation. Most importantly, these techniques determined uptake and distribution of trace amounts of short- and long-lived radionuclides in Columbia River water and biota.

In September 1957 radioactivity was measured at a station 1.6 kilometers (1.0 mile) south of the outfall of the reactor farthest downstream. At that time, the most predominant radionuclides found in river water included Na-24, P-32, Cr-51, Mn-56, Cu-64, Zn-65, As-76, and Np-239 (Table 8.1). By the time water reached this location, it was found that radionuclides with short half-lives had decayed, and many radionuclides with longer half-lives, present at lower levels in the reactor effluent, were diluted below detection limits (Davis et al. 1958).

In August 1957, aquatic organisms at the same station reflected the different rates at which individual radionuclides were bioaccumulated from water in the Hanford Reach (Table 8.2). Many radionuclides in biota were not detected in river water when only a 2-liter sample was analyzed. Algae and phytoplankton contained the highest concentrations of radionuclides, much of which was adsorbed on their surfaces. Metabo-

Table 8.1.
Concentrations of the Predominant Radionuclides in Columbia River Water 1.6 Kilometers (1 Mile) Below the Farthest Reactor Downstream in September 1957 (from Davis et al. 1958). All eight single-purpose reactors were operating.

Radionuclide	Half-life	Readioactivity, (μCi/mL)
Na-24	15.1 hours	8.6×10^{-6}
P-32	14.2 days	2.4×10^{-7}
Cr-51	27.8 days	2.0×10^{-6}
Mn-56	2.6 hours	7.4×10^{-6}
Cu-64	12.8 hours	1.7×10^{-5}
Zn-65	245.0 days	8.9×10^{-8}
As-76	26.8 hours	1.9×10^{-6}
Np-239	2.3 days	9.7×10^{-6}

Table 8.2.
Concentrations of Radionuclides in Columbia River Organisms 1.6 Kilometers (1 Mile) from the Farthest Downstream Reactor in August 1957 (from Davis et al. 1958). All eight single-purpose reactors were operating.

Concentration, pCi/g of Wet Weight						
Radio-nuclide	Green algae	Freshwater sponge	Caddis fly larvae	Snail	Crayfish	Redside shiners [a]
Na-24	5.7×10^{-4}	7.3×10^{-4}	7.1×10^{-4}	1.8×10^{-3}	2.0×10^{-3}	1.1×10^{-3}
P-32	6.6×10^{-2}	4.5×10^{-3}	2.4×10^{-2}	1.4×10^{-2}	3.0×10^{-3}	2.4×10^{-2}
Sc-46	1.7×10^{-3}	9.5×10^{-5}	7.1×10^{-5}	2.9×10^{-5}		7.1×10^{-7}
Cr-51	7.9×10^{-3}	4.6×10^{-3}	6.0×10^{-3}	5.1×10^{-3}	4.8×10^{-4}	3.7×10^{-4}
Mn-54	1.0×10^{-3}		7.9×10^{-5}			
Mn-56	8.2×10^{-2}	2.4×10^{-2}				1.9×10^{-4}
Fe-59	1.6×10^{-3}					
Co-60	1.6×10^{-4}	1.2×10^{-5}	1.7×10^{-3}	1.8×10^{-5}		
Cu-64	1.1×10^{-1}	1.7×10^{-3}	7.1×10^{-3}	9.6×10^{-3}	5.1×10^{-4}	1.6×10^{-4}
Zn-65	1.2×10^{-2}	1.5×10^{-3}	2.0×10^{-3}	1.5×10^{-3}	3.7×10^{-4}	7.6×10^{-4}
As-76	6.9×10^{-3}	6.3×10^{-3}	5.2×10^{-4}	1.0×10^{-4}	6.5×10^{-5}	1.7×10^{-4}
Zr-95/ Nb-95	1.8×10^{-3}		6.3×10^{-5}			
Ru-103	1.2×10^{-3}		1.1×10^{-4}			
Ba-140	9.0×10^{-4}		4.2×10^{-5}	2.9×10^{-5}	1.9×10^{-5}	
La-140	3.3×10^{-3}	1.2×10^{-3}	3.5×10^{-4}			
Ce-141	1.9×10^{-3}					
Np-239	2.7×10^{-3}	4.0×10^{-4}	3.1×10^{-4}			
Sr-90	2.1×10^{-4}	5.7×10^{-5}	1.2×10^{-5}			1.6×10^{-5}

[a] Contents of digestive tract not included.

lism accounted for higher amounts of Na-24 and P-32 in larger organisms, such as snails, crayfish, and shiners. Although present in algae and plankton, Fe-59 was seldom transferred to organisms at higher trophic levels. However, Zn-65 was readily transferred because zinc was an essential trace element (Davis et al. 1958).

Concentration Factors for Radionuclides

The extent to which a radionuclide was concentrated by different organisms in the Hanford Reach could be calculated as a concentration factor (CF). The CF is the ratio between the amounts of a specific radionuclide in an organism and in river water on an equivalent basis (i.e., milligram per liter). The CF represents an equilibrium condition. Con-

Table 8.3.
Observed Concentration Factors for Significant Radionuclides in Columbia River Organisms (from Foster 1959a) [a]

Radionuclide	Algae	Insect Larvae	Fish
P-32	100,000–100,000,000	100,000	100,000
Zn-65	100,000	10,000	1,000–10,000
Cs-137 [b]	1,000–5,000	1,000	5,000–10,000
Sr-90	10,000	100	1,000
Na-24	100	100	100–1,000
As-76	10,000	1,000	100
Sc-46	100,000	1,000	10
Cr-51	100–1,000	100–1,000	10
Cu-64	10,000	1,000	10

[a] In 1959, many of the listed values were not well defined, and they were subject to revision as more data were obtained.
[b] Data for Cs-137 came from a laboratory pond containing water spiked with the radionuclide and not from the Columbia River.

centration factors for the most abundant radionuclides, calculated from data in early river studies, confirmed the relative importance of P-32, Zn-65, and Cs-137 in bioaccumulation (Table 8.3).

Concentration factors for As-76, Sc-46, Cr-51, and Cu-64 in fish were low compared to CFs for the same radionuclides in algae and insects. This indicated that food chains may lower the concentrations of a radionuclide in large aquatic animals (Foster 1959a,b). As might be expected, CFs also differed seasonally in relation to water temperature and metabolism in aquatic organisms. Radioactivity levels in Columbia River fish between winter and late summer differed by a factor of 75 (Davis and Foster 1958).

In general, early studies in the Hanford Reach demonstrated that the radionuclides most likely to have high CFs were readily assimilated by aquatic organisms, were retained for relatively long periods of time, and had long radioactive half-lives (Foster 1959b). These same properties account for bioaccumulation of a radionuclide in humans.

The Food Web Concept of Radionuclide Transfer

The first 10 years of studies in the Hanford Reach showed that radionuclides were transferred through the river ecosystem from one biotic component to the other in a food chain or, more precisely, a food web. This concept provided a basis for environmental monitoring pro-

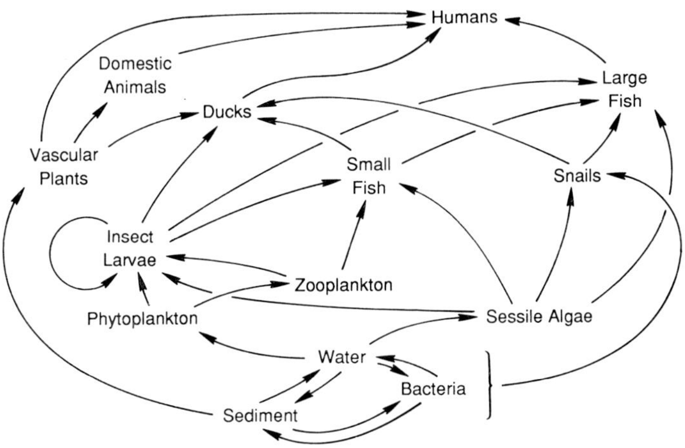

Fig. 8.11. The basic food web illustrating the main routes for transfer of radionuclides among Columbia River organisms (from Davis 1960).

grams at Hanford and, in later studies, for developing computer models to quantify the transfer of artificial radionuclides in the biosphere.

The food web illustrates how radionuclides accumulate in aquatic organisms, when their food is the predominant route of uptake (Figure 8.11). The extent that radioactivity concentrates in a consuming organism is influenced by the time a radionuclide requires to pass through links in the food web.

Any radionuclide metabolized internally or adsorbed externally by an aquatic plant or animal is available to a consumer, the next link in the food web. Metabolic processes of animals in each link select radionuclides according to the kind available and amounts assimilated. Radionuclides not assimilated are discharged as waste products (Davis 1960).

Generally, the time between transfer of radioactive material from one life form to another allowed decay of any of the short-lived radionuclides still present. Therefore, radioactivity among components of the aquatic food web was reduced downstream from the Hanford reactors (Foster and Davis 1956). From a practical viewpoint, this meant that concentration of radionuclides with short half-lives in fish and other higher life forms was not important from the standpoint of human health.

Evaluating Offsite Exposure to Radioactivity

It was recognized before the reactors were built that people who used the Columbia River downstream (e.g., in the city of Pasco) from Hanford would be exposed to small amounts of radioactivity in river water (Figure

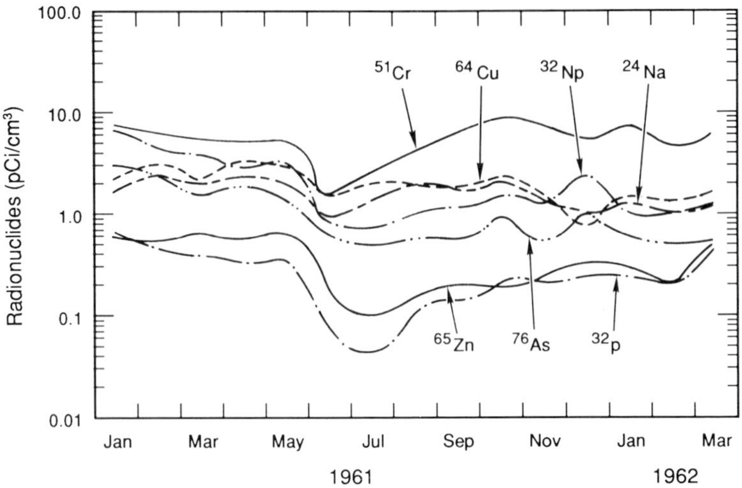

Fig. 8.12. Changes in availability of seven prominent radionuclides in the Columbia River at Pasco, Washington, 1961 to 1962 (from Foster and McConnon 1965).

8.12). As a result, radioactivity at municipal water intakes was closely monitored (Junkins 1960). In addition, considerable effort was made to identify how humans might be exposed, the radionuclides and pathways of greatest importance, and the probable magnitude of exposure. Results were referenced to recommendations of the International Commission on Radiological Protection, the National Council on Radiation Protection (NCRP) and Measurements, and the Federal Radiation Council (Parker 1956; Parker et al. 1964).

Biomedical studies at Hanford, conducted separately from aquatic studies, examined the effects of radioactivity in Columbia River water on mammals. In one study, the uptake and retention of radionuclides were evaluated by providing rats with undiluted reactor effluent as their sole source of water for periods up to one year. Average amounts of radioactivity appearing in the bone and combined soft tissues of rats were only a fraction of the maximum permissible body concentrations for humans in official recommendations. Yet the water consumed by rats contained radionuclides at levels many times higher than did river water downstream of the effluent mixing zones (Davis et al. 1958).

Local fish and produce from irrigated farms represented the most important routes of exposure to humans. The activation products P-32 and Zn-65 were of greatest concern. The short-lived radionuclides, Na-24, As-76, Np-239, and I-131, were more important in drinking water than in farm products. Sodium-32 contributed seasonally to external exposure of ardent swimmers. While Cr-51 was the most abundant radionuclide in river water, it contributed only slightly to human exposure and was not concentrated in food webs (Foster and Soldat 1966).

Individuals exposed to the most radioactivity were probably those who ate unusually large amounts of fish caught just downstream of the reactors. The estimated radiation dose for bone in these individuals was high because fish contained P-32. Later, worldwide fallout from atmospheric testing added Sr-90, which also accumulated in bone, to farm products. Drinking water was the only source of water-borne radionuclides for most people living near Hanford and their gastrointestinal tracts received the greatest exposure. Virtually all long-lived radionuclides, such as Sr-90 and Cs-137, that appeared in humans originated in the atmosphere (Foster and Soldat 1966).

In 1962, a whole-body counter first became available for environmental surveillance work at the Hanford Site. Radioactivity was measured in residents along the Columbia River at Ringold, 21 kilometers (13 miles) beyond the reactor farthest downstream, and the results were compared with their diets. The main radionuclides detected were P-40, Zn-65, Cs-137, and I-131. Phosphorus-32, a beta emitter, was measured by radiochemical analysis. Phosphorus-40 was a naturally occurring radionuclide. The most likely source of Zn-65 and P-32, both associated with reactor effluent, was irrigation water on local crops. The source of Cs-137 and I-131 was primarily worldwide fallout, which was relatively high at the time. Amounts detected by whole-body counts were approximately 2% of the maximum permissible whole-body burden established by the NCRP (Nelson and Foster 1965).

Subsequent evaluations confirmed that the total exposure of people near Hanford to radionuclides was always within applicable regulatory limits. This conclusion was supported by whole-body counting of large numbers of local residents (almost 6000 people from 1959 through 1964). Whole-body counts, in fact, showed that the actual intake of radioactivity from Hanford operations by local residents was less than estimated intake from food and water.

Radionuclide releases – later studies (1961–1971)

Battelle Memorial Institute (Battelle) assumed responsibility for onsite research and development activities at Hanford in January 1965. Many people involved in Columbia River studies were reassigned from the General Electric Company to Battelle, and their research continued.

However, additional staff members were hired, and new studies were designed to provide new perspectives.

Research on thermal effects expanded in both the laboratory and field. However, amounts of cooling water released to the Hanford Reach decreased with the closure of the single-purpose reactors. The last year that all reactors operated was 1964. Four reactors remained in production until February 1968, three until April 1969, and two until February 1970. Studies dealing directly with effects of cooling water discharges ceased after January 1971.

Radioanalyses of river organisms in the Hanford Reach in early years dealt primarily with seasonal variations, species differences, and geographic distribution of radionuclides. Before 1956, analyses measured only total beta radioactivity (Watson et al. 1970a).

In the 1960s, ecological studies with Na-24 and P-32 (primarily beta emitters) and with Cr-51 and Zn-65 (primarily gamma emitters) were undertaken. Thermal effects studies also were conducted. By this time, two hydroelectric dams (Priest Rapids and Wanapum) had been constructed upstream of Hanford that regulated flows and modified, to a degree, ecological functions in the Hanford Reach.

Reexamination of Radionuclide Cycles in Biota

Studies were initiated in 1966 to update seasonal information on variability of reactor-produced radionuclides in organisms from the Hanford Reach. Improved instruments were used to measure radioactivity. Extensive data on three radionuclides (P-32, Zn-65, and Cr-51) were collected from three organisms representing an autotroph (net plankton), a herbivore (caddis fly larvae), and an omnivorous fish (redside shiner).

The high-flow period of spring and early summer had two effects on radionuclide concentrations (Figure 8.13). First, it slowed the increasing rate of uptake in biota as light and temperature conditions for metabolism improved. Second, biota concentrations declined as a result of greater dilution in river water (Watson and Cushing 1969; Watson et al. 1970a). Essentially, these results agreed with earlier investigations in the Hanford Reach.

The radionuclides most highly concentrated in biota were P-32, Zn-65, and Cr-51. Of these, Cr-51 was most abundant in river water. Radioactivity followed a pattern of high in winter and low in summer, particularly in organisms with high surface-to-volume ratios whose mode of uptake was largely adsorption. In spring, dilution of radionuclides in river water by run-off overwhelmed the more rapid uptake brought about by sea-

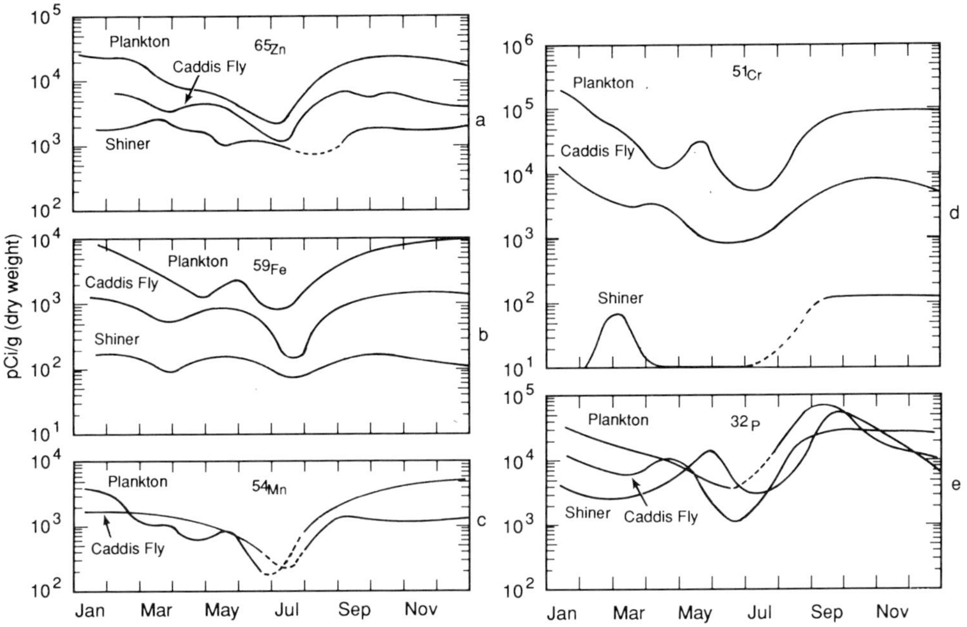

Fig. 8.13. Seasonal changes in levels of (a) Zn-65, (b) Fe-59, (c) Mn-54, (d) Sr-51, and (e) P-32 in plankton (mixed species), caddis flies, and redside shiner in the Hanford Reach, 1961–1968 (from Watson and Cushing 1969).

sonal changes in light and temperature. However, seasonal effects were less pronounced at higher trophic levels (Watson et al. 1970b).

More precise analysis of individual radionuclides in fish revealed an affinity for different body organs. Sodium-24, Mn-54, and Sr-89/90 appeared in bone; Cs-137 in muscle; P-32, Fe-59, and Cu-64 in gut contents; Sc-46 and Co-60 in spleen; Cr-51 in blood; and Zn-65 in retina (Seymour 1964).

Uptake and Transport of Radionuclides by Plankton

Accumulation and transport of P-32 and Zn-65 by net plankton (free-floating organic material retained by a No. 20 mesh plankton net) were examined in 1963 and 1964. Radionuclide input from the cooling effluent was relatively constant over this period because all reactors remained operating.

Seasonal values of P-32 and Zn-65 were lowest during the high-water period of spring and early summer. This period coincided with maximum biomass, since production of net plankton rapidly increased in the spring.

The result was a combination of more plankton per unit volume of water at a time when radionuclide levels per volume of water were reduced by dilution. Concentration factors for plankton ranged from 5000 to 118,000 for P-32, and from 300 to 19,000 for Zn-65 (Cushing and Watson 1966; Cushing 1967b). The CFs reported previously in net plankton (i.e., Coopey 1953b; Davis et al. 1956) were based only on beta radioactivity and, therefore, were much lower.

Radioactivity associated with net plankton represented 1% to 2% of the total transport of P-32 and Zn-65 from the Hanford Reach. Although lesser amounts of P-32 and Zn-65 were present per unit weight of plankton in spring and late summer, larger amounts of the radionuclides were transported downriver because the plankton biomass was much greater (Cushing and Watson 1966).

Uptake of Radionuclides by Periphyton

Periphyton readily acquired radionuclides from river water. The effect of environmental conditions on uptake of P-32 and Zn-65 by new periphyton colonies was examined next. From August 1963 to May 1964, periphyton was grown on glass slides placed in the Hanford Reach below the effluent discharges.

Concentrations of Zn-65 in periphyton were high in fall and early winter and low in March. However, concentrations of P-32 fluctuated with no apparent trend. Net production rates, starting from a bare slide surface, were highest in the spring and summer, lowest during winter. The fall-winter community of periphyton consisted mostly of diatoms. Accumulations of Zn-65 and P-32 were highly correlated with dry and ash weight, supporting adsorption as the primary means of uptake of the two radionuclides by periphyton (Cushing and Watson 1966; Cushing 1967a).

Movement of Radiotagged Fish

Zinc-65 is a tracer useful in monitoring the movement of fish because it is readily bioconcentrated and has a relatively long half-life (245 days). In 1965, fish movement in the Hanford Reach was traced in relation to the effluent outfalls. Radioactivity was measured in the eyes and gastrointestinal (GI) tract, a concentration site for Zn-65.

Levels of Zn-65 in fish from the Hanford Reach were lowest during summer because high spring flows diluted the reactor discharges. Radioactivity increased in the fall as the river's dilution capacity declined (Figure 8.14).

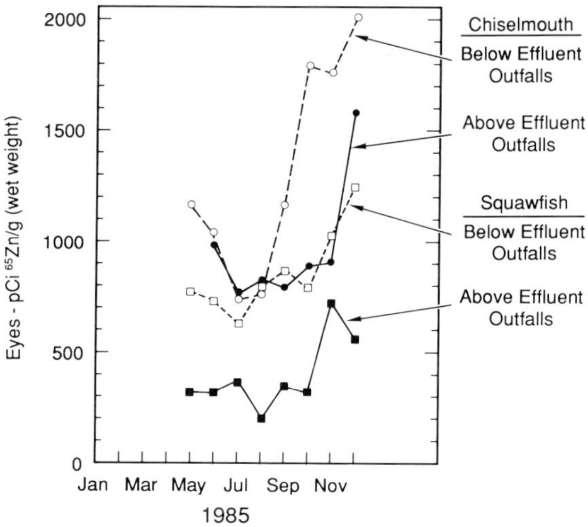

Fig. 8.14. Seasonal changes in concentration of Zn-65 from resident fish in the Hanford Reach during 1965 (from Cushing and Watson 1966.)

Radioactivity was translocated not only up the mainstem Columbia River, but into the Yakima River, a tributary downstream of Hanford, by fish exposed in the Hanford Reach. More than half the fish collected near Priest Rapids Dam, upstream of Hanford, had been exposed to radionuclides from the cooling effluent.

Mountain whitefish started moving into the Yakima River in July, but not upstream in the Hanford Reach until September. In December 1965, whitefish from the Yakima River had Zn-65 values (pCi/g wet weight) of 1300 (eye) and 680 (GI tract), compared to those from the Hanford Reach of 540 (eye) and 1200 (GI tract) (Watson 1966). Whitefish taken below Priest Rapids Dam carried Zn-65 values of 1060 (eye) and 1400 (GI tract) (Cushing and Watson 1966).

Upstream Dispersion of Radionuclides by Caddis Flies

All aquatic organisms downstream of the reactors were tagged with the radionuclides in the cooling effluent. The possibility that flights of aquatic insects emerging in the Hanford Reach might spread radioactivity upriver was briefly examined in 1966.

Levels of Zn-65 above background were detected in adult caddis flies from shoreline swarms as far as 16 kilometers (10 miles) upstream from the uppermost reactor. Radioactivity among individual caddis flies varied highly, and ranged from an average of 200 pCi/g dry weight just upstream from the uppermost reactor to 5 pCi/g farther upstream. As a radiological hazard, upstream dispersal of radioactivity by caddis flies was judged insignificant (Coutant 1967b, 1982).

Elimination of Radionuclides from Benthic Organisms

The rates at which radioactivity was lost from some benthic organisms was examined in the field. Limpet snails and colonies of algae, diatoms, and mixed periphyton were transplanted by researchers from downstream of the reactors to uncontaminated areas upstream.

Elimination had two main components. Radionuclides were first lost rapidly, then gradually. Rapid loss was due to the voidance of gut contents in limpet snails and "washing" of radioactivity from transplanted colonies. Gradual loss was due to physical decay of radionuclides and metabolic turnover (Coutant 1967a). Loss of Zn-65 from limpets was most rapid in October, slowing through winter from March to April (Coutant 1968).

Thermoluminescent Dosimetry Measurements

Radiation dose to organisms in an aquatic ecosystem was usually estimated by measuring radioactivity from composite samples of small organisms or from large specimens. Determining actual exposure to radiation from water or short-lived radionuclides was more difficult. In the late 1960s, development of thermoluminescent dosimeters (TLD) containing lithium fluoride allowed in situ measurement of radiation doses in the Hanford Reach.

The estimated radiation dose at the upper surfaces of cobblestones 7 kilometers (4 miles) downstream of the effluent outfalls was 220 millirads per day (mR/day), equivalent to 80 roentgens (R) annually. Radioactivity in the water contributed 30% to 50% of the total dose to the benthic organisms. Maximum dose to fish 19 kilometers (12 miles) downstream of the farthest reactor was 23 mR/day (8.3 R annually). During the fall, when metabolic rates and radionuclide uptake were high, a consistent inverse relationship existed between measured dose rate and weight of fish (Watson and Templeton 1973). No harmful effects from radiation in the Hanford Reach were expected because test fish, exposed to radionuclides in undiluted reactor effluent during bioassays at concentrations several times greater, were not affected.

Doses to periphyton upstream of, in, and downstream of the reactor discharges were measured with TLDs from June to July, when flows were high and the reactor effluent was well diluted. The mean dose rate upstream of the reactors was 0.1 mR/day, about the same as background radiation, regardless of water depth. The mean dose rate in the effluent was 285 mR/day near the surface of the river and declined with depth in relation to periphyton photosynthesis. The mean dose rate 24 kilometers (15 miles) downstream of the outfalls was 9.6 mR/day for TLDs shielded from light (to reduce periphyton growth) and 12.2 mR/day for unshielded TLDs, and the highest dose rates were at mid-depths (Lappenbusch et al. 1971).

At sites exposed to the effluent, TLDs indicated that P-32, Sc-46, Cr-51, and La-140 accounted for 70% of the radioactivity. The average dose rate in effluent was more than 20 times the dose rate downstream, but mean radioactivity was only four times greater. The difference was probably because the short-lived radionuclides had decayed downstream, or heat in the upstream effluent had affected metabolic assimilation of radionuclides by periphyton (Lappenbusch et al. 1971).

Radionuclides in Biota at the Columbia River Outlet

Most of the 60 radionuclides in the reactor effluent disappeared, or were reduced to trace amounts, before water passing through the Hanford Reach reached the Pacific Ocean, 595 kilometers (370 miles) downstream. Reduction was due largely to physical decay or retention of radionuclides by sediments and biota.

Near the outlet of the Columbia River, the principal gamma emitters remaining from the reactor discharges at Hanford were Cr-51, Np-239, and Zn-65. Gamma emitters found in brown algae, mussels, and oysters near the outlet included Cr-51, Mn-54, Zn-65, Zr/Nb-95, Ru-103, 106, and Ce-141, 144. Radionuclides other than Cr-51 and Np-239 were added to the river ecosystem by atmospheric fallout. Surveys along the coasts of Oregon and Washington near the river's outlet in April 1959 and 1960 generally showed the highest radioactivity in plankton and the sessile algae. Zinc-65, contributed by the Hanford reactors, was the gamma emitter of greatest biological importance, particularly in mollusks (Watson et al. 1961, 1963).

Transport and behavior of radionuclides downstream from Hanford

Extensive monitoring of water in the lower Columbia River though the 1950s showed that radioactivity introduced by the reactor effluent remained well within accepted radiological standards for drinking water (ICRP 1959) at all points of withdrawal downstream. For protection of human health, this had always been a strong environmental concern at Hanford. But much more needed to be learned about the downstream transport and behavior of specific radionuclides.

Mechanisms regulating downstream transport and association of radionuclides with sediments in the lower Columbia River were examined in the 1960s. By this time, individual radionuclides could be measured much more precisely by multidimensional gamma-ray spectrometry, a new radiological technique (Perkins 1965). Scientists realized that a complete inventory of radionuclides in the lower river ecosystem during the 1960s would include three main components: 1) long-lived activation products released in reactor effluent since 1944, 2) some fission products from irradiating trace amounts of uranium in river water and from fuel cladding ruptures, and 3) substantial amounts of fission products deposited over the Columbia River drainage basin from atmospheric tests.

Radiological monitoring to date had indicated that radionuclides tended to accumulate in the sediments of Lake Wallula (the impoundment behind McNary Dam), and in other slackwater areas downstream of Hanford (Nelson et al. 1964). However, no ecologically significant accumulations of long-lived fission products, which would have resulted from accidental fuel cladding rupture in Hanford's reactors, had been detected in the lower Columbia River.

Transport of Radionuclides in River Water

The removal of radionuclides from Columbia River water by natural processes could be calculated from measurements of radioactivity in the reactor effluent and at various points downstream. About 35% of the most abundant radionuclides disappeared in the first 64 kilometers (40 miles) from Hanford. Incorporation in sediments accounted for most of this removal (Nelson and Perkins 1962).

The relative abundance of radionuclides changed during downstream transport between Pasco and Vancouver, Washington, a distance of 595 kilometers (370 miles; Table 8.4). Radionuclides depleted from the river water were Sc-46, Mn-54, Co-58, Fe-59, Co-60, Zn-65, and Zr-95/Nb-95. These radionuclides occurred mainly as particulates in reactor effluent or were readily sorbed by particulates in river water. Radionuclides transported with little depletion were Cr-51, Ru-106, Sb-124, and Ba-140. They entered the river in soluble phase and were not strongly associated with particulates (Perkins et al. 1966).

Table 8.4.
Percent of Several Radionuclides in the Particulate Phase When Discharged in Hanford Reactor Effluent and at Locations in the Downstream Columbia River (from Perkins et al. 1966). Data are expressed as a percent.[a]

Radionuclide	Effluent Discharge	Pasco, Washington	Hood River, Oregon	Vancouver, Washington
Sc-46	36.0	74.0	85.0	89.0
Cr-51	2.4	6.4	4.0	7.6
Mn-54	2.6	20.0	88.0	88.0
Co-58	4.2	27.0	ca 83.0	ca 83.0
Fe-59	64.0	85.0	80.0	80.0
Co-60	1.8	26.0	80.0	91.0
Zn-65	1.8	14.0	64.0	76.0
Zr-95/Nb-95	ca 69.0	69.0	68.0	85.0
Ru-106	32.0	24.0	15.0	17.0
Sb-124	1.1	3.4	–	ca 5.9
Ba-140	2.3	9.0	–	37.0

[a] Samples were collected and analyzed from January to March 1965 when river flows were low.

Physical decay during downstream transport was negligible for most radionuclides other than Cr-51 and Ba-140. Decay was influenced by the time required for passage downstream to Vancouver, near the Columbia River outlet. Transport time depended on river flow volume, and varied from 3 to 14 days. During low flows, when transport time between Pasco and Vancouver was extended, decay reduced the radioactivity of Cr-51 as much as 30% and Ba-140 as much as 65% (soluble phases only). Radioactive decay of the other nine radionuclides (total transport values) was less than 10% (Nelson et al. 1966a; Perkins et al. 1966).

Further analysis indicated that, from January 1964 through September 1966, downstream transport of the eight most abundant radionuclides from Hanford's reactors averaged 9190 curies per week at Pasco and 6630 curies per week at Vancouver, Washington (Haushild et al. 1971).

Association of Radionuclides with Particulates

Net transport of radionuclides to the Pacific Ocean was associated with movement of particulate material. Net transport included activation products from Hanford's reactors and fission products from atmospheric fallout.

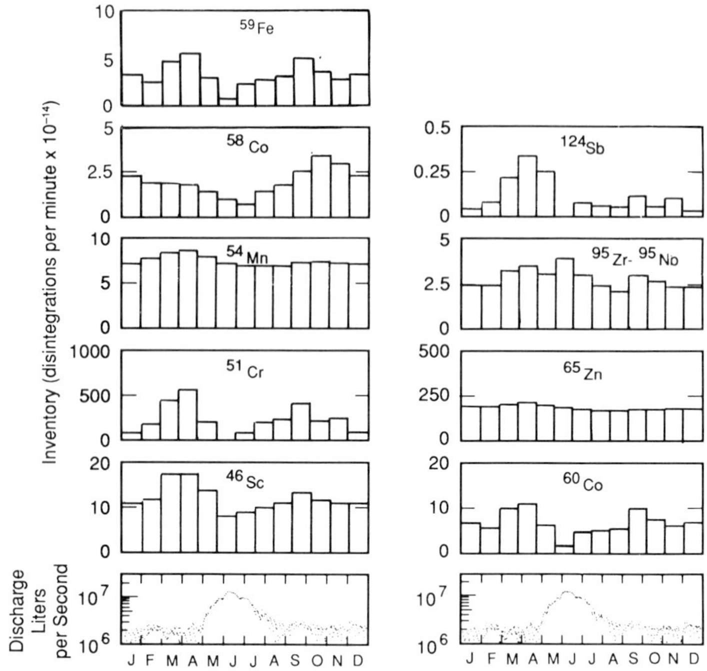

Fig. 8.15. Inventory of radionuclides in Columbia River sediments between Pasco and Vancouver, Washington, from January 1964 through January 1965 in relation to river discharge (from Perkins et al. 1966).

During low flow periods, radionuclides in particulate form or readily sorbed to particulates were depleted from river water by factors of from five to ten as the suspended particulates settled. During the spring freshet, net transport of radionuclides to the river's outlet equaled or exceeded releases upstream because sediment was resuspended at points between. Thus, the amount of radioactivity in bottom sediments between Pasco and Vancouver reflected, to a degree, current or recent discharges (Figure 8.15).

Overall, gross radioactivity was depleted from flowing river water by as much as 90% (Nelson et al. 1966a; Nelson and Haushild 1970). While the most important depletion mechanisms were sorption and assimilation by suspended particulates and deposition in sediment deposits, they also included physical decay, sorption and assimilation by aquatic life, and discharge to the Pacific Ocean.

Theoretically, if input of activation products from Hanford's reactors had remained steady, a state of equilibrium would be reached in a few months or years, depending on the half-lives of contributing radionuclides. In this situation, the amount of a radionuclide released to the Hanford Reach each day would balance the amount lost by decay, downstream transport, assimilation, and other mechanisms.

Inventories of Radionuclides in Sediments

A radionuclide inventory is an estimate of the amount of a radionuclide present in the river sediment at a specific location on a specific date. Inventories were made in the lower Columbia River soon after the amount of radioactivity from Hanford's reactors peaked in 1962, 1963, and 1964.

Through 1965, inventories of total radioactivity in the sediments downstream from Hanford varied from about 11,000 to 38,000 curies, and consisted largely of Cr-51 and Zn-65. Amounts of radionuclides in sediments were relatively low downstream as far as Richland. Much radioactivity appeared in Lake Wallula sediments, the first downstream area where particulates could be deposited. Scouring in Lake Wallula during the spring freshet passed about 30% of the radioactivity deposited each year downstream. Sediments tended to retain rather than release isotopes, and Cr-51 was reduced from the hexavalent to the trivalent state (Nelson et al. 1966b).

In October 1965, an inventory showed about 16,000 curies of gamma-emitting radionuclides in sediments between the Hanford Reactors and McNary Dam. Concentrations of Cr-51, Zn-65, Co-60, Mn-54, and Sc-46 generally were much higher in sediments upstream of McNary Dam than

Table 8.5.
Estimated Amounts of the Five Major Gamma Emitters in Sediments of the Columbia River on October 1, 1965 (from Nelson and Haushild 1970).

Radionuclide	Amount of Radionuclide Ci		
	Calculated from Bed Sediment Data		Calculated from Radionuclide Discharge Data
	Pasco	Pasco to McNary Dam	Pasco to McNary Dam
Cr-51	600	10,500	10,100
Zn-65	630	3,600	4,900
Co-60	80	250	430
Sc-46	40	330	360
Mn-54	80	130	260
Total	1,430	14,800	16,000

Table 8.6.
Estimated Amounts of the Five Major Gamma Emitters Transported in the Columbia River at Pasco, Washington, and Umatilla, Oregon, from July 1, 1965 to June 30, 1966 (from Nelson and Haushild 1970)

Radionuclide	Total Discharge, Ci		Solute Discharge,[a] as a Percent of Total Discharge	
	Pasco	Umatilla	Pasco	Umatilla
Cr-51	362,000	288,000	92	94
Zn-65	10,100	5,790	59	34
Sc-46	2,070	1,210	18	15
Co-60	170	120	34	27
Mn-543	500	350	44	28
Total	375,000	295,000	91	93

[a] The part of the total radionuclide discharge that was not solute is the particulate radionuclide discharge; a particle of 0.30 μ was used to separate solute and particulate radionuclides.

in the Hanford Reach (Table 8.5). The cause was deposition of fine-grained, suspended particulates in the slower moving waters of Lake Wallula and the radionuclides associated with these particulates (Nelson and Haushild 1970). Total downstream transport from the Hanford Reach was greatest for Cr-51 and Zn-65 (Table 8.6).

As a result of their affinity for specific radionuclides, the finest sediments, usually contained the greatest amounts of radioactivity. Total radioactivity was about 319 $\mu Ci/m^2$ in fine sediments from Lake Wallula compared with 27 $\mu Ci/m^2$ in coarse sand and gravel. Further, concentrations of Sc-40, Co-60, and Zn-65 generally decreased with depth in Lake Wallula sediments.

Physicochemical Affinity of Particulates

Starting in 1969, the physicochemical affinity of particulates for radionuclides was studied closely. Materials carried by river water were first isolated by centrifugation, and their sizes were determined. The strontium cation exchange capacity (Sr CEC) of isolated materials was then measured.

Amounts of particulates suspended in the river changed seasonally from a low of 2002 μg/L in winter to a high of 34,680 μg/L in spring. During fall, winter, and summer, the bulk of the material in suspension measured 53.0 to 2.0 microns (silt), while only 12% to 30% measured < 2.0 microns (clay). During high flows in spring, most of the suspended material was clay sized. More than 40% of the Sr CEC was in material < 0.5 microns in diameter, which was often classified as soluble (Wildung 1970). Thus, the Sr CEC in equivalent volumes of river water was 6 to 26 times higher in the spring than during other periods (Wildung et al. 1972).

Particulates in suspension consisted of mixed organic, mineral, and organomineral detritus. Electromicrographic analysis indicated that the size distribution of the particulates depended primarily on their density. Relatively large organic particles low in density were mixed with smaller mineral particles high in density. Because largely undercomposed organic particles had lower charge densities and smaller surface areas, they contributed less to ion sorption than smaller mineral particles (Wildung and Schmidt 1971a).

Seasonal changes in mixtures of primary minerals (quartz, feldspar, amphibole) passing down the Columbia River were not pronounced. But portions of layer silicate minerals (mica-illite, chlorite, and montmorillonite) increased from April to August because of input from terrestrial sources. Irrigation return water added montmorillonite to the river. The increase in Sr CEC with decreased particle size, previously observed, was caused by differences in mineral type and concentration, as well as by increased reactive area (Wildung and Schmidt 1971b)

Radioactivity in ecosystem after reactor closures

Amounts of radioactivity entering the Hanford Reach were influenced by the number of single-purpose reactors operating, which changed as a result of startups and shutdowns. While eight reactors operated in 1964, three were shut down permanently in 1965. Subsequently, shutdowns continued, and no reactor remained in operation after January 1971. The response of the river ecosystem to changes in releases of reactor-derived radioactivity were examined during shutdowns and after closure of the last reactor.

Shutdown of Three Reactors in 1965

The F, H, and DR Reactors were shut down in early 1965. Radioactivity in river water and aquatic organisms was monitored from August through September 1964 and from August through October 1965 to

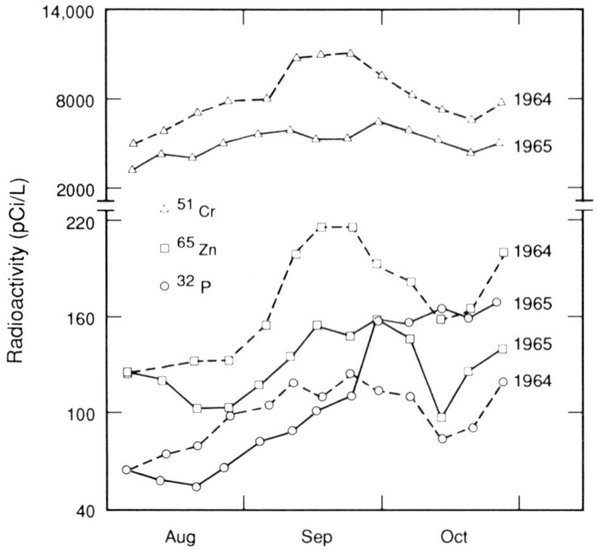

Fig. 8.16. Concentrations of P-32, Cr-51, and Zn-65 in the Hanford Reach during the fall of 1964, when eight single-purpose reactors were operating, and 1965, when only five were operating (from Cushing and Watson 1966).

detect significant changes. Radioactivity in the Hanford Reach was near an annual peak during these periods because river flows were low.

Amounts of P-32, Cr-51, and Zn-65 in river water during the two seasons were compared. In 1964, radioactivity peaked near 11,000 Ci/L for Cr-51, 220 for Zn-65, and 120 for P-32 (Figure 8.16). There was some variation from month to month. Radioactivity was lower in 1965 than in 1964, except for P-32 during October (Cushing and Watson 1966).

Measurement, using multidimensional gamma-ray spectrometry, of 18 radionuclides in river organisms downstream of the reactor areas reflected the decrease in radioactivity of the river water. The decline was more uniform in primary producers than in other organisms. Radioactivity of four radionuclides (Na-24, Mn-56, Cu-64, and Zn-65) decreased about 80% in primary producers (plankton and algae) and about 55% in herbivores and filter feeders (fish and benthic invertebrates). But radioactivity in a filter-feeding clam showed no decrease. Radioactivity decreased by about 72% in crayfish, an omnivore (Cushing and Watson 1966; Watson et al. 1966).

Temporary Shutdown of All Reactors in 1966

In July and August 1966, a labor dispute caused the five single-purpose reactors in operation to be shut down for 6 weeks. Radioactivity in river organisms was measured before and during the shutdown and again after startup. Effects from shutdown and startup were superimposed on the annual cycle of radioactivity associated with dilution of effluent by seasonal flows.

After shutdown, formation of radionuclides by neutron activation in the reactor core ceased. However, small quantities of long-lived radionuclides, such as Sc-46, Mn-54, Zn-65, and Co-60, continued to be released as films on reactor fuel cladding and piping eroded and were desorbed. Within a few days, the radionuclides still leaving the reactors dropped to very low levels. However, quantities of Sr-46, Mn-54, Zn-65, and Co-60 in the Hanford Reach did not drop as much as was expected from the effluent data. These radionuclides apparently were retained in river sediments and recycled to water by continued scouring and leaching (Hall et al. 1970).

Shutdown was followed by rapid decrease in radioactivity in plankton, periphyton, other invertebrates, and juvenile fish. In primary producers, P-32 fell below detection limits 7 days after shutdown, and Cr-51 fell below detection limits after 5 weeks (Figure 8.17). Levels of Zn-65, Mn-54, Fe-59, and other radionuclides declined less, and measurable amounts remained at all time. Fish lost P-32 rapidly, Zn-65 more slowly. After startup, radionuclides in river organisms neared equilibrium levels in 2 to 4 weeks. The loss and return of radionuclides in adult fish were not as extensive as in other trophic levels (Watson et al. 1967, 1970a).

Depletion of Radionuclides after Reactor Closure

Studies on transport and depletion of radionuclides were continued after the final 1971 closure to examine the response of a river ecosystem after an influx of artificial radionuclides stopped. In fact, the lower Columbia River was ideal for examining the depletion of radionuclides in bottom sediments and for evaluating natural cleaning processes that might take place after an accidental release of radioactive materials.

Sampling began in April 1971. After 2 months, most of the short-lived radionuclides in the Columbia River disappeared through physical decay. But some residual, long-lived radionuclides of Hanford origin remained, largely in sediments in Lake Wallula, behind McNary Dam. The most abundant radionuclides left were Fe-55, Zn-65, Eu-155, C-60, Eu-152, Mn-54, and Sc-46 (Figure 8.18). Also present, but at much lower levels, were Sb-125, Cs-137, Ce-144, and Pu-139. Concentrations of radionuclides with short half-lives decreased rapidly with depth in sediment cores. But

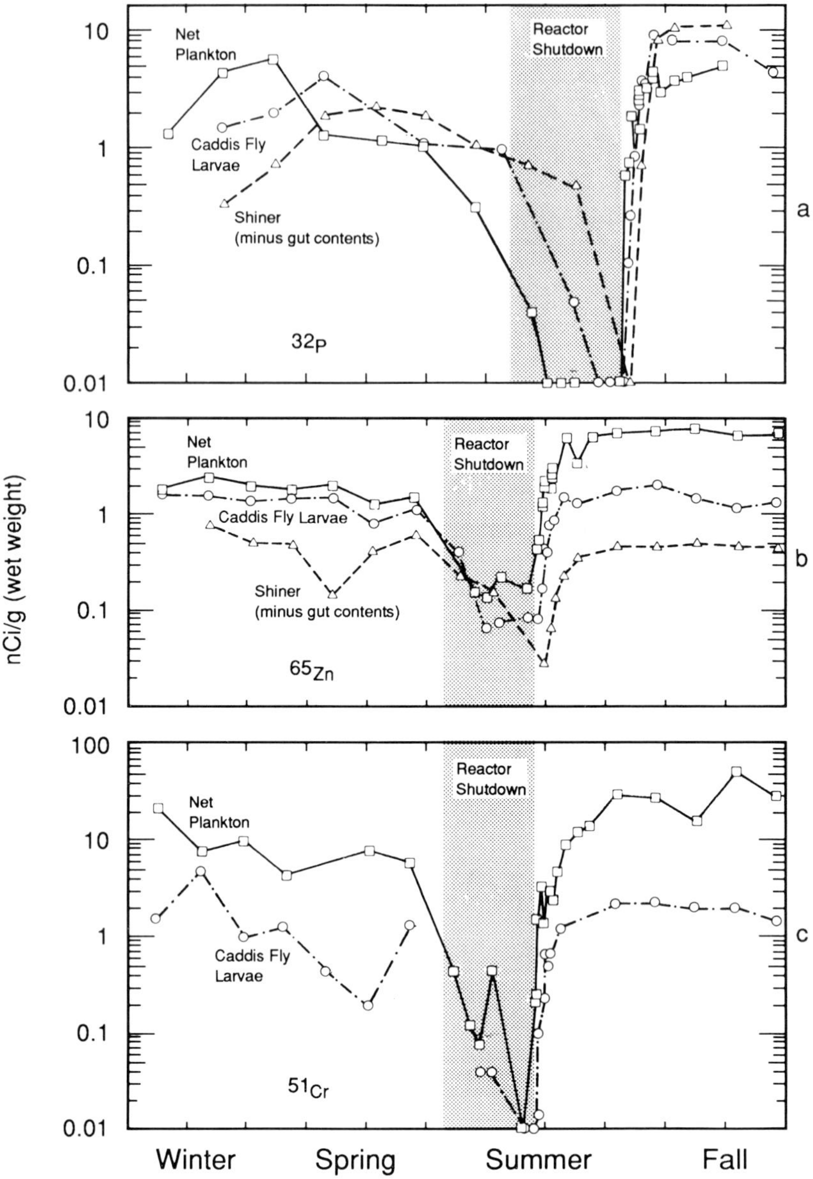

Fig. 8.17. Effect of a 6-week shutdown in 1966 of all single-purpose reactors on (a) P-32, (b) Zn-65, and (c) Cr-5 in aquatic organisms from the Hanford Reach (from Watson et al. 1967).

radionuclides with long half-lives occurred at a depth of 50 centimeters (19.5 inches) at concentrations similar to those near the surface. These radionuclides were tightly bound to sediment particles, and they returned

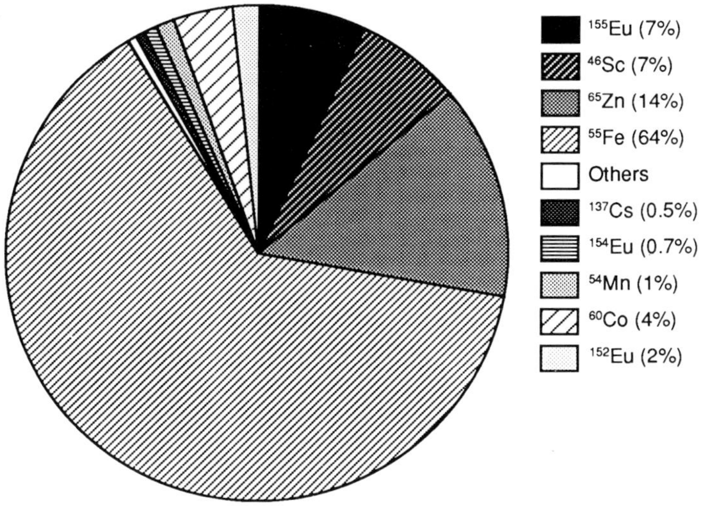

Fig. 8.18. Typical composition of proton-emitting radionuclides in surface sediments behind McNary Dam during April 1971, 3 months after shutdown of the last single-purpose reactor at Hanford. Data include radionuclides originating at Hanford and from atmospheric fallout (modified from Robertson et al. 1973). Percentages show the estimated fraction contributed by each radionuclide to the total radioactivity present at that time.

to the surface mainly by resuspension during high flows of spring and early summer (Robertson et al. 1973).

Radionuclide inventories, sediment resuspension and transport, and sedimentation rates were estimated from sediment cores taken upstream of McNary, The Dalles, and Bonneville dams. More than 98% of the radionuclides that originated at Hanford and re-entered the Columbia River from bottom sediments were in particulate form. Again, scouring was found to be the primary mechanism removing radionuclides from bottom sediments, other than radioactive decay. Flows then transported suspended particulates farther downstream. In the past, when the single-purpose reactors were operating, net deposition of radioactive sediments in Lake Wallula, behind McNary Dam, was greater than removal by scouring. After 1971, new layers of silt were expected to cover the layers containing radionuclides of Hanford origin (Robertson et al. 1973).

Net Transport of Radioactivity Before and After Final Closure

From April 1971 to April 1972, only about 500 curies of eight gamma-emitting radionuclides (reduced from 11 because 3 had nearly decayed) entered the Pacific Ocean. By comparison, the annual discharge of the same eight radionuclides was about 11,000 curies in 1964 and 13,000 curies in 1965, when most reactors were still operating. During the first year after all of Hanford's production reactors were shut down, the annual discharge of these radionuclides was reduced by a factor of 20 to 25.

All gamma emitters, which included both long-lived and short-lived radionuclides, plus the beta-emitter P-32, entering the Pacific Ocean in the mid-1960s totaled about 300,000 curies annually. This amount was about 600 times more than the amount discharged the first year after all single-purpose reactors were closed at Hanford (Haushild et al. 1973; Robertson et al. 1973).

Radionuclides Retained by Sediments in 1976

Radioactivity was monitored in Columbia River sediments through 1976, 5 years after the last reactor shutdown. At this time, a completely different spectrum of intermediate- and long-lived radionuclides was present in bottom deposits of Lake Wallula.

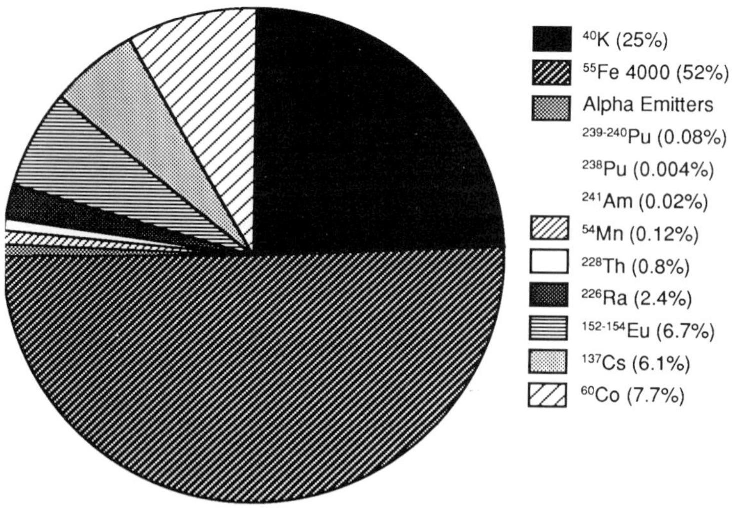

Fig. 8.19. Typical composition of radionuclides in sediments behind McNary Dam during September 1976, 5 years and 9 months after shutdown of the last single-purpose reactor at Hanford (modified from Robertson and Fix 1977). Percentages show the estimated fraction contributed by each radionuclide to the total radioactivity present at that time.

The short- and intermediate-lived radionuclides had decayed, and only a few long-lived radionuclides (Mn-54, Fe-55, Co-60, Cs-137, Eu-151–154, Pu-238, Pu-231–240, and Am-241) remained buried in the sediments at trace concentrations. Upper layers contained much lower concentrations of radionuclides than deeper layers because a cover of 40 to 80 centimeters (15 to 30 inches) of new sediment had been added since 1971.

Radioactivity in the upper layers was primarily from K-40 (48%), but included Ra-226 (4%) and Th-228 (1.8 %), radionuclides all originating from natural sources. Amounts of Cs-137, Pu-238, Pu-231–240, and Am-241 in sediments of Lake Wallula were at low levels typical of those in deposits behind Priest Rapids Dam upstream of Hanford (Robertson and Fix 1977). This pointed to atmospheric deposition as their source.

The deepest layers of sediments in Lake Wallula, deposited between 1953 and 1960, contained very low levels of radionuclides of Hanford origin. The major gamma-emitting radionuclides from Hanford present in September 1976 were Co-60 (5.3-year half-life), Cs-137 (30.1 years), and Eu-152/153 (13 and 8.6 years) (Figure 8.19). Zinc-65, Sc-46, and Mn-54, which were in relatively high concentrations after the reactors shut down, had decayed to extremely low levels (Robertson and Fix 1977). The X-ray emitter, Fe-55 (2.7 years), of minor importance in calculating radiation dose, remained the most abundant radionuclide.

Post-Facto Assessment of Radioactivity in Sediments

Considerable data had been obtained on absolute levels of radioactivity and geochemical behavior of short- and moderately long-lived radionuclides in Columbia River sediments through 1976. But information on very long-lived radionuclides in the river ecosystem, including the plutonium series and other transuranics, was limited. This gap was filled in later by an independent university researcher, T. M. Beasley of Oregon State University.

Plutonium inventories were evaluated from sediment cores taken behind McNary Dam on the Columbia River and Ice Harbor Dam on the Snake River (the control site). Samples were taken in August 1977 and analyzed by mass spectrometry and absolute radioactivity determinations. An estimated 20% to 25% of the total plutonium inventory (Pu-239, -240, -241) in Lake Wallula sediments was ascribed to the single-purpose reactor operations from 1944 to 1971 (Beasley et al. 1981). The rest probably issued from global fallout.

Only Pu-239 (24,131-year half-life) was attributed to the Hanford reactor discharges. It was believed to originate from decay of Np-139

(2.3-day half-life), one of the most abundant radionuclides released to water in the Hanford Reach when the single-purpose reactors were operating. This radionuclide was produced by slow neutron capture in uranium (U-238, 239), followed by decay of U-239 (23.5-minute half-life) to Np-239. The sequence was initiated by the natural uranium present in Columbia River water (about 1 μg/L), plus uranium occluded to the outside of aluminum-clad fuel elements ("tramp uranium") (Beasley et al. 1981).

Subsequently, inventories of Co-60, Cs-137, Am-241, and Pu-239/240 in sediments of the lower Columbia River and its estuary were estimated from 50 sediment cores taken from 1977 to 1978. Cobalt-60 radioactivity was attributed to the terminated reactor discharges at Hanford. In contrast, the majority of the radioactivity from Pu-239/240 and Cs-137 and all of the radioactivity from Am-241 were derived from global fallout. Although substantial amounts of artificial radioactivity had been released to the Columbia River from the Hanford reactors (up to 300,000 Ci/year in the mid-1960s), few radionuclides remained in the lower river ecosystem. The amount of natural radioactivity from potassium, thorium, uranium, and radium isotopes in the river sediments was nearly twice that of artificial radioactivity (Beasley and Jennings 1984).

Significance of field studies with radioactivity

With the start of operations at Hanford, a complex mixture of artificially produced radioactive elements was, for the first time, discharged to an aquatic ecosystem. Forty years ago, even the basic ecological functions of a flowing river were little known. The two phenomena, artificial radioactivity and river ecology, were interrelated and knowledge about them had to be developed together. This led to the scientific field of inquiry called "radioecology."

Natural radioactivity in Columbia River water was first measured in 1944, several months before the first single-purpose reactor began operation. Measurements of artificial radioactivity (primarily beta) in river water and aquatic biota followed in 1946. The first exploratory studies in the Hanford Reach had a broad and expansive base, to determine the distribution and fate of the artificial radionuclides that appeared in the effluent reactor. Efforts soon expanded to include routine yearly measurements at selected sites between Priest Rapids and McNary dams. Radioactivity was measured extensively in the municipal water supplies of the Tri-Cities (Richland, Kennewick, and Pasco) downstream from the Hanford Site. Additional bioenvironmental samples were collected downriver as far as the mouth of the Columbia River. Several radioecological principles emerged from these studies.

Most of the radioactivity in river water and aquatic invertebrates immediately downstream of the reactor discharges originated from radionuclides with half-lives of only a few hours. At the former Hanford townsite just downstream from F Reactor, amounts of radioactivity had diminished because of decay of short-lived radionuclides, dispersion in river water, removal by suspended particulates, assimilation by living organisms, and other factors. At Richland, about 55 kilometers (34 miles) farther downstream, the amount of radioactivity in river water had diminished to the point where no health effects were expected. Downstream from the Tri-Cities area, radioactivity was further reduced by dilution from the Yakima and Snake rivers and by deposition of suspended particulates in bottom sediments.

With respect to radioactivity and aquatic life, food chain relationships were shown to be important. The highest radioactivity usually appeared in plankton, photosynthetic microorganisms in the river. Plankton tended to accumulate relatively large amounts of the shorter-lived radionuclides directly from river water. The next higher levels of radioactivity generally appeared in invertebrates that fed on plankton, such as the aquatic larvae of certain insects, but the time needed for conversion of plant tissue to animal tissue enabled further decay of radionuclides with limited half-lives. Fish that fed directly on aquatic invertebrates generally acquired lesser amounts of radioactivity, primarily from P-32. Carnivorous fish, at the top of the food chain, were generally less radioactive than fish that fed only on invertebrates.

Variations in amounts of radioactivity appearing in aquatic organisms were also influenced by deposition of different radionuclides in specific tissues. Radioactive materials deposited largely in proportion to the metabolic requirements of an organ or tissue were considered to be "biologically active." Thus, the bones and scales of fish, normally rich in phosphorus, acquired large amounts of P-32, while the muscle and nerve tissues did not. Also, the eyes of fish acquired relatively large amounts of Zn-65. The rate of metabolism also played a role. Young, rapidly growing fish acquired more radioactivity than older, slower-growing fish, and resting stages of immature aquatic insects acquired less radioactivity than feeding stages. In addition, the form, size, and covering of aquatic organisms played a role in the absorption and adsorption of radionuclides.

Physicochemical changes in the river ecosystem influenced the types and amounts of radioactivity appearing in river water and aquatic organisms. Assuming a constant level of reactor operation, amounts of radioactivity leaving the Hanford Reach decreased with high flow and increased with low flow because of dilution in the Columbia River. Levels of radioactivity in plankton closely corresponded to those in the river water. In fish and other aquatic animals, uptake and depletion of radioactivity coincided closely with water temperature, the main factor controlling feeding activity and metabolic rates in cold-blooded organisms. The downstream distribution of radioactivity in aquatic biota followed a pattern similar to that of radioactivity in river water, except that large organisms near the upper trophic level acquired larger amounts of the longer-lived radionuclides.

Researchers at the Hanford Site could detect no significant health hazard to downstream users of river water. Estimates of probable radioactive doses received by people in various occupations were made from data on levels of radionuclides in river water, aquatic organisms, foodstuffs, people's life styles and habits, and direct measurements with whole-body counters. Exposure evaluations confirmed repeatedly that total doses of radioactivity to people living near Hanford remained well within appropriate regulatory limits. Quantitative measurements and radioecological studies in the Hanford Reach provided the scientific basis on which estimates of maximum radioactive doses were made.

Following shutdown of the last single-purpose reactor at Hanford in 1971, artificial radionuclides originating at Hanford rapidly declined in Columbia River water and aquatic organisms downstream. After 2 months, most of the short-lived radionuclides had decayed, while some of the longer-lived radionuclides persisted in particulate form, primarily in sediment deposits. Today, relatively few artificial radionuclides originating at Hanford remain in the Columbia River ecosystem. An era in which a major river was used for the disposal of large amounts of radioactive materials had passed. It will never come again.

References

Beasley, T.M., and C.D. Jennings. 1984. "Inventories of 239,240 Pu, 241 Am, 137 Cs, and 60 Co in Columbia River Sediments from Hanford to the Columbia River Estuary." *Environ. Sci. Technol.* 18:201–212.

Beasley, T.M., L.A. Ball, and J.E. Andrews III. 1981. "Hanford-Derived Plutonium in Columbia River Sediments." Science 214:911–915.

Coopey, R.W. 1948. *Preliminary Report on the Accumulation of Radioactivity as Shown by a Limnological Study of the Columbia River in the Vicinity of the Hanford Works.* HW-1162, Hanford Works, Richland, Washington.

Coopey, R.W. 1951. *Radioactive Plankton from the Columbia River.* HW-20668, Hanford Works, Richland, Washington.

Coopey, R.W. 1953a. *The Abundance of the Principal Crustacea of the Columbia River and the Radioactivity They Contain.* HW-25191, Hanford Works, Richland, Washington.

Coopey, R.W. 1953b. "Radioactive Plankton from the Columbia River." *Trans. Am. Microsc. Soc.* 62:311–327.

Coutant, C.C. 1967a. "Retention of Radionuclides in Columbia River Bottom Organisms." In: *Pacific Northwest Laboratory Annual Report for 1966 to the USAEC Division of Biology and Medicine, Vol. I Biological Sciences*, pp. 170–171. BNWL-480, Pacific Northwest Laboratory, Richland, Washington.

Coutant, C.C. 1967b. "Upstream Dispersal of Adult Caddis Flies." In: *Pacific Northwest Laboratory Annual Report for 1966 to the USAEC Division of Biology and Medicine, Vol. I Biological Sciences*, pp. 181–187. BNWL-480, Pacific Northwest Laboratory, Richland, Washington.

Coutant, C.C. 1968. "Retention of Radionuclides in Columbia River Bottom Organisms." In: *Pacific Northwest Laboratory Annual Report for 1967 to the USAEC Division of Biology and Medicine, Vol. I Biological Sciences*, pp. 9.21–9.22. BNWL-714, Pacific Northwest Laboratory, Richland, Washington.

Coutant, C.C. 1982. "Evidence for Upstream Dispersion of Adult Caddis Flies in the Columbia River." *Aquat. Insects* 4:61–66.

Cushing, C.E. 1967a. "Periphyton Productivity and Radionuclide Accumulation in the Columbia River, U.S.A." *Hydrobiologia* 24:121–139.

Cushing, C.E. 1967b. "Concentration and Transport of P-32 and Zn-65 by Columbia River Plankton." *Limnol. Oceanogr.* 12:330–332.

Cushing, C.E., and D.G. Watson. 1966. "Accumulation and Transport of Radionuclides by Columbia River Biota." In: *Disposal of Radioactive Wastes Into Seas, Oceans and Surface Waters*, ed. A. Guillon, pp. 551–570. International Atomic Energy Agency, Vienna, Austria.

Davis, J.J. 1958. "Dispersion of Radioactive Materials by Streams." *J. Am. Water Works Assoc.* 50:1501–1515.

Davis, J.J. 1960. "The Effects of Environmental Factors upon the Accumulation of Radioisotopes by Ecological Systems." In: *Proceedings Second Annual Texas Conference on Utilization of Atomic Energy*, pp. 31–41. Texas A&M, College Station, Texas.

Davis, J.J. 1965. "Accumulation of Radionuclides by Aquatic Insects." In: *Third Seminary in Biological Problems in Water Pollution*, pp. 211–215. Publ. No. 999-WP-25, R.A. Taft Sanitary Engineering Center, Cincinnati, Ohio.

Davis, J.J., and R.W. Cooper. 1951. *Effect of Hanford Pile Effluent Upon Aquatic Invertebrates in the Columbia River.* HW-20055, Hanford Works, Richland, Washington.

Davis, J.J., and R.F. Foster. 1958. "Bioaccumulation of Radioisotopes Through Aquatic Food Chains." *Ecology* 39:530–535.

Davis, J.J., R.W. Coopey, D.G. Watson, C.C. Palmiter, and C.L. Cooper. 1952. "The Radioactivity and Ecology of Aquatic Organisms of the Columbia River." In: *Biology*

Research – Annual Report 1951, pp. 11–29. HW-25021, Hanford Works, Richland, Washington.

Davis, J.J., R.W. Coopey, D.G. Watson, and C.C. Palmiter. 1953. "Radiobiological Survey of the Columbia River." In: *Biology Research – Annual Report 1952*, pp. 1–13. HW-28636, Hanford Works, Richland, Washington.

Davis, J.J., D.G. Watson, C.C. Palmiter, and R.W. Coopey. 1954. "1953 Radiobiological Survey of the Columbia River." In: *Biology Research Annual Report 1953*, pp. 1–12. HW-30437, Hanford Atomic Products Operation, Richland, Washington.

Davis, J.J., D.G. Watson, and C.C. Palmiter. 1955. "1954 Radiobiological Survey of the Columbia River." In: *Biology Research Annual Report 1954*. HW-35917, Hanford Atomic Products Operation, Richland, Washington.

Davis, J.J., D.G. Watson, and C.C. Palmiter. 1956. *Radiobiological Studies of the Columbia River Through December 1955*. HW-36074, Hanford Atomic Products Operation, Richland, Washington.

Davis, J.J., R.W. Perkins, R.F. Palmer, W. C. Hanson, and J. F. Cline. 1958. "Radioactive Materials in Aquatic and Terrestrial Organisms Exposed to Reactor Effluent Water." *Second International Conference on Peaceful Uses of Atomic Energy*, Vol. 18, pp. 421–428. United Nations, New York.

Foster, R.F. 1952. "Biological Problems Associated with the Discharge of Pile Effluent into the Columbia River." In: *Biology Research – Annual Report 1951*, pp. 11–13. HW-25021, Hanford Works, Richland, Washington.

Foster, R.F. 1959a. "Behavior of Radionuclides in Food Chains Freshwater Studies." Presented at Radiological Health Training Course, Cincinnati, Ohio, on Sept. 17, 1959.

Foster, R.F. 1959b. "The Need for Biological Monitoring of Radioactive Waste Streams." *Sewage Ind. Wastes* 31:1401–1415.

Foster, R.F., and J.J. Davis. 1956. "The Accumulation of Radioactive Substances in Aquatic Forms." In: *First International Conference on Peaceful Uses of Atomic Energy*, Vol. 13, pp. 361–367. United Nations, New York.

Foster, R.F., and D. McConnon. 1965. "Relationships Between the Concentration of Radionuclides in Columbia River Water and Fish." In: *Biological Problems in Water Pollution, Third Seminar*, pp. 211–224. PHS Document #999-WP-25, U.S. Public Health Service, Cincinnati, Ohio.

Foster, R.F., and J.K. Soldat. 1966. "Evaluation of the Exposure that Results from the Disposal of Radioactive Wastes into the Columbia River." In: *Disposal of Radioactive Wastes into Seas, Oceans and Surface Waters*, ed. A. Guillon, pp. 681–696. International Atomic Energy Agency, Vienna, Austria.

Foster, R.F., R.W. Coopey, and J.J. Davis. 1949. *A Cursory Survey of the Radioactivity in Biological Materials of the Lower Columbia River*. HW-12573, Hanford Works, Richland, Washington.

Hall, R.B., J.P. Corley, J.K. Soldat, and R.T. Jaske. 1970. "Environmental Effects of an Extended Plant Shutdown (Appendix E)." In: *Effect of Hanford Plant Operations on the Temperature of the Columbia River, 1964 to Present*, eds. R.T. Jaske and M. O. Synoground. PNL-1345, Pacific Northwest Laboratory, Richland, Washington.

Haushild, W.L., H.H. Stevens, Jr., J.L. Nelson, and G.R. Dempster, Jr. 1971. *Radionuclides in Transport in the Columbia River from Pasco to Vancouver, Washington*. Professional Paper 433-N, U.S. Geological Survey, U.S. Government Printing Office, Washington, D.C.

Healy, J.W. 1946. *Accumulation of Radioactive Elements in Fish Immersed in Pile Effluent Water*. Doc. # 1–3442, Hanford Works, Richland, Washington.

Henderson, C., and R.F. Foster. 1957. "Studies of Smallmouth Black Bass (*Micropterus dolomieui*) in the Columbia River near Richland, Washington." *Trans. Am. Fish Soc.* 86:111–127.

Herde, K.E. 1946. *Studies in the Accumulation of Radioactive Elements in Oncorhynchus tshawytscha, Chinook Salmon, Exposed to a Medium of Pile Effluent.* HW 1–5064, Hanford Works, Richland, Washington.

Herde, K.E. 1947. *Radioactivity in Various Species of Fish from the Columbia and Yakima Rivers.* Doc. # 1–5501, Hanford Works, Richland, Washington.

Herde, K.E. 1948. *A One-Year Study of Radioactivity in Columbia River Fish.* HW-11344, Hanford Works, Richland, Washington.

International Commission on Radiological Protection (ICRP). 1959. *Report of Committee II on Permissible Dose for Internal Emitters.* ICRP Publication No. 2.

Junkins, R.L. 1960. "Removal of Radionuclides from the Pasco Supply by Conventional Treatment." *J. Am. Water Works Assoc.* 52:831–840.

Lappenbusch, W.L., D.G. Watson, and W.L. Templeton. 1971. "In Situ Measurement of Radiation Dose in the Columbia River." *Health Phys.* 21:247–251.

Nelson, I.C., and R.F. Foster. 1965. "Ringold Farms – a Hanford Environmental Study." *Health Phys.* 11:391–401.

Nelson, J.L., and W.L. Haushild. 1970. "Accumulation of Radionuclides in Bed Sediments of the Columbia River Between the Hanford Reactors and McNary Dam." *Water Resour. Res.* 6:130–137.

Nelson, I.C., R.W. Perkins, and J.M. Nielsen. 1964. *Progress in Studies of Radionuclides in Columbia River Sediments. A Summary of Hanford Achievements in This Program Under General Electric 1961–1964.* HW-83614, Hanford Atomic Products Operation, Richland, Washington.

Nelson, J.M., and R.W. Perkins. 1962. *The Removal of Radioisotopes from the Columbia River by Natural Processes.* HWSA-2411, Hanford Works, Richland, Washington.

Nelson, I.C., R.W. Perkins, and W.L. Haushild. 1966a. "Flow Time Measurements of the Columbia River Using Radioactive Tracers Introduced by the Hanford Reactors." *Water Resour. Res.* 2:31–39.

Nelson, J.L., R.W. Perkins, J.M. Nielsen, and W.L. Hauschild. 1966b. "Reactions of Radionuclides from the Hanford Reactors with Columbia River Sediments." In: *Disposal of Radioactive Wastes into Seas, Oceans and Surface Waters*, pp. 131–161. Proceedings of the 1966 Symposium, International Atomic Energy Agency, Vienna, Austria.

Olson, P.A., and R.F. Foster. 1952. *Accumulation of Radioactivity in Columbia River Fish in the Vicinity of the Hanford Works.* HW-23093, Hanford Works, Richland, Washington.

Parker, H.M. 1956. "Radiation Exposure from Environmental Hazards." In: *First International Conference on Peaceful Uses of Atomic Energy*, Vol. 13, pp. 301–310. United Nations, New York.

Parker, H.M., R.F. Foster, I.L. Ophel, F.L. Parker, and W.C. Reinig. 1964. "North American Experience in the Release of Low-Level Wastes to Lakes and Rivers." In: *Third International Conference on Peaceful Uses of Atomic Energy*, Vol. 14, pp. 61–69. United Nations, New York.

Perkins, R.W. 1965. "An Anticoincidence-Shielded Multidimensional Analyzer." *Nucl. Instr. Meth.* 33:71.

Perkins, R.W., J.L. Nelson, and W .L Haushild. 1966. "Behavior and Transport of Radionuclides in the Columbia River Between Hanford and Vancouver, Washington." *Limnol. Oceanogr.* 11:231–248.

Pruter, A.T., and D.L. Alverson eds. 1972. *The Columbia River Estuary and Adjacent Ocean Waters*. University of Washington Press, Seattle, Washington.

Robeck, G.G., C. Hendersen, and R.C. Palange. 1954. *Water Quality Studies on the Columbia River*. U.S. Public Health Service, Robert A. Taft Sanitary Engineering Center, Cincinnati, Ohio.

Robertson, D.E., and J.J. Fix. 1977. *Association of Hanford Origin Radionuclides with Columbia River Sediments*. BNWL-2305, Pacific Northwest Laboratory, Richland, Washington.

Robertson, D.E., W.B. Silker, J.C. Langford, M.R. Peterson, and R.W. Perkins. 1973. "Transport and Depletion of Radionuclides in the Columbia River." In: *Proceedings of Symposium on Radioactive Contamination of the Marine Environment*, pp. 141–158. International Atomic Energy Agency, Vienna, Austria.

Seymour, A.H. 1964. "Contributions of Radionuclides to Our Understanding of Aquatic Ecosystems." *Verh. Int. Verein. Limno.* 15:221–236.

Soldat, J.K. 1962. *A Compilation of Basic Data Relating to the Columbia River. Section 8, Dispersion of Reactor Effluent in the Columbia River*. HW-69369, Hanford Works, Richland, Washington.

Watson, D.G. 1952. "Observations on Spawning and Migration of Chinook Salmon, *Oncorhynchus tshawytscha* (Walbaum) in the Columbia River in the Vicinity of Hanford Works." In: *Biology Research Annual Report 1951*, pp. 14–18. HW-25021, Hanford Works, Richland, Washington.

Watson, D.G. 1966. "Migration of Columbia River Fish." In: *Pacific Northwest Laboratory Annual Report for 1965 in the Biological Sciences*, pp. 121–126. BNWL-280, Pacific Northwest Laboratory, Richland, Washington.

Watson, D.G., and C.E. Cushing. 1969. "Seasonal Variation in Radionuclides in Columbia River Organisms." In: *Pacific Northwest Laboratory Annual Report for 1968 to the USAEC Division of Biology and Medicine*, Vol. I Life Sciences, Part 2 Ecological Sciences, pp. 2.11–2.15. BNWL-1050, PT2, Battelle, Pacific Northwest Laboratories, Richland, Washington.

Watson, D.G., and J.J. Davis. 1957. *Concentrations of Radioisotopes in Columbia River Whitefish in the Vicinity of the Hanford Atomic Products Operation*. HW-48523, Hanford Works, Richland, Washington.

Watson, D.G., and W.L. Templeton. 1973. "Thermoluminescent Dosimetry of Aquatic Organisms." In: *Radionuclides in Ecosystems*, Proceedings of the Third National Symposium on Radioecology, ed. D.J. Nelson, pp. 1121–1129. CONF-710501-P2, National Technical Information Service, Springfield, Virginia.

Watson, D.G., J.J. Davis, and W.C. Hanson. 1961. "Zinc-65 in Marine Organisms Along the Oregon and Washington Coasts." *Science* 133:1821–1828.

Watson, D.B., J.J. Davis, and W.C. Hanson. 1963. "Interspecies Differences in Accumulation of Gamma Emitters by Marine Organisms near the Columbia River Mouth." *Limnol. Oceanogr.* 8:301–309.

Watson, D.G., C.E. Cushing, and R.W. Perkins. 1966. "Radionuclides in Columbia River Biota." In: *Pacific Northwest Laboratory Annual Report for 1965 in the Biological Sciences*, pp. 121–127. BNWL-280, Pacific Northwest Laboratory, Richland, Washington.

Watson, D.G., C.E. Cushing, C.C. Coutant, and W.L. Templeton. 1967. "Radionuclides in Columbia River Organisms." In: *Pacific Northwest Laboratory Annual Report for 1966 to the USAEC Division of Biology and Medicine, Vol. I, Biological Sciences*, pp. 161–169. BNWL-480, Pacific Northwest Laboratory, Richland, Washington.

Watson, D.G., C.E. Cushing, C.C. Coutant, and W.L. Templeton. 1970a. *Radiological Studies on the Columbia River. Part I.* BNWL-1377, Pacific Northwest Laboratory, Richland, Washington.

Watson, D.G., C.E. Cushing, C.C. Coutant, and W.L. Templeton. 1970b. "Cycling of Radionuclides in Columbia River Biota." In: *Trace Substances in Environmental Health IV*, ed. D.D. Hemphill, pp. 141–157. Environmental Health Center and Extension Division, University of Missouri, Columbia, Missouri.

Wildung, R.E. 1970. "Isolation and Measurement of the Physiochemical Properties of Particulate Matter Suspended in a River System." In: *Pacific Northwest Laboratory Annual Report for 1969 to the USAEC Division of Biology and Medicine, Vol. I Life Sciences*, Part 2 Ecological Sciences, pp. 3.1–3.4. BNWL-1306, Battelle, Pacific Northwest Laboratories, Richland, Washington.

Wildung, R.E., and R.L. Schmidt. 1971a. "Electron Micrographic Observations of Particulate Matter Suspended in the Columbia River." In: *Pacific Northwest Laboratory Annual Report for 1970 to the USAEC Division of Biology and Medicine, Vol. I Life Sciences, Part 2 Ecological Sciences*, pp. 2.1–2.2. BNWL-1550, Pacific Northwest Laboratories, Richland, Washington.

Wildung, R.E., and R.L. Schmidt. 1971b. "Mineral Composition of Particulate Matter Suspended in the Columbia River as Influenced by Watershed Characteristics and Particle Size." In: *Pacific Northwest Laboratory Annual Report for 1970 to the USAEC Division of Biology and Medicine, Vol. I Life Sciences, Part 2 Ecological Sciences*, pp. 2.1–2.4. BNWL-1550, Battelle, Pacific Northwest Laboratories, Richland, Washington.

Wildung, R.E., R.C. Routson, and R.L. Schmidt. 1972. *Seasonal Changes in Particle Size Distribution, Composition, and Strontium Exchange Capacity of Particulate Matter Suspended in the Columbia River*. BNWL-1638, Pacific Northwest Laboratory, Richland, Washington.

Chapter 9

LABORATORY STUDIES WITH RADIOACTIVITY AND AQUATIC ORGANISMS, 1945 TO 1971

During the early years of Hanford operations, particularly the 1950s, fish were exposed to specific radionuclides in the aquatic laboratory to determine factors such as exposures causing death, concentrations appearing in tissues and organs, gross damage produced by radioactivity, fraction of radioactivity retained, and uptake and retention rates. Other studies involved partitioning of radioactive materials in experimental aquatic microcosms.

Of the 60 or more artificial radionuclides appearing in the reactor cooling discharges, relatively few were sufficiently biologically active or persistent in the river ecosystem to warrant intensive investigation in the aquatic laboratory. Experimental studies also were conducted with radionuclides from other sources such as nuclear detonations in the atmosphere. Some studies provided data for comparing the uptake and effect of a fission product (strontium, cesium, or plutonium) by a cold-blooded vertebrate (fish) with those by a warm-blooded vertebrate (mammal). (These comparisons were made by the Biomedical Group and are reported elsewhere.)

Direct exposure of organisms to radionuclides

In the aquatic laboratories, fish and other river organisms were exposed to radionuclides by different routes, including ingestion (direct feeding) and contamination of media (uptake from water).

The activation products P-32, Zn-65, and Cr-51 were the most important radionuclides found in Hanford reactor effluent in terms of their ability to transfer radioactivity to aquatic life. The fission products Sr-90 and Cs-137 also were used in Hanford studies. These radionuclides gained national attention as a result of atmospheric testing in the 1950s, and small amounts appeared in the river ecosystem from atmospheric fallout. Only trace amounts of Sr-90 and Cs-137 were present in the cooling water discharges of Hanford's reactors.

Radionuclides used in experiments with fish included

- Phosphorus-32 (14.3-day half-life) – Phosphorus is essential to metabolic processes in aquatic organisms; therefore, P-32 was taken up readily by aquatic organisms and tends to have relatively high concentration ratios and slow biological turnover.
- Zinc-65 (245-day half-life) – Zinc is an essential trace element; therefore, Zn-65 was readily accumulated in aquatic organisms, persisted in their bodies, and passed through the food web.
- Chromium-51 (27.8-day half-life) – Chromium was abundant in reactor effluents; however, it had limited biological function. It did not accumulate to any extent in aquatic organisms; therefore, it was not studied in the laboratory.
- Strontium-90 (28.1-year half-life) – Strontium behaved similarly to the essential element calcium in aquatic organisms; hence, it was a bone-seeking radionuclide. As a fission product, trace amounts were bioaccumulative.
- Cesium-137 (30-year half-life) – Cesium behaved similarly to the element potassium in aquatic organisms; hence, it occurred primarily in body muscle. As a fission product, trace amounts were bioaccumulative.

In field and laboratory studies, words used to describe radioactivity have precise meaning. The term "half-life," commonly used in reference to radioactive materials, is the time required for the activity of a specific radionuclide to be reduced by half through physical or radioactive decay. In successive half-lives, radioactivity is reduced to 1/2, 1/4, 1/8, 1/16, and so on of its initial value. This phenomenon is also called "physical half-life." In contrast, "biological half-life" is the time required for half of the activity of a specific radionuclide to be lost from an organism as a result of biological processes. "Effective half-life" is derived from both physical and biological half-lives, and is best measured experimentally. Radioactivity is usually reported as picocuries (pCi) or microcuries (μCi), where 1 pCi equals 1,000,000 μCi.

Feeding P-32 to Rainbow Trout

To determine the amounts harmful to fish, P-32 was fed to yearling rainbow trout in doses of 600, 3000 and 3800 μCi weekly for 12 weeks

Fig. 9.1. Yearling rainbow trout were fed small prey fish into which P-32 was injected. Scientist Donald G. Watson examined how large doses of P-32, taken internally, affected trout.

(Figure 9.1). Control fish received stable phosphorus at rates equivalent to those associated with the highest dose of P-32. To control exposure rates and avoid handling stress, P-32 was first injected into small trout, which were then fed to the test yearlings (Watson 1957).

Trout fed the two higher doses of P-32 died of radiation damage within 7 weeks after each had received about 12,000 μCi. The lowest dose did not cause mortality over the same time period, but did cause extensive vascular tissue breakdown. About 55% of the ingested P-32 was retained by trout. Concentrations were greatest in bone, scales, and liver (Figure 9.2).

In a follow-up experiment, lesser amounts of P-32 were fed to rainbow trout at rates of 0.006, 0.06 and 0.6 μCi per gram (μCi/g) of body weight for 5 days weekly over 6 months. At 0.006 μCi, P-32 produced no abnormalities. At 0.06 μCi, growth was reduced in 17 weeks, and about 62% of the radionuclide was retained. At 0.6 μCi, the dose was lethal, reducing growth at 11 weeks and damaging the gastrointestinal tract,

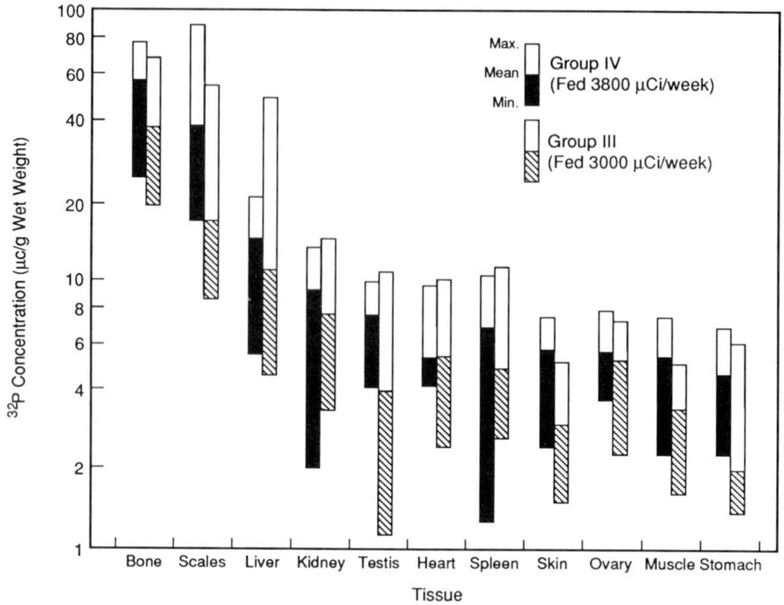

Fig. 9.2. Radioactivity in different tissues and organs of rainbow trout after they had been fed large, lethal doses of P-32 (from Watson 1957).

anterior kidney, and leucocyte production at 3 weeks. Phosphorus-32 approached equilibrium levels in hard tissues and muscles of trout in 30 to 40 days, and in other soft tissues in 20 to 30 days (Figure 9.3). Effective half-lives extended from 8.4 days in the liver to 13 days in muscle and 14 days in bone (Watson et al. 1959). [a]

Uptake of P-32 by fish in water and food was also compared experimentally. Warm-water cichlids (a semitropical fish) were placed in aquaria in which the water was spiked with P-32 to provide 100 μCi per milliliter (μCi/mL) and by dosing the food ration with P-32 at 0.5 μCi/g. Exposure via water was not feasible because P-32 adsorbed to the aquaria walls and on debris, leading to inconsistent exposures among groups of fish. Dosing the food ration with P-32 gave reasonably consistent ex-

[a] At the time these studies were conducted, the International Committee on Radiological Protection had recommended a maximum permissible concentration (MPC) for P-32 in drinking water of 2×20^{-4} μCi, which is equivalent to an intake of about 3 μCi of P-32 each week. If a person consumed 1 pound of fish per week and a safety factor of 10 were applied, the MPC for edible parts (flesh) of fish would be 7×10^{-4} μCi/g. This was about 1% of the concentration previously reported (i.e., Watson 1957) as sublethal to trout fed P-32 for 12 weeks (Donaldson and Foster 1957).

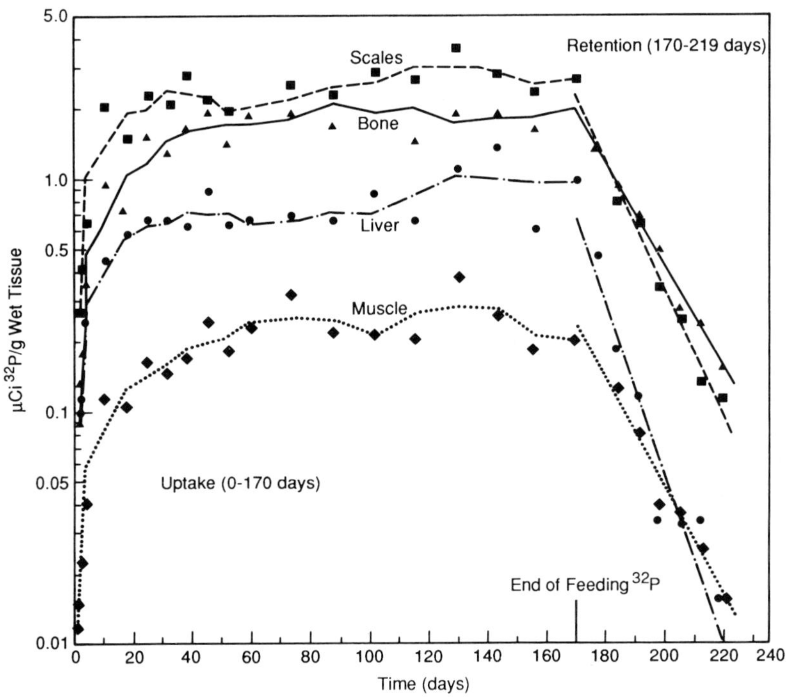

Fig. 9.3. Uptake and retention activity by rainbow trout while they were fed chronic, sublethal doses of P-32 for 6 months, followed by depuration (from Watson et al. 1959).

posures. Levels of P-32 in fish reached equilibrium with levels in food in about 2 weeks (Olson and Foster 1960).

Effect of Zn-65 Fed to Rainbow Trout

Zinc-65 was readily taken up by fish in the Hanford Reach. Furthermore, the radionuclide was present in marine fish and shellfish near the outlet of the Columbia River. During 1963 and 1964, the period of maximum reactor operation at Hanford, Zn-65 averaged about 37 pCi/g in the muscle of mountain whitefish and reached as high as 100 pCi/g. Yet, because body burdens of Zn-65 were low, radiation damage had never been observed in fish from the Hanford Reach.

A series of tests in the laboratory were conducted to examine uptake, metabolism, and retention of Zn-65. Fish were fed Zn-65 in capsules because they acquired it primarily through food organisms in the Hanford Reach.

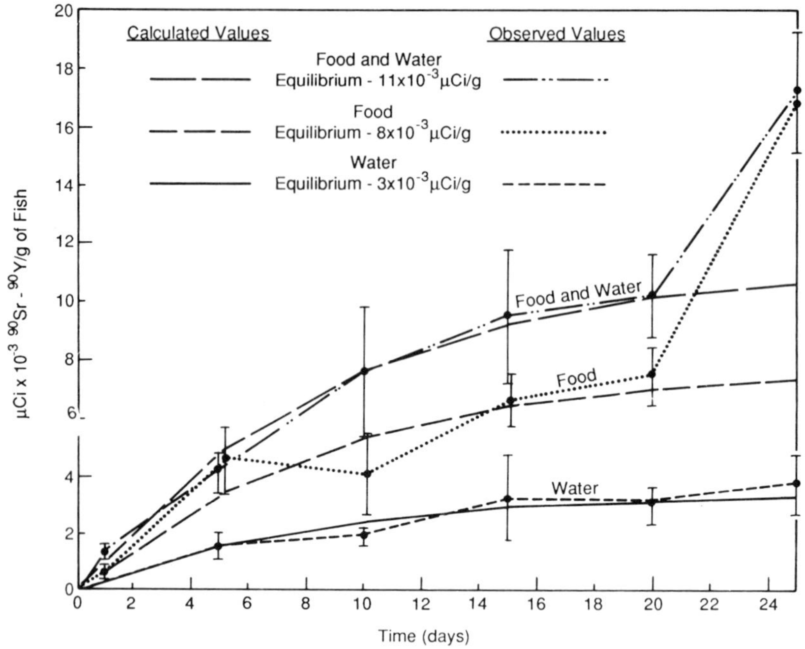

Fig. 9.4. Uptake of Sr-90 by rainbow trout from spiked food, water, and both food and water (from Schiffman 1959).

After one feeding of 8.8 μCi Zn-65 (low oral dose), yearling rainbow trout excreted the radionuclide for several days, primarily through the gastrointestinal tract and gills. Zinc-65 appeared initially in the blood,

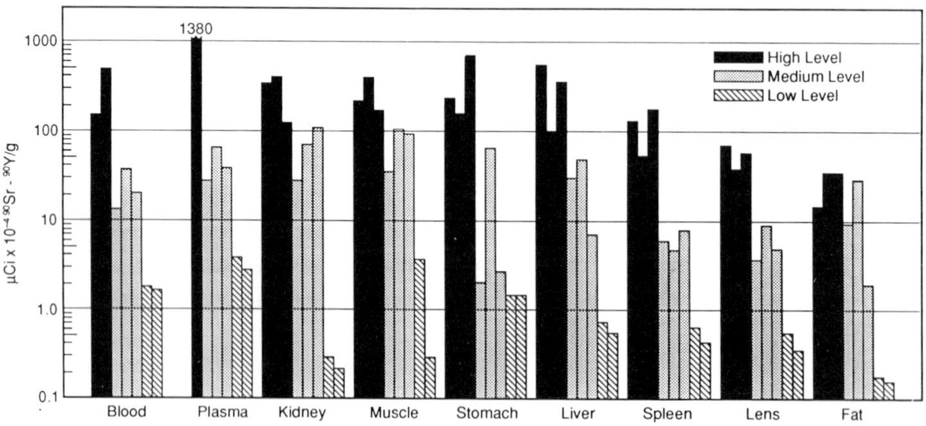

Fig. 9.5. Concentrations of Sr-90 at equilibrium in tissues and organs of rainbow trout after 21 weeks of feeding sublethal doses (from Nakatani and Foster 1961).

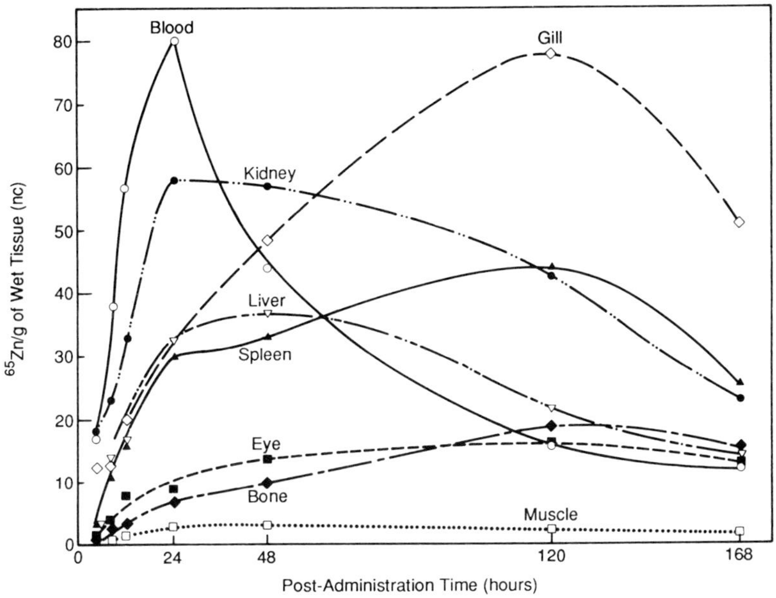

Fig. 9.6. Concentrations of Zn-65 in tissues and organs of rainbow trout over a 7-day period following a single oral dose of 8.8 μCi (from Nakatani and Miller 1963).

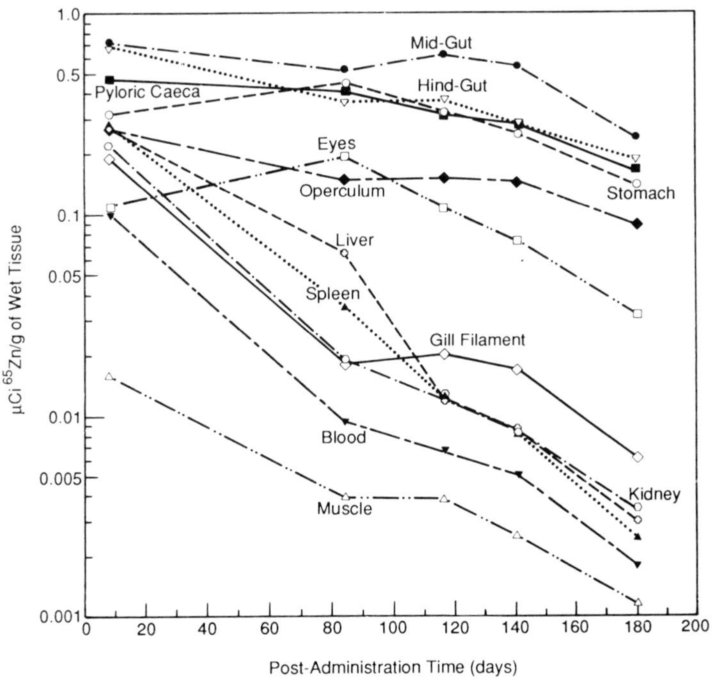

Fig. 9.7. Concentrations of Zn-65 in tissues and organs of rainbow trout over a 200-day period following a single oral dose of 200 μCi (from Nakatani and Liu 1964).

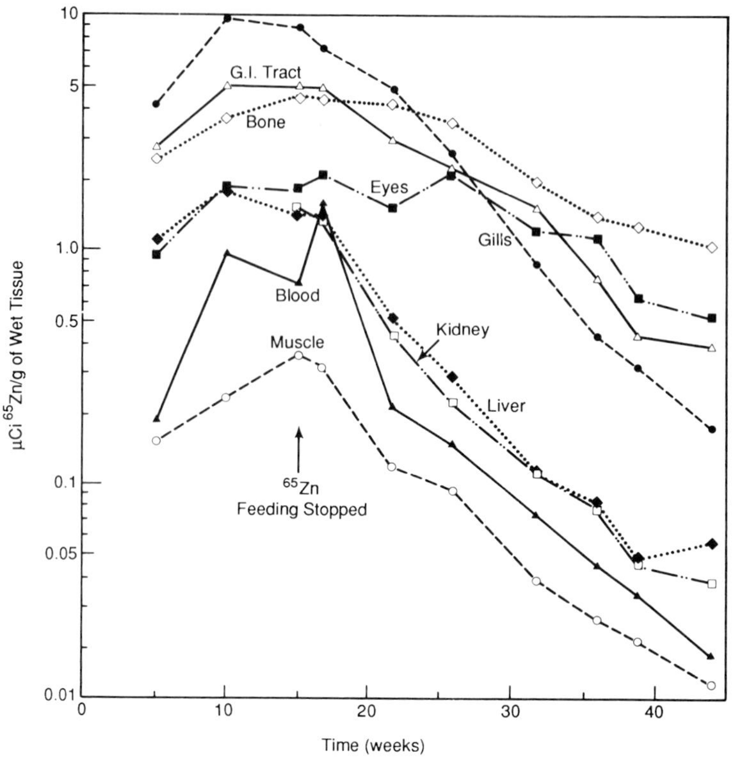

Fig. 9.8. Concentrations of Zn-65 in tissues and organs of rainbow trout fed 0.1 μCi/g of weight daily for 17 weeks (from Nakatani 1966).

peaking in 24 hours (Figure 9.6). Appearance was slower in the eye, bone, and muscle. After 7 days, Zn-65 concentrated most in the gills, spleen, and kidney (Nakatani and Miller 1963).

After one feeding of 200 μCi Zn-65 (high oral dose), body burdens in yearling rainbow trout declined from 25 μCi on day 8 to about 10 μCi on day 182. Concentrations were highest in the gastrointestinal tract of the fish, which retained more than 50% of the body burden from day 85 to day 182 (Figure 9.7). Apparently, zinc was needed by certain enzymes in the gastrointestinal tract (Nakatani and Liu 1964).

In another test, rainbow trout were fed low levels of Zn-65, 0.1 μCi/g of fish daily for 17 weeks. Activity became relatively high in the gastrointestinal tract, bone, and gill filaments and relatively low in the muscle (Figure 9.8). Concentrations of Zn-65 in various tissues generally declined, even though the fish were still fed the radionuclide daily. Little Zn-65 was excreted in urine. Levels of Zn-65 in gill tissues declined rapidly after feeding ceased, indicating that the radionuclide was excreted by the gills (Nakatani 1966).

The effects of feeding small amounts of Zn-65 (chronic ingestion) was examined further in groups of yearling trout fed Zn-65 each day at rates of 0.01, 0.1, and 1.0 μCi/g of fish for 17 weeks. Growth, mortality, blood composition, tissue structure, and swimming performance were compared between fed and control trout. Trout fed Zn-65 did not differ significantly from controls, except that they grew faster. However, when trout were fed at 10 μCi/g for 10 weeks, blood indicated some leukopenia and the gill filaments were slightly damaged (Nakatani 1966; Nakatani et al. 1965).

In these tests, trout fed Zn-65 acquired 10,000 times greater burdens than did river fish but still showed no adverse effects. Therefore, the much lower levels of Zn-65 in the reactor effluents were considered to be too low to impact Columbia River fish populations.

Binding of Zn-65 in Fish and Invertebrates

Marked interspecies differences in tissue levels of Zn-65 were noted in fish from the Hanford Reach during field studies. The biochemical factors that might influence the binding of radionuclides in different fish and invertebrate tissues were evaluated experimentally.

Fish accumulated Zn-65 primarily through their food, with uptake occurring in the intestinal tract. The eyes acquired the highest concentrations of Zn-65, corresponding to their high content of stable zinc. But specific activities in eyes were comparable with those in muscle and blood (Buhler 1967). Specifically, Zn-65 (and Fe-59) concentrated in melanin pigment of the choroid, iris, and also the peritoneal membrane of the eye. Melanin appeared to be weakly cationic, and could bind many heavy metals and radionuclides by simple ion exchange processes (Buhler 1968).

The tissues of freshwater mussels held for 36 days in water taken from downstream of the reactors also accumulated Zn-65. Activity reached about 100 μCi in soft tissues and 300 μCi in shells. In order of descending concentration, Zn-65 accumulated in the gills, mantle and palps, body mass (digestive gland, digestive tract, and gonads), adductor muscles, and foot (muscle). Uptake was proportional to levels in the water over a range of 1 to 100 μCi Zn-65 per liter (Pauley and Nakatani 1967).

Uptake of Sr-90 by Rainbow Trout

One study examined how Sr-90 entered fish from aquatic habitats. Three groups of rainbow trout were exposed to Sr-90 (as Sr-90/Y-90) irradiation. One group was held in water spiked with 2×10^{-3} μCi/mL. Another group was held in similarly spiked water, but each fish was fed gelatin capsules containing 0.24 μCi Sr-90 daily. The third group was held in unspiked water and fed Sr-90 in capsules as above. Radionuclide uptake was assessed for up to 25 days. In addition, natural food organisms (aquatic insects and small fish) were exposed to Sr-90 in water, homogenized, and fed to the fish.

Activity of Sr-90 in trout peaked after 3 weeks at 11×10^{-3} μCi/g of tissue, about 1.5 times the level in spiked water (Figure 9.4). Equilibrium was reached at the same rate whether the source was food or water. About 21% of Sr-90 administered in capsules was retained in trout after 1 day. But, when Sr-90 was incorporated in natural food, only about 7% was retained. Feedings of 0.24 μCi Sr-90/Y-90 each day to the fish damaged tissues lining the gut (Schiffman 1959). Strontium-90 from spiked water was taken up by trout primarily through their gills. The 0.24 μCi of Sr-90 fed in capsules was about 5 times greater than the level taken up from contaminated food organisms.

Damage to Trout Tissues from Sr-90

A subsequent study quantified the amount of Sr-90 that produced radiation damage when fed to yearling rainbow trout and identified the accompanying pathological symptoms. One control and three study groups of trout were fed capsules containing known amounts of Sr-90/Y-90 daily for 21 weeks.

Growth was slightly depressed, and higher mortalities occurred among trout fed the maximum dose, 0.5 μCi of Sr-90 per gram of tissue, for 21 weeks. However, fish fed 0.05 and 0.005 μCi/g daily showed no effects. Pronounced leukopenia had appeared in the high-dose (0.5-μCi) group when feeding ended, and in the medium-dose (0.05-μCi) group 6 months after feeding ended (Nakatani and Foster 1961). At secular equilibrium, after 21 weeks, concentrations of Sr-90/Y-90 varied among tissues (Figure 9.5).

Subsequent analyses showed that about 25% of the Sr-90 activity in all treatments was retained. The total body burden of one surviving trout (309 grams) in the maximum-dose group (0.5 μCi) when feeding ended was about 2200 μCi. Total body burdens of fish in the medium-dose group (0.05 μCi) were about 310 μCi, but the fish showed no apparent damage; leukopenia was indicated 6 months after treatment. Symptoms of Sr-90 exposure among fish in the maximum-dose group were loss of appetite and weight, listlessness, and lower response to stimuli (Nakatani and Foster 1963).

Injection of Rainbow Trout with Sr-90

Intramuscular injection was another method used to irradiate rainbow trout with Sr-90 and to determine toxic limits. A control and three study groups of fish were injected twice weekly for 17 weeks with up to 1.5×10^{-2} μCi Sr-90 (as Sr-90/Y-90) per gram of weight.

Growth and mortality were not affected in trout injected with Sr-90 at the levels used. Further, there was no apparent effect on pathology of the gastrointestinal tract, red blood cell counts, hemoglobin content, or hematocrit values. Overall, radionuclide burdens attained in fish from injection were higher than those attained from feeding Sr-90 (Schiffman 1960).

Elimination of Sr-90 by Rainbow Trout

Excretion of Sr-90 by rainbow trout was examined to quantify elimination mechanisms. In the preceding studies, fish readily took up Sr-90 from water and eliminated the radionuclide from their gills against a concentration gradient. Strontium-90 also was excreted via the kidney and lower intestine.

About 50% to 70% of the activity from Sr-90 injected into trout remained after 22 hours. Of this, about 6% to 7% remained in the urine and 3% to 4% in the gut. Apparently, 15% to 40% of the injected dose was removed by diffusion through the gills and/or the skin (Schiffman 1961).

Distribution and Retention of Sr-90 in Rainbow Trout

Sublethal exposures to Sr-90 were examined in yearling rainbow trout, with emphasis on distribution and retention of the radionuclide. The higher levels of Sr-90 required to induce radiation damage had been determined earlier. Test fish were fed a single, 15-μCi dose of encapsulated Sr-90 (as Sr-90/Y-90). Concentrations of the radionuclide in various tissues were monitored for 100 days.

Strontium-90 increased sharply in all tissues during the initial 24 hours. Concentrations then declined rapidly, with the exception of those in bone. Activity in bone averaged 0.2 μCi/g at 16 days and 0.05 μCi/g at 100 days. Trout retained an average of 25% of the administered dose at 24 hours, which gradually decreased to about 4.4% at 100 days (Nakatani 1962).

Cycling of Sr-90 in Crayfish

Aquatic crustacea take up calcium from the water and deposit it in their exoskeleton. Concentrations of both calcium and strontium, two elements with similar properties, in aquatic crustacea will change when the exoskeleton is replaced during the molt cycle. Activity levels were examined during molting in the crayfish common to the Columbia River.

Crayfish injected with Sr-90/Y-90 while in premolt took up Sr-90 in structures that concentrated calcium (gastroliths) and fixed the radionuclide in the exoskeleton. At molting, Sr-90 in the gastroliths was mobilized and deposited in the postmolt exoskeleton. Premolt crayfish retained about 91% of the Sr-90 injected. About 10% of the activity was lost with the exoskeletons during the first and second molts (Dean 1965).

Metabolism of Cs-137 in Trout

Cesium-137 readily enters aquatic organisms through exposure to contaminated water. One study examined metabolism and retention of known amounts of Cs{171}137 in yearling rainbow trout. Test fish were injected intravenously with 10 pCi of Cs-137, held in 18°C water, and examined at intervals up to 14 days.

Cesium-137, which is similar chemically to potassium, was soon distributed uniformly through the soft tissues of trout, with the exception of white muscle (Figure 9.9). In contrast to Sr-90, no measurable Cs-137 accumulated in bone. After 6 hours, activity declined in all soft tissues but white muscle. The effective half-life of Cs-137 was 1–1/2 days in red muscle and 13 days in white muscle (Dean et al. 1965).

Temperature and Metabolism of Cs-137 in Trout

The previous study was repeated with rainbow trout held in 5°C water to examine the effect of temperature on metabolism and retention of Cs-137. Again, yearling trout were injected intravenously with 10 pCi of Cs-137 and examined at intervals afterwards.

The half-life of Cs-137 in soft tissues (whole animal) of trout held at 5°C was 20 days. This was twice the half-life of Cs-137 (about 10 days) in trout held at 18°C. Uptake at the lower temperature was rapid in nerve tissue, but little turnover took place after uptake peaked. Levels of Cs-137 were still increasing in muscle tissue 28 days postinjection when observations ceased. Hence, fish in river ecosystems might acquire higher

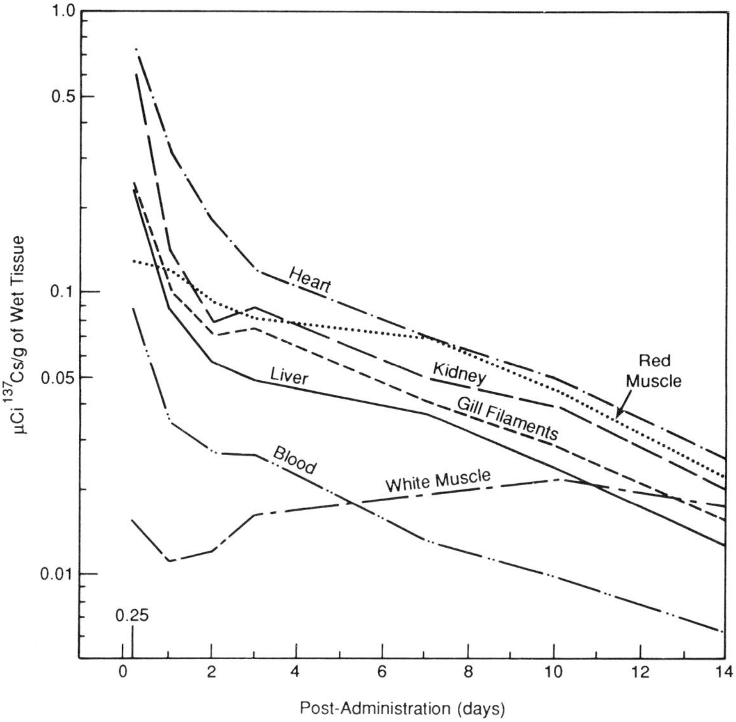

Fig. 9.9. Concentrations of Cs-137 in tissues and organs of rainbow trout following a single intravenous injection of 10 pCi (from Dean et al. 1965).

levels of Cs-137 in colder waters, enhancing transfer of fallout radionuclides in the food web (Dean and Nakatani 1966).

Uptake and transfer of radionuclides in microcosms

Laboratory studies on the uptake and cycling of biologically important radionuclides continued in the 1970s. While studies took a variety of forms, they were intended to clarify lingering questions related to the fate of radionuclides in river ecosystems.

Many experiments involved miniature ecosystems, or microcosms. Laboratory aquaria, tanks, or troughs were stocked with aquatic organisms in which ecological processes took place under conditions that were, for the most part, controlled. Measured amounts of radionuclides were added to these microcosms to follow partitioning among occupants and to identify features affecting the uptake and fate.

Partitioning of P-32 in an Oligotrophic System

In the early 1950s, accumulation and transfer of P-32 was studied in an oligotrophic system containing limited amounts of nutrients (i.e., low in phosphate). Three 60-gallon aquaria were filled with uncontaminated river water, seeded with plankton, and allowed to develop for 6 weeks. At that time, snails and small fish were added, and the water was spiked with a P-32 tracer at average doses of 170.6 μCi per aquaria.

Phosphorous-32 was effectively removed from aquaria water by biological processes. Algae concentrated the radionuclide to levels 300,000 times greater than in the water. Removal of P-32 from the spiked water showed three phases (Figure 9.10). In the first and shortest phase, suspended algae (planktonic) rapidly absorbed P-32. In the intermediate phase, most of the remaining P-32 was gradually taken up by attached algae (periphytic) and mud, while the activity acquired by planktonic algae decreased. In the final, and longest phase, 18 days after maximum levels of P-32 were reached in algae, the radionuclide continued to be removed from the water by sedimentation and binding to surfaces.

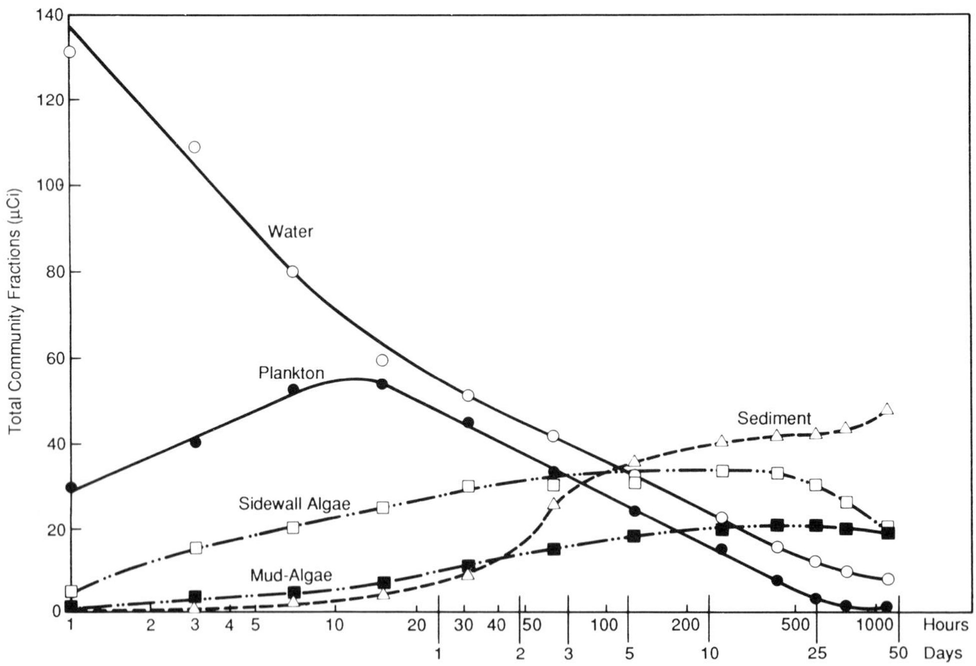

Fig. 9.10. Distribution of P-32 in an aquatic system, showing partitioning among biotic components as a function of time (from Whittaker 1953).

Uptake by larger organisms was irregular, but maximum densities were reached in fish after 11 days and in snails after 18 days (Whittaker 1953, 1961).

Effects of Phosphate on Uptake of P-32

Because natural phosphate is essential for life processes, amounts in the Columbia River influenced the uptake of P-32 by aquatic organisms. Generally, biological uptake of P-32 was rapid in the Hanford Reach because levels of natural phosphate were low, usually below 0.05 part per million (ppm).

One early study examined the effects of phosphate on uptake of P-32 in aquaria microcosms. Tracer P-32 was introduced in aquaria containing natural phosphate at levels from 0.05 to 525 ppm. Only small amounts of P-32 were removed from the water by organisms when natural phosphate levels were high (5 to 525 ppm). However, P-32 was rapidly removed from the water (50% in 2.5 days) and concentrated in organisms when natural phosphate levels were low (0.05 to 0.5 ppm). Removal and concentration of P-32 could be predicted from the amount of natural phosphate present, but the relationship was complex and not linear (Whittaker 1954).

Other Implications from Microcosm Studies with P-32

Eventually, seven studies on the uptake of P-32 in microcosms were completed during the 1950s. This effort added to the understanding of what happened to P-32 entering the Hanford Reach in reactor effluent (Whittaker 1961).

The relationship between absorption and adsorption of P-32 was puzzling. Experiments with a small, aquatic crustacean (Daphnia) suggested rapid uptake and turnover on surfaces (adsorption), so that equilibrium with water concentration was reached in 1 hour. After this, slower uptake by ingestion (absorption) occurred at a near linear rate until the adsorbed P-32 was a small fraction of total uptake. The effect of adsorption on uptake rate was inversely related to the organism's size.

Uptake rates were affected by water temperature. As a rule, aquatic communities develop more rapidly at warmer water temperatures. However, one test suggested that raising the temperature in the microcosm from 10°C to 25°C did not produce a threefold to fourfold increase in uptake and turnover of P-32, as indicated by the van't Hoff equation. Rather, a slower, fractional rate of increase occurred at 25°C.

Another study examined how P-32 moved between community components. When contaminated organisms were transferred from spiked to unspiked aquaria, P-32 rapidly passed to the water and, hence, to all components of the new community. The "back-and-forth movement" of P-32 spread the radionuclide through the ecosystem and gave insight to a multidimensional flow model (Whittaker 1961).

Uptake of Zn-65 by Periphyton in Closed System

A closed recirculating system was developed in 1968 to provide continuous measurement of uptake and cycling of radionuclides by stream periphyton. This system provided control of environmental conditions that often compromised field studies on radionuclides (Cushing and Porter 1969).

The flowing system (a type of microcosm) was initially spiked with Zn-65, and exposure conditions were varied according to design. Uptake of radioactivity by periphyton was considerable during continuous light, continuous darkness, a 12-hour photoperiod, or by killed communities (boiled or immersed in formalin). Thus, activity was transferred to periphyton primarily by adsorption, and was largely independent of photosynthesis. Uptake was proportional to the initial concentration of Zn-65 in the water; Zn-65 subsequently decreased during the test. Increasing the concentration of stable zinc or magnesium decreased the uptake of Zn-65 proportionately, suggesting competition of elements for cation binding sites and supporting the adsorptive uptake theory. At least some zinc, an essential micronutrient for plants, must be assimilated by periphyton (Cushing and Rose 1970).

Mechanisms for radionuclide uptake by periphyton were examined macroscopically by use of autoradiographic techniques. A natural, mat-like growth of periphyton is on the upper surface of most stream beds. Zinc-65 was sorbed largely on this layer; therefore, a diffusion gradient existed and the entire periphyton mass was not exposed to the water transporting radioactivity. Data from short-term spiking experiments with radionuclides might have more meaning if expressed by area rather than on a gravimetric basis (Rose and Cushing 1970). Use of individual radionuclides as tracers in microcosm studies could improve efforts to model the cycling of minerals by periphyton and lead to more appropriate models of the dynamics of radionuclide transfer. Simple uptake and retention models were not adequate for this purpose (Cushing et al. 1975).

Cycling of Zn-65 in Lotic Microcosms

Cycling of Zn-65 was investigated in simulated streams containing a simple water-periphyton-fish food web. The streams were developed from laboratory troughs where flow, light, and radionuclide levels were controlled. Water temperature in the troughs paralleled those of the Hanford Reach from January to June. The troughs, stocked with periphyton and young carp, were supplied with uncontaminated river water (control) or river water with Zn-65 added continuously at levels near 1 pCi/mL and 10 pCi/mL. Test concentrations were 20 and 200 times above levels in the Hanford Reach downstream from the reactors.

Amounts of Zn-65 neared equilibrium in periphyton after about 28 days at levels of 17,000 and 90,000 pCi/g wet weight in the low- and high-level streams, respectively (Figure 9.11). These represented levels approximately 23 and 150 times those in control periphyton, and were close to differential spike levels. Thus, Zn-65 concentrated in periphyton

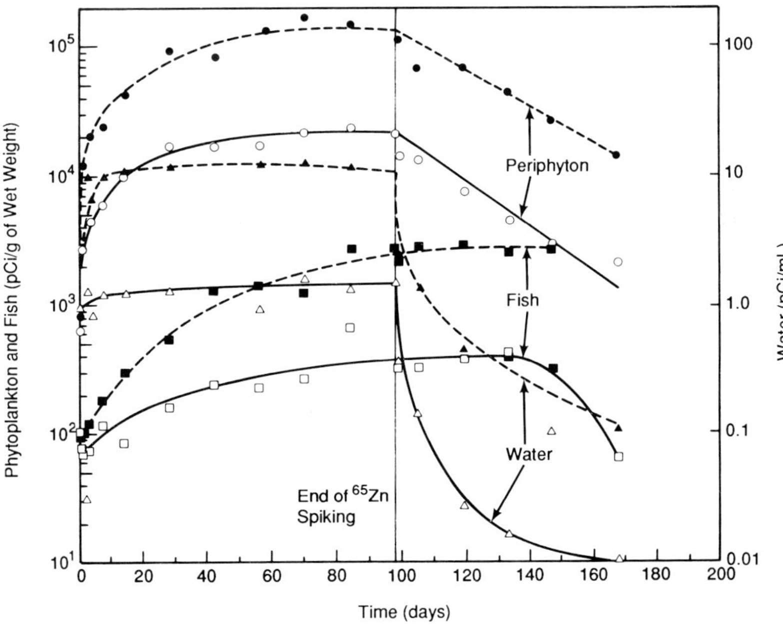

Fig. 9.11. Uptake and retention of Zn-65 by periphyton and fish in laboratory troughs containing recirculating water spiked with either 1 (dashed line) or 10 (solid line) pCi/mL of the radionuclide (from Watson and Cushing 1971).

at amounts nearly proportionate to levels available in the troughs. Specific amounts at equilibrium were 35 and 210 pCi/mg in the low- and high-level streams, respectively. Effective half-life was about 15 days.

Uptake of Zn-65 by fish in the troughs was variable because their metabolic rate was influenced by water temperature (6° to 18°C). Equilibrium was approached after about 43 days at levels of 250 and 1300 pCi/g wet weight in the low- and high-level streams, respectively. These levels were about 6 and 29 times those in control fish. Specific amounts at this time were 3 and 20 pCi/mg. These data suggested that true equilibrium had not been reached (Cushing and Watson 1973).

Uptake of Zn-65 by Tubificid Worms

Zinc-65 and other radionuclides were readily accumulated by tubificid worms held in containers. Tubificids, a food source for fish, live in bottom sediments of lakes and streams where they feed on organic detritus. Uptake of Zn-65 by tubificids was examined in both static and flowing systems for 9 days.

Tubificids took up radioactivity when Zn-65 was dissolved in water but not when bound to sediments. Bioaccumulation of dissolved Zn-65 depended on temperature and concentration of the radionuclide in water. Uptake was fastest, and reached greater concentrations (15 pCi/g dry weight), in worms exposed to 12.5 μCi/L at 25°C (Dean 1974). Zinc-65 had not reached equilibrium in tubificids after 9 days exposure.

Bioaccumulation of Cs-137

Bioaccumulation of Cs-137 was studied in a large, outdoor concrete pond containing a community of organisms typical of a farm fish pond, including emergent plants. Pond water was spiked with Cs-137 to provide 6×10^{-3} μCi/mL of radioactivity. Organisms representing potential pathways to humans were sampled for 17 months.

All organisms bioaccumulated Cs-137. Concentration factors ranged from 50 to more than 10,000 times that in the pond water. Bioaccumulation in aquatic plants was at least 500 times those reported in other studies for terrestrial plants. Levels of Cs-137 changed in algae, submerged seed plants, grass, fish, and frogs with temperature changes, which regulated metabolic rates in the pond. Shading reduced uptake of Cs-137 by submerged plants. Emergent plants rooted in gravel accumulated more of the radionuclide than plants rooted in mud. Radioactivity decreased in all organisms by transfer to bottom sediment and by partitioning among increasing biomass. Generally, the highest activity occurred at the highest trophic levels. Aquatic plants acted as reservoirs of Cs-137, enabling transfer to terrestrial grazers (Pendleton 1958, 1959).

Subsequent efforts examined the availability of Cs-137 in different types of aquatic ecosystems. Uptake and partitioning were compared between the large concrete pond and a dirt pond containing mixed fission products from chemical separation facilities (in the 200 Areas) at Hanford.

Again, all aquatic organisms exposed to Cs-137 accumulated the radionuclide in amounts higher than were in the water. Results indicated that terrestrial animals, such as ducks, feeding on aquatic plants would become highly contaminated. Furthermore, all aquatic animals or animals used for human food could take up hazardous amounts of Cs-137 from water containing low levels of Cs-137 (Pendleton and Hanson 1958).

Significance of laboratory studies with radioactivity

Ecologists recognize that the data obtained from laboratory experiments, while more qualitative, are apt to differ from the data obtained in field studies. The principal reason is that natural, uncontrolled variables are present in field situations, and these variables strongly influence the results. Laboratory studies not only permit control of the experiment, but allow certain variables to be altered to examine mitigating factors such as temperature, light, and size/age of the organism tested. The broadest and deepest understanding of ecological phenomena is usually obtained from both laboratory and field studies in related areas. Researchers at Hanford initiated such an approach with radionuclide studies in the early years, and continued with this dual approach when examining other ecological phenomena.

In general, laboratory experiments provide the best quantitative data from sublethal exposures of aquatic organisms to radionuclides because exposure conditions, particularly dose rates, can be controlled. Effluent monitoring (Chapter 7) and field studies (Chapter 8) indicated that kinds and amounts of radioactivity in the cooling effluent of the single-purpose reactors were not likely to be lethal to aquatic life. Furthermore, they indicated that the amounts of radioactivity encountered in the Hanford Reach were not likely to cause sublethal damage. This assessment could be confirmed only by laboratory studies that 1) identified the routes from which fish and other aquatic organisms took up radionuclides from water, partitioned the radionuclides in their bodies, and eliminated them; 2) determined the actual radiation doses causing sublethal effects, particularly for the most biologically active radionuclides; and 3) examined the transfer of radionuclides among the components and trophic levels of aquatic ecosystems.

Laboratory studies at Hanford were among the first to examine the relationships between specific radionuclides and fish. However, the number of radionuclides studied and the scope of effort were somewhat limited. Interest at Hanford focused on a few activation products (i.e., P-32, Zn-65) that appeared in Hanford's reactor effluent, and a few fission products (i.e., Sr-90 and Cs-137) that were introduced as a result of nuclear material production and nuclear energy use. The radionuclide P-32, as a tracer, proved beneficial for aquatic ecological studies elsewhere (as did C-14). Partitioning of the element phosphorus, believed to be a limiting factor in primary productivity of phytoplankton in aquatic ecosystems, could be followed simply by spiking water with its radioactive isotope (P-32).

We now know that only radionuclides with a specific biological function can bioaccumulate significantly in the body of fish. Conversely, radionuclides with no specific biological function have low bioaccumulation potential. Furthermore, the potential for bioaccumulation in fish is reduced by partitioning among other components of the aquatic environment. Models capable of predicting chemical speciation and partitioning of radionuclides in aquatic habits in general, and fish in particular, are weakened by many assumptions because the different biotic and abiotic factors that influence speciation and partitioning are not yet completely understood. In recent years, research in this aspect of radioecology has declined.

Radiological aspects were emphasized at Hanford in early years not only because more than 60 radioactive materials appeared in the cooling effluent of the single-purpose reactors, but because so little radiological information was available. Laboratory studies, when integrated with field work, played a major role in explaining the fate, distribution, and effects of radionuclides in aquatic environments.

References

Buhler, D.R. 1967. "Tissue Binding of Zinc-65 in Fish and Other Vertebrates." In: *Pacific Northwest Laboratory Annual Report for 1966 to the USAEC Division of Biology and*

Medicine, Vol. I Biological Sciences, pp. 171–174. BNWL-480, Pacific Northwest Laboratory, Richland, Washington.

Buhler, D.R. 1968. "Some Sites for Radionuclide Binding in Fish." In: *Pacific Northwest Laboratory Annual Report for 1967 to the USAEC Division of Biology and Medicine, Vol. I Biological Sciences*, pp. 9.21–9.29. BNWL-714, Pacific Northwest Laboratory, Richland, Washington.

Cushing, C.E., and N.S. Porter. 1969. "Radionuclide Cycling by Periphyton: An Apparatus for Continuous In Situ Measurements and Initial Data on Zinc-65 Cycling," In: *Symposium on Radioecology*, eds. D.J. Nelson and F.C. Evans. (AEC CONF-670503), National Technical Information Service, Springfield, Virginia.

Cushing, C.E., and F.L. Rose. 1970. "Cycling of Zinc-65 by Columbia River Periphyton in a Closed Lotic Microcosm." *Limnol. Oceanogr.* 15:761–767.

Cushing, C.E., and D.G. Watson. 1973. "Cycling of Zinc-65 in a Simple Food Web." In: *Radionuclides in Ecosystems, Third National Symposium on Radioecology*, ed. D. J. Nelson, pp. 211–322. CONF-710501-P1, National Technical Information Service, Springfield, Virginia.

Cushing, C.E., J.M. Thomas, and L.L. Eberhardt. 1975. "Modeling Mineral Cycling by Periphyton in a Simulated Stream System." *Verh. Int. Verein.* Limnol. 19:1591–1598.

Dean, J.M. 1965. "Cycling of Sr-90 in Molting Crayfish." In: *Hanford Biology Research Annual Report for 1964*, pp. 41–48. BNWL-122, Pacific Northwest Laboratory, Richland, Washington.

Dean, J.M. 1974. "The Accumulation of Zn-65 and Other Radionuclides by Tubificid Worms." *Hydrobiol.* 45:31–38.

Dean, J.M., and R.E. Nakatani. 1966. "Temperature Effects on Cesium Metabolism in Trout." In: *Pacific Northwest Laboratory Annual Report for 1965 in the Biological Sciences*, pp. 111–113. BNWL-280, Pacific Northwest Laboratory, Richland, Washington.

Dean, J.M., J. Eapen, and R.E. Nakatani. 1965. "Metabolism of Cs-137 in Trout." In: *Hanford Biology Research Annual Report for 1964*, pp. 71–101. BNWL-122, Pacific Northwest Laboratory, Richland, Washington.

Donaldson, L.R., and R.F. Foster. 1957. "Effects of Radiation on Aquatic Organisms." In: *The Effects of Atomic Radiation in Oceanography and Fisheries*, pp. 91–102. Publishing No. 551, National Academy of Sciences, National Research Council, Washington, D.C.

Nakatani, R.E. 1962. "Distribution and Retention of Sr-90-Y-90 in Trout"' In: *Hanford Biology Research Annual Report for 1961*, pp. 21–29. HW-72500, Hanford Atomic Products Operation, Richland, Washington.

Nakatani, R.E. 1966. "Biological Response of Rainbow Trout (Salmo gairdneri) Ingesting Zinc-65." In: *Disposal of Radioactive Wastes into Seas, Oceans and Surface Waters*, ed. A. Guillon, International Atomic Energy Commission, Vienna.

Nakatani, R.E., and R.F. Foster. 1961. "Damage to Rainbow Trout from Repetitive Feeding of Sr-90/Y-90." In: *Hanford Biology Research Annual Report for 1960*, pp. 1–7. HW-69500, Hanford Atomic Products Operation, Richland, Washington.

Nakatani, R.E., and R.F. Foster. 1963. "Effects of Chronic Feeding of Sr-90-Y-90 on Rainbow Trout." In: *Radioecology*, eds. V. Schultz and A.W. Klement, pp. 351–362. Reinhold Publishing Corp., New York and AIBS, Washington, D.C.

Nakatani, R.E., and D.H.W. Liu. 1964. "Distribution and Retention of Zn-65 in Trout." In: *Hanford Biology Research Annual Report for 1963*, pp. 101–194. HW-80500, Hanford Atomic Products Operation, Richland, Washington.

Nakatani, R.E., D.H.W. Liu, and W.J. Clarke. 1965. "Effect of Chronic Ingestion of Zn-65 in Trout." In: *Hanford Biology Research Annual Report for 1964*, pp. 101–186. BNWL-122, Pacific Northwest Laboratory, Richland, Washington.

Nakatani, R.E., and W.P. Miller. 1963. "Distribution of Zn-65 after Ingestion by Trout." In: *Hanford Biology Research Annual Report for 1962*, pp. 111–113. HW-76000, Hanford Atomic Products Operation, Richland, Washington.

Olson, P.A., and R.F. Foster. 1960. "Effect of Mode of Administering P-32 to Fish." In: *Hanford Biology Research Annual Report for 1959*, pp. 91–98. HW-65500, Hanford Atomic Products Operation, Richland, Washington.

Pauley, G.B., and R.E. Nakatani. 1967. "The Uptake of the Radioisotope Zn-65 by Various Tissues of the Freshwater Mussel, *Anodonta californiensis* Lea." *Proc. Nat. Shellfish Assoc.* 57:1–8.

Pendleton, R.C. 1958. "Absorption of Cs-137 by an Aquatic Community." In: *Hanford Biology Research Annual Report 1957*, pp. 31–34. HW-53500, Hanford Atomic Products Operation, Richland, Washington.

Pendleton, R.C. 1959. "Effects of Some Environmental Factors on Bioaccumulation of Cesium-137 in an Aquatic Community." In: *Hanford Biology Research Annual Report for 1958*, pp. 41–46. HW-59500, Hanford Atomic Products Operation, Richland, Washington.

Pendleton, R.C., and W.C. Hanson. 1958. "Absorption of Cesium-137 by Components of an Aquatic Community." In: *Second International Conference on Peaceful Uses of Atomic Energy*, Geneva, Switzerland, Vol. 18, P/392, pp. 411–422. United Nations, New York.

Rose, F.L., and C.E. Cushing. 1970. "Periphyton: Autoradiography of Zinc-65 Adsorption." *Science* 168:571–577.

Schiffman, R.H. 1959. "The Uptake of Strontium from Diet and Water by Rainbow Trout." In: *Hanford Biology Research Annual Report for 1958*, pp. 16–19. HW-59500, Hanford Atomic Products Operation, Richland, Washington.

Schiffman, R.H. 1960. "Effects of Intramuscular Injections of Sr-9-Y-90 on Rainbow Trout." In: *Hanford Biology Research Annual Report for 1959*, pp. 56–58. HW-65500, Hanford Atomic Products Operation, Richland, Washington.

Schiffman, R.H. 1961. "Preliminary Studies on the Elimination of Strontium by Trout." In: *Hanford Biology Research Annual Report for 1960*, pp. 81–85. HW-69500, Hanford Atomic Products Operation, Richland, Washington.

Watson, D.G. 1957. "Effect of Massive Doses of P-32 on Trout." In: *Biology Research Annual Report 1956*, pp. 228–233. HW-47500, Office of Technical Services, Washington, D.C.

Watson, D.G., and C.E. Cushing. 1971. "Cycling of Zinc-65 in a Simple Food-Web." In: *Pacific Northwest Laboratory Annual Report for 1970 to the USAEC Division of Biology and Medicine, Vol. I Life Sciences, Part 2 Ecological Sciences*, pp. 2.1–2.9. BNWL-1550 PT2, Battelle, Pacific Northwest Laboratories, Richland, Washington.

Watson, D.G., L.A. George, and P. L. Hackett. 1959. "Effects of Chronic Feeding of Phosphorus-32 on Rainbow Trout." In: *Hanford Biology Research Annual Report for*

1958, pp. 71–77. HW-59500, Hanford Atomic Products Operation, Richland, Washington.

Whittaker, R.H. 1953. "Removal of Radiophosphorus Contaminant from Water in an Aquarium Community." In: *Biology Research – Annual Report 1952*, pp. 14–19. HW-28636, Hanford Atomic Products Operation, Richland, Washington.

Whittaker, R.H. 1954. An Experiment on the Relation of Phosphate Level in Water to the Removal and Concentration of Radiophosphorus." In: *Biology Research – Annual Report 1953*, pp. 11–23. HW-30437, Hanford Atomic Products Operation, Richland, Washington.

Whittaker, R.H. 1961. "Experiments with Radiophosphorus Tracer in Aquarium Communities." *Ecol. Monogr.* 31:151–188.

Chapter 10

THERMAL EFFECTS STUDIES IN THE HANFORD REACH, 1960 TO 1971

Through the 1950s, aquatic studies focused on the radioactivity in reactor cooling effluent at Hanford, with supporting studies on chemicals and heat. Around 1965, most power plants generating electricity from nuclear and fossil fuels in the United States were cooled by the "once-through" method, in which heated water was returned to the source environment. Hence, the possibility of widespread "thermal pollution" in the nation's lakes and streams came to be viewed with alarm. Concern focused on facilities at locations where thermal effects might be harmful to aquatic biota. Criteria useful for assessing potential damage from thermal effects were largely unavailable.

The Water Quality Act of 1965 directed attention to temperature as a pollutant. Water quality standards developed by the states of Oregon and Washington then prompted the U.S. Atomic Energy Commission to calculate the impact of heat from once-through cooling of Hanford's eight single-purpose reactors on the Columbia River. From that point on, potential thermal effects were closely examined in the Hanford Reach until the last reactor was shut down in January 1971. The interagency Columbia River Thermal Effects Study (CRTES), driven by the Environmental Protection Act of 1969 and by the concerns of fisheries agencies, concluded that no significant thermal effects could be attributed to Hanford's reactors (EPA 1971a). However, information from thermal studies at Hanford proved valuable for thermal effect evaluations at the nation's power plants long after the last single-purpose reactor was shut down.

One aspect of this effort, the relationship between thermal discharges and *Flexibacter columnaris* – a bacterial disease endemic among salmonids and other Columbia River fish, was studied from 1961 to 1973. Highly virulent strains of columnaris, capable of causing severe disease outbreaks at relatively low water temperatures, appeared in the 1940s and 1950s. Many adult salmon became infected during upstream migration and died before spawning. Some people speculated that heated effluent discharged to the river from the single-purpose reactors at Hanford since 1944 had, in some way, influenced outbreaks of the disease (Pacha and Ordal 1970).

Field studies: thermal releases to the Columbia River

Features of cooling water discharges unique to the single-purpose reactors and the mainstem Columbia River were the focus of field studies (Coutant 1969a; Nakatani 1969; Templeton and Coutant 1971; Becker 1973b). In general, thermal effects were sought on migratory salmonids, resident fish and invertebrates, and ecosystem functions. Part of the effort was planned and conducted under the CRTES from 1968 to 1970 (EPA 1971a).

The heated water released from each single-purpose reactor at Hanford entered the river primarily through a single large outlet above the river bottom near midstream. Effluent temperatures were nearly always greater than 80°C as they left the outfall. At some locations, retention basins along the shoreline and pipeline breaks created shoreline seeps and secondary discharges of heated water. All sources were of regulatory concern.

Imposition of effluent temperatures on seasonal river temperatures result in a thermal increment (delta T). This increment is lethal only if it exceeds the temperature tolerated by a fish that is acclimated to ambient temperatures above the reactors. Thermal effects are largely specific to time, site, and species because river temperatures vary seasonally, the temperature of a heated plume varies with plant power level, the mixing zone varies with river discharge volume, and the thermal tolerance of a species varies with acclimation.

Effects of Reactor Cooling Water Intakes

Impingement and entrainment of small fish were common wherever water was diverted from streams, lakes, and estuaries in the 1960s. Large amounts of water from the Columbia River water were required for once-through cooling of Hanford's reactors. Microscopic aquatic organisms carried in the reactor cooling water were probably lost early in the water treatment process, especially during filtration and chlorination.

Records are lacking on potential destruction of small fish at the cooling water intakes of Hanford's single-purpose reactors. Losses would depend, in part, on the attraction of the intake structures to fish and on the velocity of the water entering the intake well. All intakes were indented in the shoreline away from the directional, downstream flow of the Columbia River. Currents in each indentation were slowed to a clockwise rotation of about 0.1 to 1.0 foot per second (Soldat 1962). Intake velocities were probably sufficiently low that only the smallest immature fish were impinged or entrained.

Some observations were made on passage of fish through the 100-F pumphouse screens in the early 1950s. River water pumped from the intake well spilled into the reservoir of a treatment station in such a way that part of the flow could be directed into a fish trap. A few dozen small salmon (probably chinook salmon fry) from the Hanford Reach were collected each spring. Large numbers of larval lampreys were also caught. Preserved specimens, stored in jars in the 146-FR Building, were lost in the 1964 fire (R.F. Foster, written communication). Thus, the species and a number of fish entrained at the intakes of the single-purpose reactors at various seasons were never quantified.

Effect of Thermal Loading

Beginning in 1962, physical changes in river water caused by the added heat were assessed at Hanford. For several years, Hanford scientists monitored water temperatures at points along the mainstem Columbia River upstream and downstream from Hanford with automatic recording thermographs. These data were analyzed by state-of-the-art computer models that simulated the dispersal of thermal plumes (the first such simulations ever done). Temperature records kept since 1933 at Rock Island Dam, upstream of Hanford, were used to provide historical background.

Assessments indicated that all low-head dams built on the mainstem of the Columbia River had no significant effect on average water temperatures. However, storage and release of water since 1941 from behind Grand Coulee Dam (a high-head dam below the Canadian border) had delayed the transport of water through mainstem reservoirs by about 30 days. Releases from Grand Coulee Dam also moderated temperature extremes in the mainstem Columbia River at all points below. Thus, temperatures in the Hanford Reach when the first single-purpose reactor began operating in 1944 were slightly lower in the summer and slightly higher in the winter than before Grand Coulee began operations (Jaske 1969; Jaske and Goebel 1967).

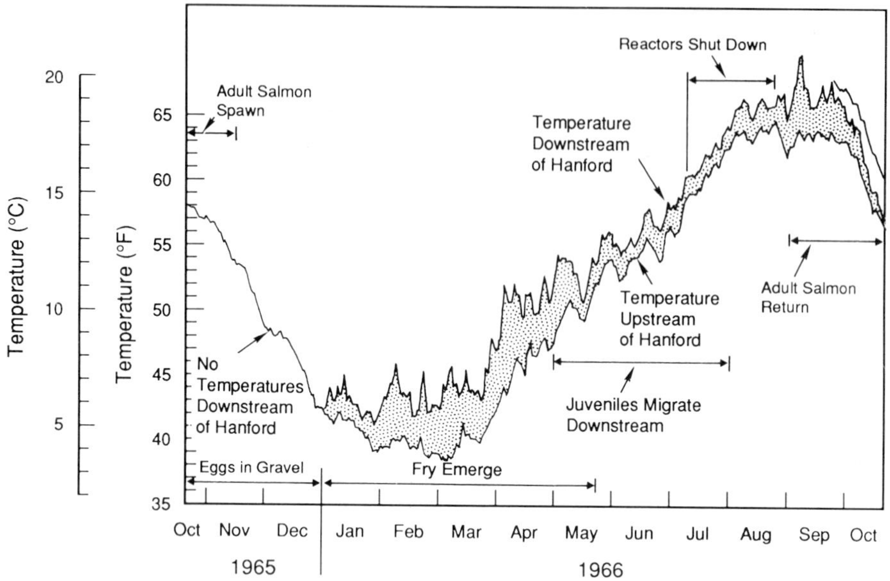

Fig. 10.1. Temperatures of the Columbia River upstream and downstream of Hanford 1965 to 1966. The effect of thermal increments from the Hanford reactor discharges before, during, and after a total shutdown in July and August are shown in relation to the life cycle of fall chinook salmon (modified from Templeton and Coutant 1971).

Water in the Hanford Reach, the last flowing section of the mainstem Columbia River, responded more rapidly to weather changes than water in reservoirs. Conductive heat transfers were higher and evaporative rates were lower in the Hanford Reach than in impoundments. The effect of these phenomena was clear during July and August 1966 when the reactors shut down during a labor strike. At this time, natural insolation alone increased temperatures in the Hanford Reach from 0.5° to 0.75°C (Figure 10.1). An opposite cooling effect occurred in winter (Jaske and Synoground 1970).

As long as the single-purpose reactors operated, incremental temperatures in the Hanford Reach decreased as water flowed downstream to the Snake River because much of the added heat dissipated to the atmosphere. An average of 35% (5% to 40%) of the heat added by the reactor effluents was retained at the Oregon-Washington border. The contribution of the reactor discharges to increasing Columbia River temperatures downstream from McNary Dam from 1965 to 1969 was about equal to that added from the warmer Snake River (Jaske and Synoground 1970).

Discharge Plume Characteristics

Effluents from mid-river releases extended downstream in buoyant plumes that first surfaced as turbulent upwellings. The highest temperatures were near the discharge points, but they rapidly declined as the plumes passed downstream and mixed with river water. About 80% of the incremental temperature from a plume was lost in 5 seconds. The surface width of each plume and its mixing efficiency varied with river flow. High river flows narrowed plume width and dispersed heat more effectively than low flows.

Temperatures in various plumes were examined sporadically from 1946 to 1961, along with flow velocity and radioactivity. Initial measurements were direct. For example, continuous measurements were taken during a traverse of the river with a thermometer (accurate to $< 0.1°C$) fastened to a torpedo-shaped weight and towed behind a boat. In 1950 and 1951, the surface distribution of the plumes was examined when either a flocculant or fluorescein dye was added to the effluent before its discharge (Soldat 1962). Physical dimensions of the effluent plume from KE Reactor, the last single-purpose unit still in operation, were defined in the late 1960s (Figure 10.2). Surface temperatures were quantified by dye releases and a new technique, infrared imagery. Vertical temperatures were measured by sensitive thermocouples trailed behind a boat at various depths. Furthermore, plume measurements were obtained at different stages of river flow, ranging from 35,000 cubic feet per second (ft^3/s) (the regulated minimum) to 160,000 ft^3/s (Jaske et al. 1969, 1970; Jaske and Synoground 1970).

The KE Reactor plume included a zone of prompt mixing immediately downstream of the release point, a zone of transition 183 meters (200 yards) to 1.6 kilometer (1 mile) downstream, and a zone of complete mixing. The zone of prompt mixing was the area where juvenile salmonids migrating downriver might, in theory, encounter lethally high temperatures. Although temperatures lethal to fish did exist near the outfall, the heated effluent was completely mixed from top to bottom when it reached a point 366 meters (400 yards) farther down. Plume boundaries varied with river flow, being widest at low flow (Jaske et al. 1970).

A thermal death model for salmonid outmigrants was developed from field data on plume discharge temperatures and laboratory data on the resistance of juvenile fish to thermal shock. The model not only predicted the mortality of fish encountering lethal temperatures by chance, but conditions that would impose thermal stress and change fish behavior (Jaske et al. 1970). No losses were predicted when salmonid outmigrants

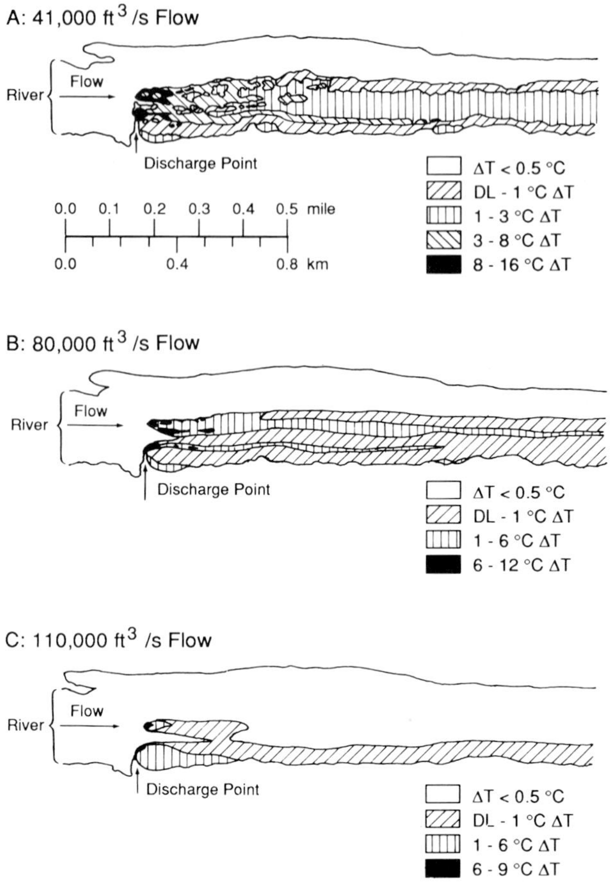

Fig. 10.2. Surface distribution of incremental temperatures from the KE Reactor's effluent plume under three discharge conditions: 41,000, 80,000, and 110,000 ft^3/s (modified from Templeton and Coutant 1971).

passed 18 meters (20 yards) from the plume center line, when river flow was 80,000 ft^3/s or more, or when ambient river temperatures were less than 16°C (Templeton and Olson 1973). Ecological conditions leading to possible loss of outmigrants were infrequent, because flows in the Hanford Reach ranged from 80,000 to 200,000 ft^3/s during the low-flow season (July to September).

Thermal Effects on Benthic Organisms

Most aquatic animals are cold-blooded, and their metabolism and growth respond to water temperature. Thus, thermal increments may affect development of benthic organisms in the Hanford Reach downstream from the reactor discharges.

One study showed that adult caddis flies emerged about 2 weeks earlier downstream from the reactors than upstream. This effect was attributed to slightly warmer water during winter, which hastened the development and maturation of larval caddis flies on the river bed (Coutant 1968a).

The size and growth rates of limpet snails did not differ upstream and downstream from the reactor discharges 1969 through 1970, even though temperatures downstream were 0.5° to 1.5°C higher. Most limpets had a1-year life cycle, undergoing high mortality after depositing egg cases during summer. Adult limpets held in warmed laboratory water reproduced successfully (Coutant and Becker 1973).

The thermal resistance of caddis fly larvae, one of the most abundant food organisms for fish in the Hanford Reach, was examined after acclimation to near maximum river temperatures of 19.5°C during late summer and fall. Larvae abruptly exposed to a 10°C higher temperature underwent 50% mortality in 68 hours, but most larvae survived a 7.5°C increment (Becker 1971b). The lethal thermal limit for most river invertebrates appeared to be well above temperatures in effluent mixing zones.

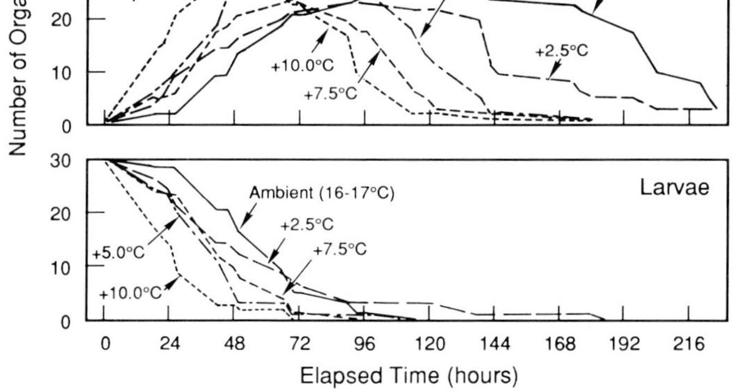

Fig. 10.3. Development of black fly larvae from the Columbia River at incremental temperatures differing by 2.5°C (from Becker 1973c).

Larvae of the black fly were exposed to increments ranging from 17° to 27°C greater than seasonal water temperatures. Increments of 2.5°C proportionately increased rates of development through the pupal stage until adults emerged (Figure 10.3). A rise in temperature of 10°C reduced the mean development span by about half, from 8 days at 17°C to 4 days at 27°C. Temperatures of 27°C were tolerated by black fly larvae (Becker 1973c).

Thermal Effects on Returning Adult Salmonids

Theoretically, the spawning of fall chinook salmon in the Hanford Reach could be influenced by the heated reactor discharges. Most salmon spawned upstream of the reactor discharges to Priest Rapids Dam, but some dug redds (cavities in the streambed to deposit eggs) downstream of the reactors. Most redds below the reactors were placed on either side of the midstream plumes. However, some redds were dug along the margins of islands some distance downstream from the outfalls. Because the plumes were initially buoyed upward, they probably did not interact with the redds until mixing had occurred (Watson 1970, 1976).

During the winter of 1954 to 1955, temperatures were measured 15 centimeters (6 inches) above the gravel, and 15, 38, 53, and 76 centimeters (6, 15, 21, and 30 inches) in the gravel of a simulated redd about 0.8 kilometer (0.5 mile) downstream of the 100-F effluent outfall. Intergravel water was actually 1° to 4°C warmer during the winter, when salmonid eggs were developing in redds, than in the river (Nakatani 1969). Temperatures in the upper layers of gravel varied up to 4.5°C daily, but they remained isothermal at a depth of 76 centimeters (30 inches). Also, beta radioactivity was less than 1/10 that in the overlying water, at an intergravel depth of 30 centimeters (12 inches) and less than 1/50 at a depth of 16 centimeters (30 inches). Upwelling groundwater apparently provided dilution (Honstead and Clark 1953) and restricted the penetration of river water containing radioactivity.

In some years, temperatures in the Hanford Reach were above optimum for egg survival (> 15°C) when fall chinook salmon began spawning in October. Incremental heat from the effluent plumes might impair spawning activity or cause egg mortality before water temperatures declined during winter. However, less than 1% of the salmon run to the Hanford Reach spawned where their eggs might have been affected by the plumes (Foster et al. 1972).

In the late 1960s, it was believed by some that the Hanford discharges represented a block to upstream migration of salmonids, similar to the block occurring at the outlets of some warm tributaries above Hanford. Upstream migration of summer chinook salmon and steelhead through the Hanford Reach in relation to the discharge plumes was monitored in 1967 and 1968. In a cooperative effort with the Bureau of Commercial Fisheries (now the National Marine Fisheries Service), ultrasound transmitters were inserted in adult salmon captured at McNary Dam. The fish were tracked electronically as they continued upstream.

Returning salmonids generally swam near the shorelines in less than 3 meters (10 feet) of water, and did not come near the midstream plumes. Upstream migrants usually paralleled the shoreline opposite the reactors (north shore) in the Hanford Reach, most notably during the late summer and fall when river temperatures approached 20°C. In fact, most of the upstream migrants had passed through the left fish ladder at McNary Dam, and they continued along the same shore as they approached the Hanford Reach. A few migrants along the reactor side (south shore) did pause and cross the river when encountering warm water seepages (Coutant 1968b, 1969b, 1970a). Most importantly, the discharge plumes did not block fish migration, nor did they slow the upstream passage of fish.

During the same period (1967 and 1968), fisheries agencies noted that a portion of the summer steelhead run entering the Hanford Reach each year did not continue upstream migration but disappeared from further observation. The reactor effluents were considered as a possible reason for this "loss." The cooperative tagging studies showed that these fish remained in viable condition, overwintered, and spawned the following spring in the vicinity of the Ringold Hatchery (C.C. Coutant, personal communication).

Thermal Effects on Outmigrant Salmonids

Juvenile salmonids move seaward through the Hanford Reach each year during the spring and summer. River temperatures are cool and favorable to outmigrants during spring, but increase to suboptimum levels through summer and early fall. During reactor operations, some seasonal outmigrants might enter the heated effluent plumes and be thermally shocked.

Of the total outmigration of juvenile salmonids through the Hanford Reach, only a few fish could potentially have entered the heated effluent soon after it was discharged. Only about 5% of the river's cross section was heated enough to be a hazard. Furthermore, available evidence indicated that most of the fish passed near the shore rather than in the

Fig. 10.4. Crew preparing to drift young salmonids in liveboxes through the reactor effluent plumes in the spring of 1970. The livebox could be lowered to different depths down the pole by rope and pulley, while the temperatures fish encountered were recorded from the front of the boat.

center of the river where the effluent outfalls were located (Foster et al. 1972). Some possible thermal effects were examined in the late 1960s by 1) drifting juvenile salmon through the discharge plumes in liveboxes, 2) anchoring floating traps in effluent mixing areas, and 3) examining predation by gulls near the plumes.

The livebox drifts were conducted from March to September of 1968 and 1969. Young salmonids restrained in cages were drifted through effluent plumes while temperature increase and exposure time were monitored (Figure 10.4). Results of two drift series showed that caged fish could be killed by direct thermal shock along the south reactor shoreline when they entered areas of limited mixing heated by shoreline seeps or pipeline leaks. However, fish that were drifted through the midriver plumes usually survived passage except in late summer when river flows were low and ambient temperatures were high (Becker et al. 1971).

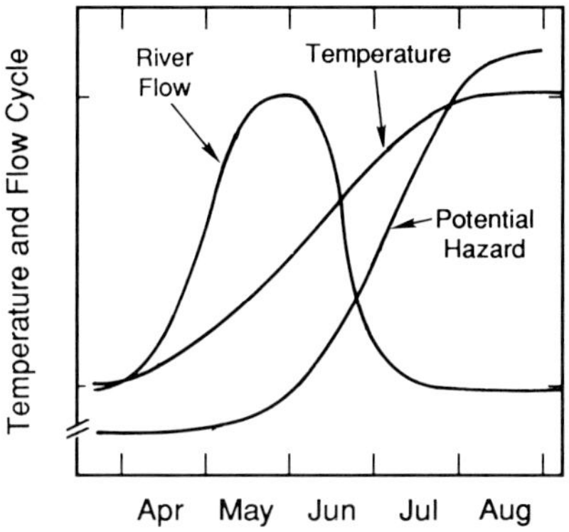

Fig. 10.5. Assessment of potential thermal hazard to juvenile salmonids passing seaward through the Hanford Reach in relation to the annual temperature and river flow cycles (from Becker et al. 1973).

Behavioral avoidance of warmed water, which was possible in most field situations, was prevented by the cages.

Further assessment revealed four seasonally changing combinations that held increasing potential for adverse effects: 1) low temperature-low flow in early spring, 2) low temperature-high flow in late spring, 3) high temperature-high flow in early summer, and 4) high temperature-low flow in late summer (Figure 10.5). Generally, environmental conditions favored survival of juvenile salmonids that migrated seaward from the Hanford Reach from April to July. But conditions were less favorable for migrants originating upstream of Priest Rapids Dam that passed downstream in July and August (Becker et al. 1973).

Floating, traveling screen traps were anchored in the KE Reactor's effluent plume to catch salmonid outmigrants in the spring of 1968. The few fish collected were not injured and showed no evidence of thermal stress (Coutant et al. 1969). Seagulls occasionally fed downstream of reactor outfalls in the Hanford Reach, suggesting that their prey might be disorganized, thermally shocked fish. The diet of 32 seagulls selectively harvested while feeding near the plume in the fall of 1968 (an off-season for outmigrants) included no juvenile salmonids (Prentice 1969).

Laboratory studies: lethal, sublethal, and physiological effects of temperature

Laboratory studies at Hanford provided a deeper understanding of potential effects of heat added to an aquatic ecosystem. Experiments with Columbia River salmonids were emphasized because, as valued resources, these anadromous fish drew the attention of federal and state agencies. Responses to warmed water may vary among species of salmonids, and may change during different phases of their life cycle. Many unknown lethal, sublethal, and physiological effects needed exploration.

Early in these thermal studies, it became clear that the effects of temperature on aquatic organisms could be indirect as well as direct. Most indirect effects have a physiological basis because they are regulated by internal functions. Therefore, physiological mechanisms leading to fish deaths at high and low temperatures were also explored at Hanford.

Temperature Regimes and Rearing of Juvenile Salmonids

In the late 1960s, several groups of salmonids were reared from eggs at different temperatures. The main goals were to determine the optimum and tolerable temperature regimes for eggs and young of Columbia River salmon, and provide input to decisions on how to draft temperature standards for the Columbia River. The new rearings used raw water from the Hanford Reach, heated or cooled beyond ambient temperatures, according to test design. The effort paralleled earlier studies that monitored fish reared from eggs in warmed, diluted reactor effluent.

Initially, a rearing was used to examine the development of chinook salmon eggs for 6 months as temperatures in the Hanford Reach declined from 13.9°C (57°F) to 2.2°C (36°F) during winter, then increased to 8.3°C (47°F) [(a)]. The control lot was held at ambient river temperatures, while experimental lots averaged about 2.2°C (4°F) colder and 1.1°, 2.2°, and 4,4°C (2°, 4°, and 8°F) warmer. Mortality was significant only in the warmest lot and occurred during the last rearing period. Results indicated that chinook salmon eggs in the Hanford Reach could develop successfully at an average increment of 2.2°C above ambient river temperatures, but not at 4.5°C above (Olson and Foster 1955).

A second rearing, started in the fall of 1966, again examined the development of chinook salmon eggs in relation to changing temperatures in the Hanford Reach. However, this time eggs were stripped from adult fish that were collected on four different dates as river temperatures dropped from about 14.4°C (58°F) to 3.3°C (38°F). The four groups of

[(a)] These studies were done using degrees Fahrenheit, hence the conversion to celsius.

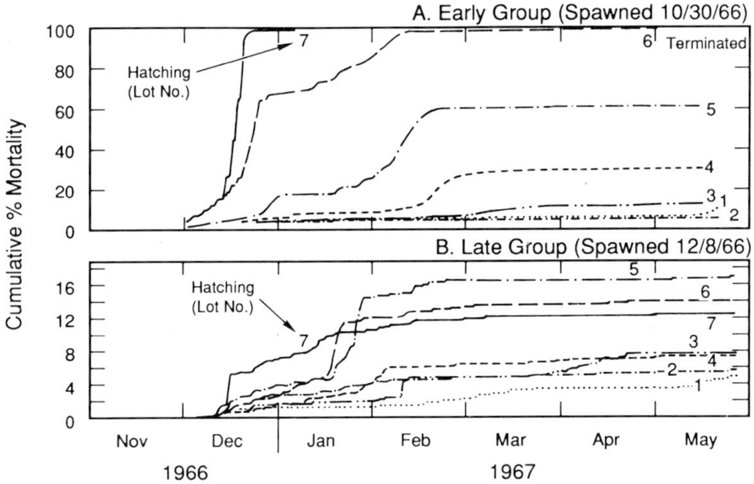

Fig. 10.6. Cumulative mortality of chinook salmon reared from (a) early and (b) late groups of eggs taken in fall of 1966 when reared at thermal increments of 1.1°C (2°F) (modified from Olson et al. 1970).

eggs were each separated into seven lots. Lots were reared at ambient river temperature (control) and six incremental temperatures (1.1°, 2.2°, 3.4°, 4.5°, 5.6°, and 6.7°C) for 177 days.

In the early (first) group of eggs, 2.2°C (4°F) increments caused excessive mortality after hatching, and 6.7°C (12°F) increments caused total mortality. However, the late (last) group tolerated increments as high as 6.7°C with no significant losses after hatching (Figure 10.6). Temperature increased the weight of fish in each incremental lot by a factor of 1.4. Results indicated that tolerance of developing chinook salmon eggs to thermal increments (i.e., the safety margin) was limited if the eggs were deposited in early fall when river temperatures were still warm. However, tolerance to thermal increments increased for eggs deposited during late fall after ambient river temperatures dropped (Olson et al. 1970).

A third rearing, started in the fall of 1967, examined the effect of periodic thermal increments on chinook salmon eggs as temperatures in the Hanford Reach declined as winter approached. One group of eggs was separated into seven lots, all held initially at ambient levels. Later, some lots were exposed to temperature increases as high as 5.0°C greater than ambient even though river temperatures had declined. A constant 2.5°C thermal increment slightly increased rearing losses. When mortality occurred in other lots, it depended on the time of year (season) and how the water was warmed. Most lots underwent no mortality when temperatures were raised to 5°C by 0.25°C stages each week. Fish in all lots grew faster in warmer water (Olson and Nakatani 1969).

A fourth rearing, started in July 1969, examined thermal tolerance of steelhead eggs reared for 19 months (i.e., through one warm season) to a time when the resultant smolts would likely migrate to sea. Eggs were divided into four lots. The control lot was held at ambient, seasonally changing river temperature, and the other lots were held at 1.7°, 2.2°, and 2.6°C above ambient. River temperatures peaked at 21°C to 22°C from mid-August to mid-September each year. The thermal increments caused no mortalities. However, growth was affected by seasonal temperature extremes. Growth in the warmest lot was depressed in the summer but highest in the winter (Olson et al. 1973).

After 2 years, steelhead from the fourth rearing were abruptly exposed to lethal temperatures of 25.0°, 27.0°, and 29.0°C. The length of time until death differed significantly among the four lots, but not length of time until loss of equilibrium. When some fish from each lot were acclimated to 12°C for 3 weeks, differences in time to death disappeared. Even though rearing temperatures did affect the thermal resistance of young steelhead, differences were soon lost when all fish were held at similar temperatures (Schneider and Templeton 1973).

A fifth rearing was started in November 1969 to examine the effect of fluctuating temperatures, such as those that might occur in the Hanford Reach when river flows vary from discharges at an upriver dam. Two groups of chinook salmon eggs were each held at a seasonally changing, ambient river temperature (control), and at temperatures ranging 0.6°, 1.1°, and 1.7°C greater and less than ambient on alternate days. Mortality and growth were similar under fluctuating and nonfluctuating conditions. Generally, the warmer temperatures produced larger fish (Olson 1970).

Thermal Resistance of Adult Salmonids

Sexually mature salmonids enter the Hanford Reach each year when river temperatures approach maximum. At these times, elevation of river temperatures by a few degrees in the Hanford Reach might cause thermal

---

Fig. 10.7. Times to loss of equilibrium (LE) and death (D) of adult salmonids taken from the Hanford Reach and held at elevated temperatures, with 95% confidence limits (horizontal lines): (a) jack chinook salmon in 1968 and 1969; (b) steelhead in 1969 (1969 chinook data included for comparison); and (c) coho salmon in 1968 (1968 chinook data included for comparison) (modified from Becker 1973b).

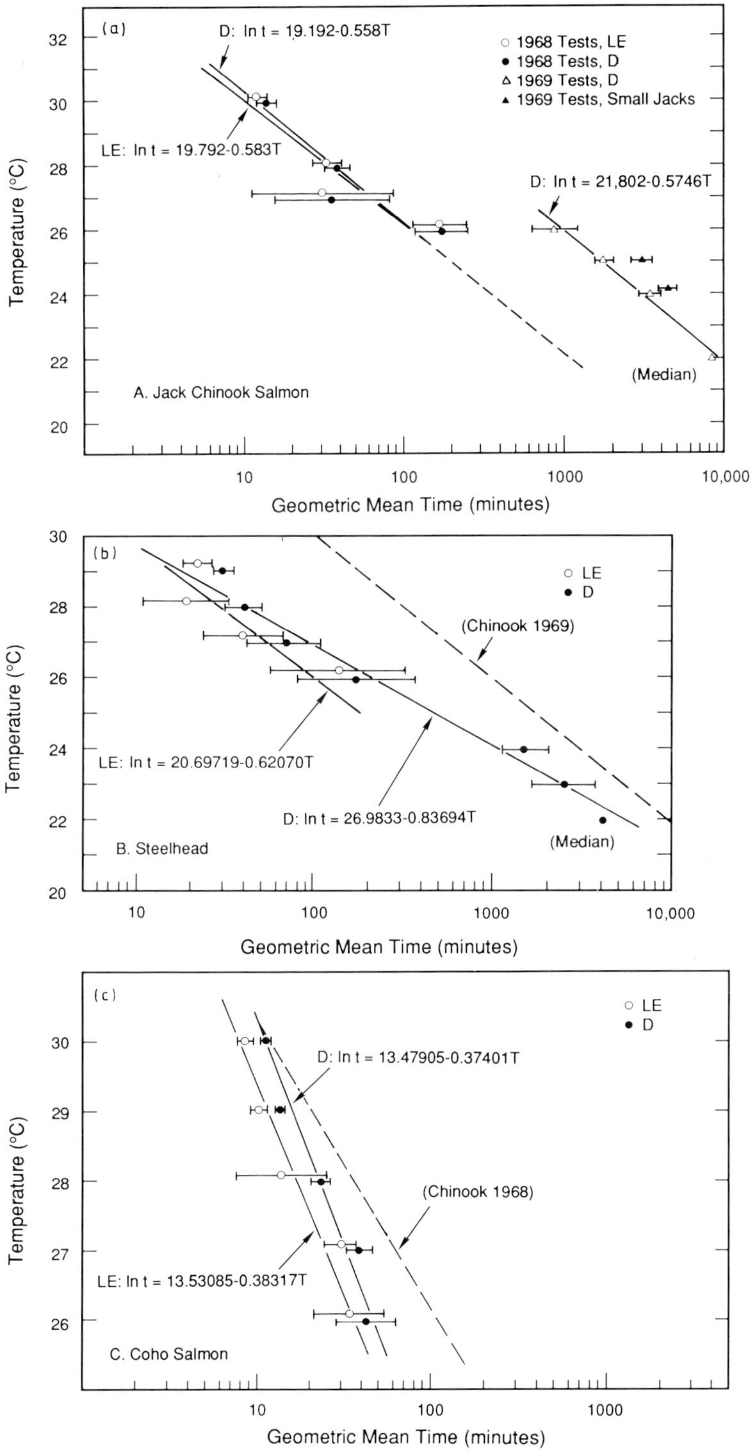
(a)
D: ln t = 19.192-0.558T
○ 1968 Tests, LE
● 1968 Tests, D
△ 1969 Tests, D
▲ 1969 Tests, Small Jacks
LE: ln t = 19.792-0.583T
D: ln t = 21,802-0.5746T
(Median)
A. Jack Chinook Salmon
Temperature (°C)
Geometric Mean Time (minutes)
(b)
○ LE
● D
(Chinook 1969)
LE: ln t = 20.69719-0.62070T
D: ln t = 26.9833-0.83694T
(Median)
B. Steelhead
Temperature (°C)
Geometric Mean Time (minutes)
(c)
○ LE
● D
D: ln t = 13.47905-0.37401T
(Chinook 1968)
LE: ln t = 13.53085-0.38317T
C. Coho Salmon
Temperature (°C)
Geometric Mean Time (minutes)

stress among any adult fish present. Accordingly, the resistance times of adult coho salmon, steelhead, and jack chinook salmon (precocious males) to temperatures above ambient were examined for 3 years, starting in 1968. Adult fish were collected at Priest Rapids Dam, held in large outdoor tanks supplied with river water, and exposed to warmer temperatures for 1 week.

Effects were quantified by calculating geometric mean times to loss of equilibrium (LE) and death (D) of fish at each test temperature (Figure 10.7). Direct comparison of data was limited because the tests were conducted yearly under different conditions. Incipient lethal temperatures (7-day exposures causing 50% mortality) for adult steelhead and jack chinook salmon were near 21° to 22°C. When tests were conducted together, adult coho salmon were less resistant than jack chinook salmon to temperatures between 26° and 30°C, and adult steelhead were less resistant than jack chinook salmon to temperatures between 22° and 26°C (Coutant 1970b; Templeton and Coutant 1971).

The experimental temperatures causing mortality of adult salmonids over a 7-day period were only slightly greater than peak seasonal temperatures in the mid-Columbia River during late summer and early fall. Because of this, responsible agencies agreed to lower temperatures in the Hanford Reach during critical periods by increasing releases of water at Grand Coulee Dam. At the same time, operating levels of the reactors were often lowered at Hanford.

Thermal Resistance of Juvenile Salmonids

Questions about the thermal resistance of juvenile salmonids were examined experimentally to assess possible site-specific effects both at Hanford and at offsite locations (Dean 1973). Of particular concern was the chance that juvenile salmonids might encounter lethal or sublethal conditions in the mixing plumes during their annual outmigrations.

Laboratory studies with juvenile salmonids at Hanford consistently emphasized thermal resistance times where the fish were experimentally exposed to an elevated temperature. The temperature of undiluted reactor effluent was so high that exposure times in the zone of initial mixing were critical factors leading to LE and D. The main experimental (and field) variables were ambient river temperature and encountered circumstances, particularly the thermal shock temperature in relation to duration of exposure.

The first experiments in 1968 examined thermal resistance among groups of juvenile chinook salmon reared from eggs at ambient river

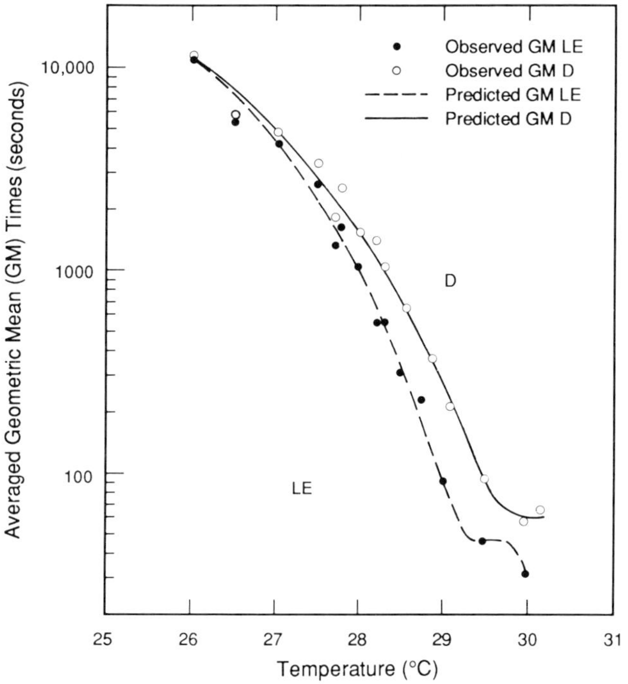

Fig. 10.8. Cubic models of geometric mean times to loss of equilibrium (LE) and death (D) of juvenile chinook salmon with similar rearing histories (modified from Coutant and Dean 1972).

temperatures and 1.1°C (2.0°F) increments. [a] Fingerlings were acclimated to 15°C for 3 weeks and abruptly exposed to lethal temperatures. Median resistance times were not affected by rearing temperature, and length of fish [50- to 97-millimeter (2- to 4-inch) range] did not influence times to death. Furthermore, times to LE were correlated with time to D. Of the fingerlings that underwent LE, 85% died when returned to nonlethal temperatures. The length of time a fish was exposed to a lethal temperature (27°C) was crucial for its survival (Dean and Coutant 1968).

In ecological assessments of thermal effects, LE may be more important than D when juvenile salmonids are abruptly exposed to potentially lethal temperatures. A fish unable to control its equilibrium is unable to avoid predators and other stresses in an aquatic ecosystem. In this relationship, a cubic model provided the best fit of geometric mean resistant times (Figure 10.8). At high lethal temperatures, LE times were about 30% to 60% of D times, and they were well separated. At low lethal

[a] These studies were done using degrees Fahrenheit, hence the conversion to celsius.

temperatures, LE times were about 80% of D times, and they were not well separated (Coutant and Dean 1972; Templeton and Coutant 1971). The tests established that LE occurred at a predictable fraction of the exposure time that caused thermal death.

Recovery of fish that underwent thermal LE and were returned to a lower, non-lethal temperature was examined further. Groups of juvenile chinook and coho salmon and rainbow trout acclimated to 15°C were abruptly exposed to lethal temperatures of 27°, 28°, 29°, and 30°C. The interaction of temperature and exposure time was crucial. When exposed to 27°C, 80% of the chinook salmon died after reaching LE, compared to 25% of the coho salmon. About 55% of the rainbow exposed to 28°C died. Fish recovered from LE when returned to acclimation temperature but died later, showing delayed mortality (Dean 1973).

Relationships between exposure time and mortality were subsequently examined. Juvenile chinook salmon and rainbow trout acclimated to 15°C

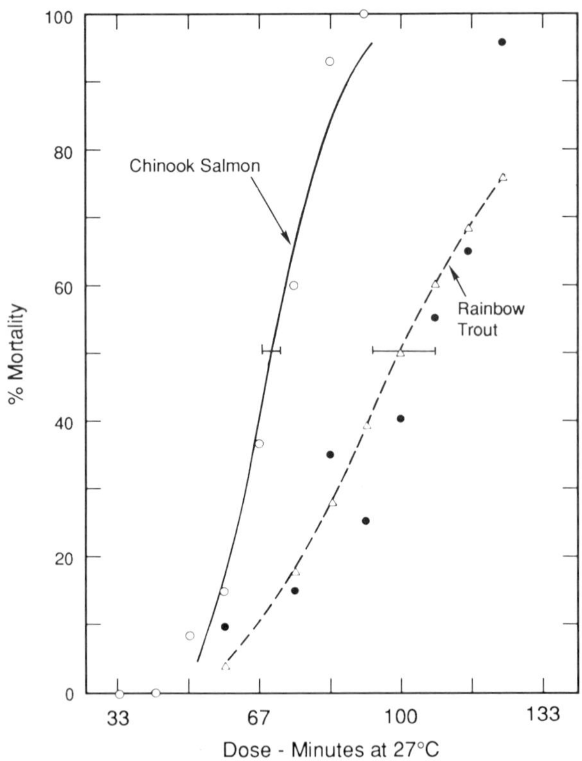

Fig. 10.9. Mortality of juvenile chinook salmon and rainbow trout when acclimated to 15°C and abruptly exposed to 27°C for different times (modified from Dean 1973).

were abruptly exposed to a lethal 27°C and returned to acclimation temperature. A characteristic 48-hour dose-response pattern emerged, involving a lethal exposure and mortality (Figure 10.9). Juvenile chinook salmon were more susceptible to thermal shock than rainbow trout (Dean 1973).

Vulnerability of Juvenile Salmonids to Predation After Thermal Shock

Loss of equilibrium, a behavioral change resulting from thermal shock, may increase the vulnerability of downstream migrants to large fish predators. Such changes may result from sublethal exposures to heated discharge plumes.

Experiments began in 1969 to examine predation rates as a sublethal component of thermal shock. Two groups of prey fish, one thermally shocked and the other unshocked, were offered simultaneously to predatory trout held at ambient river temperatures. Consumption was compared between the two prey groups when about half were eaten. Test fish were juvenile rainbow trout and chinook salmon acclimated to 15°C and shocked at 30.5°C and 28°C, respectively.

Predation was selective after a single thermal shock. Vulnerability of shocked fish generally increased in proportion to length of exposure. Predation rates became higher at exposures a fraction of those causing complete LE; 20% for rainbow and 10% for chinook. Results were partially reversed when prey were allowed 30 minutes to 1 hour to recover (Coutant 1972). In additional studies, juvenile rainbow trout acclimated to 15°C were more vulnerable to predation when thermally shocked at 26°, 28°, and 30°C than when unshocked. The relationship depended on shock temperature and exposure duration in a dose-response pattern typical of LE and D. Selective predation began at exposures about 10% of the median time causing D (Coutant 1973; Becker 1973b).

The cubic death model (Figure 10.8), developed with thermal resistance times, was used as a predictive tool to identify conditions imparting thermal shock among outmigrating salmonds. Field data indicated that sublethal shock was more likely to occur in the discharges of Hanford's reactors than lethal shock. However, no concentrations of the most likely predator were detected in midriver below the discharge plumes.

Temperature and Energy Reserves in Trout

Fish, being cold-blooded, respond metabolically to changes in water temperature. Initially, juvenile rainbow trout were acclimated to water temperatures of 5° and 18°C to allow comparison of physiological and ultrastructural features related to energy reserves at different temperatures.

Lipid levels were slightly higher in the cardiac muscle of trout held at 5° and 18°C, but not in the red muscle, white muscle, or liver. Glycogen levels were higher in the cardiac muscle and liver of trout at 5°C, but not in the red and white muscle. Theoretically, this slight increase in energy reserves provides greater stamina to some fish at lower temperatures (Dean 1966). Proteinaceous material, one of which may be plasma albumin, was more evident under an electronmicroscope in liver hepatocytes of trout held at 5°C than at 18°C (Berlin and Dean 1967).

Metabolism of Acetate and Palmitate in Trout

The metabolism of acetate and palmitate (labeled with C-14) was also examined in tissues of rainbow trout acclimated to low (5°C) and high (18°C) temperatures. The production of CO_2 was measured in homogenates of liver, red muscle, and white muscle incubated in the presence of acetate and palmitate. Movement of radioactivity from acetate into total liver lipids was also measured.

Acetate and palmitate oxidized to CO_2 more rapidly in tissues from trout adapted to cold. Further, more acetate was incorporated into the liver lipids of cold-adapted fish. Tissues of warm-adapted fish seemed to have reduced oxidation capacity (Dean 1969). However, a subsequent study showed a general increase in oxidation of acetate as temperatures rose from 18° to 38°C. Energy production was probably still not limited when the fish died from exposure to the warmed water (Caldwell 1968).

In a related study, the activity of acid phosphatase was examined in liver tissues of rainbow trout acclimated to both low (5°C) and high (18°C) temperatures. Livers from cold-adapted fish contained twice as much activity as those from warm-adapted fish (Berlin 1967; Dean and Berlin 1969).

Response of Intestinal Epithelium to Temperature and Irradiation

Epithelial cells in the intestine of fish renew continuously. They may reflect any change in aquatic habitats that influences cellular dynamics. On this basis, the effects of temperature (5°, 10°, and 18°C) and whole-body, 250 kilovolt peak X-rays [1000, 2000, and 4000 roentgens (R)] were sought on DNA-synthesis, cell kinetics, and structural changes in the intestine of juvenile silver salmon.

The time required for intestinal epithelial cells to renew depended on the temperature. Renewal took approximately 13 to 15 days at 18°C, 23 to 25 days at 10°C, and more than 35 days at 5°C. Both temperature and irradiation affected DNA synthesis. Cell damage from irradiation at 10°C was minimal at exposures below 1000 R, but severe at 8000 R and higher (Johnson et al. 1967).

Internal Temperatures of Freshwater Fish

The assumption that deep-body temperatures of fish correspond to the surrounding water needed to be applied to thermal tolerance studies with fish. In 1966, temperatures of the deep muscle and intestinal tract were measured in five species common to the Hanford Reach.

Temperatures of the intestinal tract corresponded to water temperature, but those in the deep muscle were 0.1° to 1.1°C higher. Exercise from induced swimming in rainbow trout lowered the deep muscle temperature to that of the surrounding water, presumably by increasing the flow of blood through the gills (Dean 1976).

The effect of increasing and decreasing water temperature at specific rates on internal temperature of juvenile salmonids was also studied. When water containing fish acclimated to 15°C was raised to 25°, 27°,

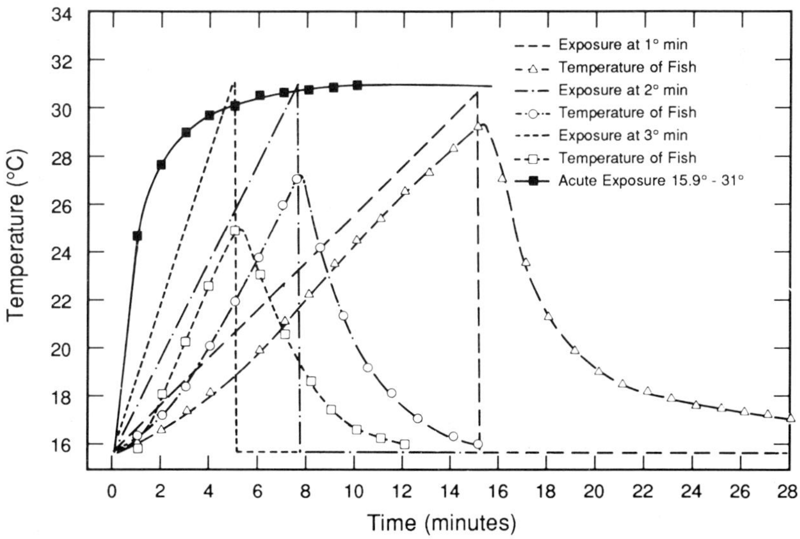

Fig. 10.10. Response of internal temperatures in juvenile salmonids acclimated to 15°C when the water temperature increases at different rates (modified from Dean 1973).

29°, and 30°C, their muscle mass increased to about 90% of the new temperature in 2 minutes (Figure 10.10). A similar time lag in internal temperature occurred when water temperatures were lowered (Dean 1973).

Investigations with the fish pathogen columnaris

Field studies with the fish pathogen *Flexibacter columnaris* dealt with 1) coarse (nongame) fish as reservoir hosts, 2) infections among coarse fish at warmed locations, 3) fish ladders as the focus for transmission of infections to returning anadromous salmonids, and 4) pathogenesis of columnaris among juvenile and adult sockeye salmon, adult chinook salmon, and adult coho salmon.

Laboratory investigations were conducted with columnaris for more than a decade to determine 1) survival of columnaris cells in river water and river mud, 2) relative susceptibility of resident fish and juvenile salmonids, 3) effects of temperature and crowding on disease outbreaks, 4) the immune response and carrier phenomena, 5) artificial immunization of juvenile salmonids, and 6) effect of tritium irradiation on immune response in rainbow trout.

Results from field and laboratory studies on columnaris at Hanford were summarized in a monograph on epizootiology of the disease in the Columbia River (Becker and Fujihara 1978). Years of study provided no evidence that releases of heated effluent at Hanford from 1944 to 1971 had an effect on outbreaks of columnaris in anadromous or resident fish.

Columnaris Disease in the River Ecosystem

Infections in resident coarse fish, particularly suckers, provided an index to seasonal status of columnaris disease in the Columbia River. However, infection criteria varied between sites and from year to year. Based on antibody production, the incidence of disease in coarse fish usually reached 70% to 90% during the fall and winter after they had been exposed the previous summer. Some coarse fish overwintered with residual antibodies. Columnaris cells were rarely isolated from coarse fish during winter but appeared in early spring when river temperatures neared 10°C (Fujihara and Hungate 1972).

Infections in coarse fish at dams in fish ladders were usually more extensive than at river locations. Large numbers of columnaris cells could be isolated from water leaving the fish ladders, identifying the ladders as sites where adult salmonids moving upriver in the summer and fall could be infected. Bonneville Dam on the lower Columbia River was probably the first site where adult salmonids were exposed to columnaris. Infection incidence and residual antibodies in returning salmonids generally increased at each dam as the fish passed upriver (Fujihara and Hungate 1971).

Coarse fish developed agglutinins and were relatively immune to columnaris in most years. Significantly, infections in coarse fish in the Hanford Reach were not consistently higher or lower than those in coarse fish in the Wenatchee River, upstream of Hanford, or downstream in the lower Snake River (Fujihara and Hungate 1972). But moribund specimens were found in June and July 1973, indicating their natural defense mechanisms had been overwhelmed. Severe outbreaks occurred in resident fish in the Columbia River in 1970 and 1973. These outbreaks corresponded with abnormally high spring temperatures, extended warm summers, and unusually low flows (Becker and Fujihara 1978).

Outbreaks and Pathogenicity of Columnaris Disease

Initial laboratory tests in the late 1950s demonstrated that irradiation of columnaris cells with X-rays had no apparent effect on their mutation rate. The experimental doses were high enough to inactivate the pathogen (Fujihara et al. 1960).

Further study showed that survival of free columnaris cells was related to temperature. After 148 hours in sterile water, survival was about 37% at 10°C, 33% at 14°C, and 28% at 18°C. After 16 days in sterile river mud, survival was far greater at 25°C than at 5°C, and few cells remained viable after 3 days at −15°C. The virulence of cells surviving at 10° and 20°C was similar to that of the parent strain. The pathogen probably overwintered in resident fish.

Acute effects of columnaris disease were examined by exposing mixed species of young resident fish and yearling rainbow trout to large numbers of cultured columnaris cells in water. The fish usually died in 1 to 2 days. Of the species tested, suckers appeared to be the most susceptible, and trout least susceptible. Smaller, younger rainbow trout and chinook salmon died in 24 hours when exposed to large numbers of columnaris cells but larger, older fish of both species survived for up to a week (Figure 10.11). Trout weighing 1.2 grams were more resistant to the disease than salmon of the same size (Fujihara and Olson 1961; Fujihara et al. 1971). Other tests showed greater resistance to disease in young

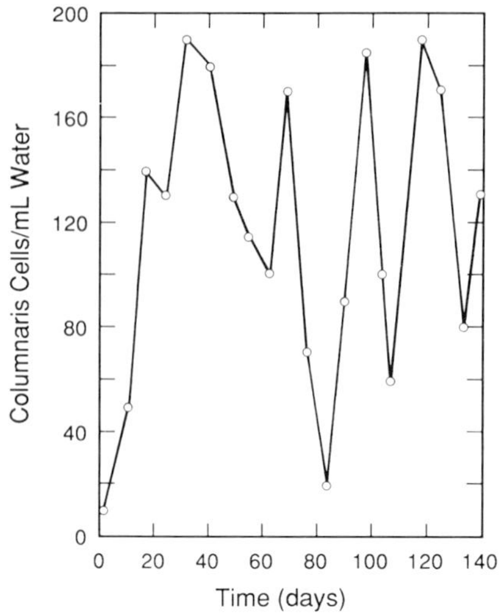

Fig. 10.11. Relative susceptibility of young rainbow trout and chinook salmon to columnaris from acute exposure (ca. 2.5×10^{-5} cells per liter of sterile water) for 30 minutes at 20°C on the basis of progressive size increases (from Fujihara et al. 1971).

chinook salmon averaging 8.5 grams (large) than those averaging 3.9 grams (small), and more rapid mortality in warmer water.

Outbreaks of disease among fish exposed to columnaris via river water were also examined in 1966. Groups of 300 young rainbow trout were held from June to October in troughs supplied with raw water from the Hanford Reach. Almost 9% of the trout died during the third and fourth weeks when temperatures reached 17°C, but few losses occurred during late summer even though temperatures climbed to 20.6°C (Fujihara et al. 1971). Survivors apparently developed resistance to the disease. Later, the effect of crowding on disease outbreak was examined in different-sized groups of young rainbow trout: 50, 150, 450, and 900 fish per trough. Losses corresponded to crowding, reaching 0%, 2.0%, 10.0%, and 11.9%, respectively, among the exposed groups.

Immune Response of Fish Exposed to Columnaris

In 1964, laboratory studies began on the immune response of salmonids exposed to columnaris, and subsequent release of cells by infected fish.

First, a group of 1600 juvenile rainbow trout were held in river water at seasonal regimes from July through December. Disease caused mortality at temperatures greater than 15°C. Peak losses (28%) occurred during the fourth week as columnaris cells in the outlet water abruptly in-

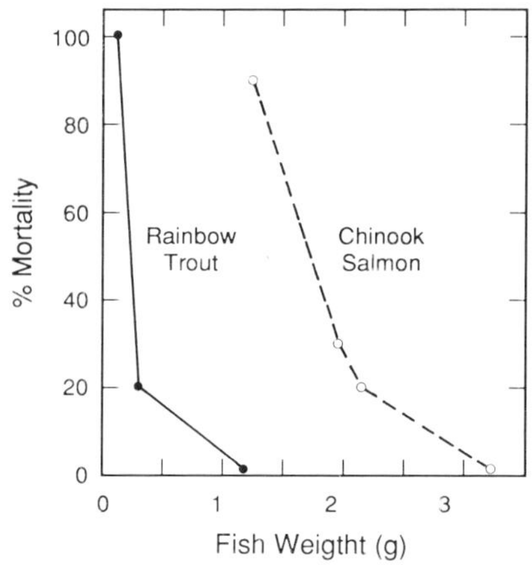

Fig. 10.12. Release patterns of columnaris cells from a trough containing 25 infected rainbow trout. Fish were retained from July through November at ambient river temperatures (from Fujihara and Nakatani 1971).

creased. Mortality then declined, even though release of cells from the fish and temperatures remained high. Apparently, immunity developed rapidly in exposed fish, and survivors were refractory carriers (Fujihara and Nakatani 1971).

Another test examined the release of columnaris cells from infected fish. Ten yearling rainbow trout, survivors of infection the previous year, were periodically isolated in jars of sterile river water for 1 hour at 20°C. Five fish did not shed the pathogen, two had three consecutive release cycles, and three had one release cycle. The gills were the primary sites of pathogen release, in the absence of body lesions.

Release of columnaris cells was next examined in a group of 25 yearling rainbow trout held in a trough supplied with river water at ambient temperatures from July through November. Samples at the trough outlet showed at least four major peak times of columnaris release (Figure 10.12). When the test ended, 10 fish were still releasing cells, and 15 were refractory. The fish initially contained some residual agglutinins, indicating prior exposure to columnaris. Effective immunity was indicated by development of high agglutinin levels.

Artificial Immunization Against Columnaris

By 1965, the development of immunity in fish exposed seasonally to columnaris in river water had been demonstrated to be a natural protective mechanism. Efforts began to experimentally induce antibody production (measured by agglutinin titers) peritoneally and orally (Fujihara and Nakatani 1971; Becker and Fujihara 1978).

In one test, uninfected yearling rainbow trout were vaccinated with heat-killed columnaris cells weekly for 4 weeks and biweekly for 8 weeks. Antibodies, detected by agglutinin titers, appeared within 2 weeks and continued to increase. Later, injection of virulent cells to their bodies killed all nonimmunized fish. Among immunized fish, subcutaneous injection of virulent cells caused 37% (no adjuvant) and 28% (adjuvant) mortality, but intraperitoneal injection killed all fish.

In a companion test, uninfected yearling rainbow trout were vaccinated intraperitoneally and exposed to virulent columnaris cells shed from infected fish. Initial agglutinin titers decreased, despite continuous exposures, then increased over a 1-year span. Vaccinated fish had no mortalities, and their agglutinin values were consistently higher than in control fish.

Subsequently, sonically disrupted columnaris cells were fed to juvenile rainbow trout, and heat-killed cells were fed to juvenile coho salmon. Fish were then exposed to river water during the warm summer. Trout developed no apparent resistance against infection. Coho salmon developed resistance, and eventual mortality was 8%, compared to 48% for controls.

Ecological functions in the Hanford Reach

Most studies in the Hanford Reach from 1945 to 1971 addressed the potential effects of releasing radioactivity, heat, and chemicals to the Columbia River. However, a number of investigations were conducted on ecosystem components and ecosystem functioning. In a holistic sense, all parts of an ecosystem are interrelated. Thus, any effect from a human-induced stress might affect all parts of the ecosystem in subtle ways. The better each component of the Hanford Reach was understood, the greater the understanding of its ecological responses.

Gas-Bubble Disease in Fish

During the 1960s, the Columbia River contained elevated levels of dissolved gases (supersaturation) at certain seasons from entrainment of air in plunge basins of hydroelectric dams. In areas where the river was warmed, such as in the discharges plumes from the Hanford reactors, supersaturation of the water theoretically might increase by about 2% for each increase of 1°C. Fish exposed to high supersaturations could develop what is called "gas-bubble" disease and die.

In exploratory studies, juvenile salmonids prestressed with high nitrogen levels representing 115% saturation suffered higher losses from thermal shock than those held at normal saturations. However, exposure of salmonid outmigrants to potentially lethal conditions in the discharge plumes at Hanford would be less than a few minutes because entrainment was so brief (0.5 to 1.0 meters per second). In most cases, no mortality would be expected.

The probable cause of death in fish afflicted with gas bubbles from supersaturation in these studies was blockage of blood circulation by emboli. Bubbles appearing in fish contained air (a mixture of N_2 and O_2) rather than nitrogen alone (Schneider 1970, 1971).

Smallmouth Black Bass Populations

In the 1950s, biological and radiological data were sought on smallmouth bass caught in sloughs near the 100-F Area with sport gear. These bass congregate in early spring in areas bordering the Hanford Reach where circulation of water is limited and insolation adds heat. Bass start to spawn when the water temperature becomes sufficiently warm, usually before the annual spring freshet.

In these collections, the average smallmouth bass was 39.4 centimeters (15.5 inches) long and weighed 1244 grams (40 oz). Bass ranged in age from 3 to 11 years, but about half the sport catch consisted of 7- to 8-year-olds. Some bass entered the sloughs and spawned as early as April, but the eggs were lost when cold water from the freshet flooded the nests. Eggs spawned in July and August, when water temperatures were 15.6° to 23.9°C, survived to hatch. Radioactivity in bass, primarily from P-32, peaked in September at levels too low to be a hazard to consumers (Henderson and Foster 1957).

Trace Elements in River Water and Phytoplankton

The seasonal abundance of primary producers, especially plankton, greatly influences the cycling of trace elements in aquatic ecosystems. Studies in the 1960s showed that plankton in the Hanford Reach consisted largely of diatoms, principally of the genera *Asterionella, Melosira, Fragilaria, Tabellaria*, and *Synedra*. The predominant diatom, Asterionella, increased in abundance twice each year, while the other genera increased once. The spring "bloom" was related to increased river flow, a decrease of nitrate and phosphate in river water, and an increase of dissolved silica, water temperature, and effective light penetration (Cushing 1963, 1964).

Cycling of trace elements in the Columbia River was examined in 1968 with the aid of a new technique, neutron activation analysis. The first step was to obtain baseline data on the elemental composition of river water and phytoplankton from samples collected upstream of the reactors. Comparison with data obtained previously by conventional methods (Silker 1964; Cushing 1964) revealed certain discrepancies. All in all, neutron activation was superior in detecting microquantities of trace elements in phytoplankton (Cushing and Rancitelli 1972).

Subsequently, trace elements were measured in caddis fly larvae that feed on phytoplankton and in whitefish that feed on caddis fly larvae. The data were compared with earlier analyses of water and phytoplankton to help establish trophic-level relationships. Concentrations of most trace elements (Ag, Co, Cr, Cs, Fe, Na, Sb, Sc, and Zn) decreased through the food web. Potassium was the only element to increase, and four elements (Br, Hg, Rb, and Se) remained relatively constant (Cushing 1979).

Production of Periphyton at Elevated Temperatures

The effect of heat on periphyton communities was examined in flowing, once-through, outdoor channels from 1971 through 1972. Six miniature experimental streams, representing outdoor microcosms, were provided with river water heated to 0 (control), 2.5°, 5.0°, 7.5°, and 10.0°C above the seasonal regimes in the Hanford Reach. The periphyton that developed on glass slides in the channels consisted of attached algae and associated organisms (Coutant and Owen 1970).

Standing crop and pigment contents of mature periphyton did not vary significantly among the six streams regardless of temperature. Periphyton accrued on clean slides more rapidly at warmer temperatures, but the excesses eroded and were exported from the troughs. Export of periphyton varied seasonally, ranging from 0.3 g/m^2 daily in midwinter (cold) to 8.0 g/m^2 daily in early fall (warm). The total biomass exported over one year varied among troughs by only 12%.

Species composition did differ among channels, apparently in relation to temperature regime. *Melosira varians* dominated among all fall and winter communities. By mid-winter, cooler streams were dominated by *Fragilara* and *Nitzschia*. *Rhopaloidia gibba* dominated the community heated 10°C above ambient in late spring, and this species became important in cooler channels by midsummer. *Melosira* regained dominance in the control channel by midsummer, and in the warmed channels later.

Apparently, periphyton communities in the Hanford Reach are maintained in a subclimax condition, in an almost constant state of recolonization and growth, as a result of grazing by aquatic animals and erosion from substrate (Owen 1971, 1973).

Feeding and Growth of Juvenile Chinook Salmon

Fall chinook salmon emerge from redds in the Hanford Reach each spring after overwintering in gravel. Some small fish appear in shoreline areas for a brief period to feed before passing downstream. The feeding and growth of zero-age fall chinook salmon were examined during the spring outmigration in 1968 and 1969. Prey selection was analyzed in relation to reactor discharges and other environmental features.

The young fall chinook salmon fed almost entirely on adult and larval stages of aquatic insects, primarily midges (Chironomidae). By numbers, adult midges provided 64% and 58% of their diet, and larval midges 17% and 18% of their diet, respectively, over the two spring seasons. Small chinook salmon appeared to be opportunistic feeders that selected insects drifting, floating, or swimming. Feeding, growth, and outmigration were influenced by seasonal changes in river temperature and flow (Becker 1970a, 1973a). However, discharges from Hanford's reactors had no apparent effect on prey selection or feeding activity. The length-weight relationships of fish collected above and below the discharges were similar.

Parasites of Fish in Hanford Reach

As part of site characterization efforts beginning in the late 1960s, some parasites were identified from fish in the Hanford Reach. Parasites acquired by fish often reflect the unique features of an aquatic habitat. The Hanford Reach retains the historical features of the once free-flowing, mainstem Columbia River. Two species of digenetic trematodes were found in the digestive tract of white sturgeon. One, described as a new species, was *Cestrahelmins rivularis* (Deropristiidae) (Becker 1971a). The other, *Tubulovesicula lindbergi* (Hemiuridae) was a species normally found in marine fish, which was probably acquired when the host sturgeon scavenged the carcass of an infected adult salmon (Becker 1970b).

Five species of monogenetic trematodes were found on the gills of juvenile largemouth bass and smallmouth bass taken in sloughs along the Hanford Reach. Infection intensities on smallmouth bass were low. The parasites were probably present when these fish were introduced decades ago to the Columbia River from the eastern United States (Becker 1972).

The blood of some fish of the Hanford Reach contained low intensities of a biflagellate parasite. The red blood cells of suckers also held low intensities of an intracellular hemogregarine (Becker 1979), but there was no apparent pathogenicity. Piscivorous leeches, four of which occur in the Hanford Reach (Becker and Dauble 1979), may have been vectors of these blood parasites.

Spawning of Fall Chinook Salmon

Enumeration of fall chinook salmon redds (deposits of eggs in gravel) by aerial surveys was initiated at Hanford in 1947 and continued through 1987. The purpose of the surveys was to detect any possible effects of the reactor discharges on spawning activity of adult salmon, survival of intragravel eggs, or return of downstream migrating young (Watson 1952).

The life cycle of anadromous fall chinook salmon requires 4 or 5 years to complete. Further, other factors operating in the river and ocean (including mortalities at dams, losses in irrigation channels, commercial and sport harvest, and release of hatchery fish) strongly influence the number of adult salmon returning to spawning areas in fresh water. Therefore, the establishment of long-term trends in the abundance of spawning salmon at Hanford was important. The effort required yearly counts as long as the single-purpose reactors remained in operation, and they were continued through 1984.

Initial redd counts were low, reflecting limited spawning activity in the Hanford Reach, until Priest Rapids Dam was completed upstream of Hanford in 1959. The number of redds increased dramatically from about 300 in 1960 to 4500 in 1969 (Figure 10.13). The initial increase probably resulted from the barrier to upstream migration created by Priest Rapids Dam, and from the upstream translocation of fish whose spawning grounds were inundated by other dams below Hanford (Watson 1970, 1976).

The initial increase in numbers of spawning fall chinook salmon took place during the time that eight all reactors were discharging effluents in the 1960s. Furthermore, the increase in adult spawners in the Hanford Reach proportionately exceeded the composite return run of fall chinook

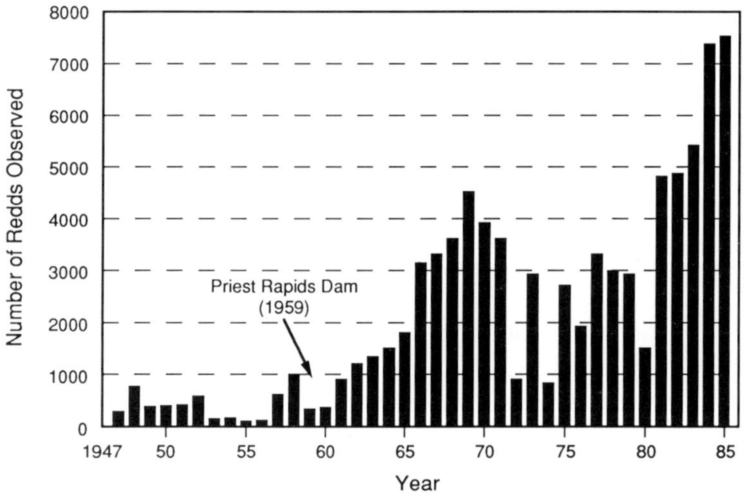

Fig. 10.13. Estimated number of fall chinook salmon redds deposited annually in the Hanford Reach from 1947 to 1985 based on aerial surveys (from Becker 1985).

salmon to the Columbia River at a time when other upriver stocks of salmonids were declining. In other words, the Hanford Reach proved to be suitable for a chinook salmon run that faced environmental alteration elsewhere.

In the 1970s, after all single-purpose reactors were shut down, the number of chinook salmon redds in the Hanford Reach fluctuated from year to year. Redd counts again increased in the 1980s, suggesting that recent efforts by fishery managers to increase the number of salmon returning to spawn in the upper Columbia River had met with some success (Becker 1985). No indication was ever obtained that operations at Hanford adversely effected the spawning of fall chinook salmon in the Hanford Reach, nor was any impact identified on returning adults or outmigrant juveniles of any anadromous fish.

Abundance of Steelhead

Steelhead is the anadromous form of rainbow trout. Like Pacific salmon, they originate in freshwater, migrate to the ocean for feeding and growth, and return as adults to spawn in natal rivers. Production of steelhead in the Hanford Reach has always been a mystery.

The number of steelhead entering the Hanford Reach from 1962 to 1971 averaged about 36,000 fish yearly. This number represents the count of steelhead moving through the fish ladders at McNary Dam, minus fish ladder counts at Ice Harbor and Priest Rapids dams. When this value was corrected for migration into the Yakima River and other tributaries, and for the legal sport catch, the spawning population was estimated to average about 16,700. An assumed 20% prespawning mortality lowered the estimate to about 10,000 fish spawning in the Hanford Reach annually over this period (Watson 1973).

Significance of thermal studies in the Hanford Reach

Field and laboratory studies to determine any direct and indirect effects of heated discharges from the single-purpose reactors in the Hanford Reach were emphasized over a relatively brief span from 1960 to 1971. However, the effect of heat was one of three areas examined during effluent monitoring since 1945 (Chapter 7). Research efforts increased in response to the Water Quality Act of 1965, leading to participation in the 2-year CRTES, which ended after closure of the last operating reactor.

Before 1960, field work in the Hanford Reach gave no evidence that resident or anadromous fish were killed by elevated temperatures in the discharges of up to eight reactors. Each effluent stream, when it left the reactor core, had been heated to such an extent that direct exposure would be almost immediately lethal to cold-blooded aquatic organisms. The discharges had been designed to prevent fish kills through release of heated effluent near midstream, the rapid mixing and dilution in the Hanford Reach, and the buoyant nature of the discharge plumes. Once all effluent had mixed downstream, river water temperatures were elevated slightly, up to 1° to 2°C during some seasons. However, effluent monitoring (Chapter 7) showed that slightly warmed water accelerated the growth of young salmon and trout rather than harmed them.

The increased emphasis on thermal effects during the 1960s was initiated largely by federal and state regulations, which directed the attention of regulatory agencies at the Hanford Site. In effect, the assumption that the heated effluent from the single-purpose reactors neither harmed any fish or exerted any significant impact on aquatic communities had to be proven.

The concerns about thermal effects led to examination of a much broader range of potential problems from discharge of reactor effluent to the Hanford Reach, culminating in the cooperative CRTES. Eventually, thermal studies at Hanford encompassed four general areas: 1) the release of heated water to the Columbia River, 2) laboratory studies on sublethal and lethal effects, 3) laboratory studies on physiological effects, and 4) ecological functions in the Hanford Reach.

Much attention centered on the salmonid resources of the middle Columbia River, specifically the fall chinook salmon and steelhead trout spawning in or migrating through the Hanford Reach. Applicable studies involved

- behavior of adult salmon and steelhead trout migrating upstream past the effluent plumes
- survival of juvenile salmonids that were drifted through an effluent plume
- effect of temperature increments on survival and growth of eggs and juvenile salmon
- upper lethal temperature limits of juvenile and adult salmonids
- vulnerability of juvenile salmonids to predation after thermal shock
- effect of the thermal discharges on the fish disease, columnaris
- status of fall chinook salmon that spawned each fall in the Hanford Reach.

In retrospect, thermal studies in the Hanford Reach were unique in three key aspects. First, the Columbia River, fed with snowmelt and precipitation from high mountain ranges bordering the Columbia Basin, is a cool and relatively large river system. It resembles no other river system in the United States. Second, the single-purpose reactors at Hanford were designed with "single-pass" cooling systems in which essentially all heat produced in the reactor cores was discharged to the Hanford Reach. Third, no other river in the world supported valuable runs of anadromous salmonids while receiving effluent containing large amounts of waste heat. The fact that river temperatures historically peaked near, and sometimes exceeded 20°C, complicated the picture. Temperatures a few degrees greater than this point were known to be lethal for juvenile salmonids.

For these reasons, thermal effects studies at Hanford were largely site-specific. They identified upper temperatures and exposure conditions that could, under certain circumstances, impact the life cycles of salmonids and some aquatic invertebrates or destroy them, and they identified many of the controlling factors. From this standpoint, the objectives of the 2-year CRTES were attained. However, no significant ecological effect from the heated reactor discharges at Hanford was revealed by this effort.

References

Becker, C.D. 1970a. "Feeding Bionomics of Juvenile Chinook Salmon in the Central Columbia River." *Northwest Sci.* 44:71–81.

Becker, C.D. 1970b. "Marine Trematode, *Tubulovesicula lindbergi* (Digenea: Deropristiidae) from Resident White Sturgeon in the Columbia River." *J. Fish. Res. Board Can.* 27:1311–1316.

Becker, C.D. 1971a. "*Cestrahelmins rivularis* sp. n. (Digenea: Deropristiidae) from White Sturgeon, *Acipenser transmontanus*, in the Columbia River, Washington." In: *Proc. Helminthol. Soc. Wash.* 38:21–26.

Becker, C.D. 1971b. "Response of Columbia River Invertebrates to Thermal Stress," In: *Pacific Northwest Laboratory Annual Report for 1970 to the USAEC Division of Biology and Medicine, Vol. I Life Sciences, Part 2 Ecological Sciences*, pp. 2.11–2.18. BNWL-1550 PT1, Pacific Northwest Laboratory, Richland, Washington.

Becker, C.D. 1972. "Initial Records of Monogenea from Juvenile Bass in the Central Columbia River, Washington." *Trans. Am. Fish. Soc.* 101:721–716.

Becker, C.D. 1973a. "Food and Growth Parameters of Juvenile Chinook Salmon, *Oncorhynchus tshawytscha*, in Central Columbia River." Fish. Bull. 71:381–400.

Becker, C.D. 1973b. "Columbia River Thermal Effects Study: Reactor Effluent Problems." *J. Water Poll. Control Fed.* 45:850–869.

Becker, C.D. 1973c. "Development of *Simulium* (*Psilozia*) *vittatum* Zett. (Diptera: Simuliidae) from Larvae to Adults at Thermal Increments from 17.0° to 27.0°C." *Am. Mid. Nat.* 89:241–251.

Becker, C.D. 1979. "Haematozoa from Resident and Anadromous Fishes of the Central Columbia River: A Survey." *Can. J. Zool.* 58:351–362.

Becker, C.D. 1985. *Anadromous Salmonids of the Hanford Reach, Columbia River: 1984 Status*. PNL-5371, Pacific Northwest Laboratory, Richland, Washington.

Becker, C.D., and D.D. Dauble. 1979. "Records of Piscivorus Leeches (Hirudinea) from the Central Columbia River, Washington State." *Fish. Bull.* 76:921–931.

Becker, C.D., and M.P. Fujihara. 1978. *The Bacterial Pathogen Flexibacter columnaris and Its Epizootiology Among Columbia River Fish. A Review and Synthesis*. Monograph No. 2, American Fisheries Society, Bethesda, Maryland.

Becker, C.D., C.C. Coutant, and E.F. Prentice. 1971. *Experimental Drifts of Juvenile Salmonids Through Effluent Discharges at Hanford. Part II. 1969 Drifts and Conclusions*. BNWL-1527, Battelle, Pacific Northwest Laboratories, Richland, Washington.

Becker, C.D., C.C. Coutant, and E.F. Prentice. 1973. "Ecological Evaluation: Migration of Juvenile Salmon in Relation to Heated Effluents in the Central Columbia River." In: *Radionuclides in Ecosystems, Third National Symposium on Radioecology*, ed. D. J. Nelson, pp. 521–536. CONF-710501-P1, National Technical Information Service, Springfield, Virginia.

Berlin, J.D. 1967. "Temperature-Induced Differences in Acid Phosphatase Levels of Rainbow Trout Livers." In: *Pacific Northwest Laboratory Annual Report for 1966 to the USAEC Division of Biology and Medicine, Vol. I Biological Sciences*, pp. 151–159. BNWL-480, Pacific Northwest Laboratory, Richland, Washington.

Berlin, J.D., and J.M. Dean. 1967. "Temperature-Induced Alterations in Hepatocyte Structure of Rainbow Trout (*Salmo gairdneri*)." *J. Exp. Zool.* 164:111–132.

Caldwell, R.S. 1968. "Effect of Acclimation Temperature on Mitochondrial Enzymes in Fish." In: *Pacific Northwest Laboratory Annual Report for 1967 to the USAEC Division of Biology and Medicine, Vol. I Biological Sciences*, pp. 7.21–7.26. BNWL-714, Pacific Northwest Laboratory, Richland, Washington.

Coutant, C.C. 1968a. "Effect of Temperature on the Development Rate of Bottom Organisms." In: *Pacific Northwest Laboratory Annual Report for 1967 to the USAEC*

Division of Biology and Medicine, Vol. I Biological Sciences, pp. 9.11–9.14. BNWL-714, Pacific Northwest Laboratory, Richland, Washington.

Coutant, C.C. 1968b. "Behavior of Adult Salmon and Steelhead Trout Migrating Past Hanford Thermal Discharges." In: *Pacific Northwest Laboratory Annual Report for 1967 to the USAEC Division of Biology and Medicine, Vol. I Biological Sciences*, pp. 9.10–9.13. BNWL-714, Pacific Northwest Laboratory, Richland, Washington.

Coutant, C.C. 1969a. "Temperature, Reproduction and Behavior." *Chesapeake Sci.* 10:261–279.

Coutant, C.C. 1969b. "Behavior of Sonic-Tagged Chinook Salmon and Steelhead Trout Migrating Past Hanford Thermal Discharges." In: *Pacific Northwest Laboratory Annual Report for 1968 to the USAEC Division of Biology and Medicine, Vol. I Life Sciences*, Part 2 Ecological Sciences, pp. 2.31–2.44. BNWL-1050 PT2, Battelle, Pacific Northwest Laboratories, Richland, Washington.

Coutant, C.C. 1970a. *Behavior of Sonic-Tagged Chinook Salmon and Steelhead Trout Migrating Past Hanford Thermal Discharges (1967)*. BNWL-1531, Battelle, Pacific Northwest Laboratories, Richland, Washington.

Coutant, C.C. 1970b. *Thermal Resistance of Adult Coho (Oncorhynchus kisutch), and Jack Chinook (O. tshawytscha) Salmon and Adult Steelhead Trout (Salmo gairdneri) from the Columbia River*. BNWL-1508, Battelle, Pacific Northwest Laboratories, Richland, Washington.

Coutant, C.C. 1972. *Effect of Thermal Shock on Vulnerability to Predation in Juvenile Salmonids. I. Single Shock Temperature*. BNWL-1521, Battelle, Pacific Northwest Laboratories, Richland, Washington.

Coutant, C.C. 1973. "Effect of Thermal Shock on Vulnerability of Juvenile Salmonids to Predation." *J. Fish. Res. Board Can.* 30:961–973.

Coutant, C.C., and C.D. Becker. 1973. "Growth of the Columbia River Limpet, *Fisherola nuttalli* (Haldeman), in Normal and Reactor-Warmed Water." In: *Radionuclides in Ecosystems, Third National Symposium on Radioecology*, ed. D. J. Nelson, pp. 561–568. CONF-710501-P1, National Technical Information Services, Springfield, Virginia.

Coutant, C.C., and J.M. Dean. 1972. *Relationships Between Equilibrium Loss and Death as Responses of Juvenile Chinook Salmon and Rainbow Trout to Acute Thermal Shock*. BNWL-1520, Battelle, Pacific Northwest Laboratories, Richland, Washington.

Coutant, C.C., and B.B. Owen, Jr. 1970. "Productivity of Periphyton Communities under Thermal Stress." In: *Pacific Northwest Laboratory Annual Report for 1969 to the USAEC Division of Biology and Medicine, Vol. I Life Sciences, Part 2 Ecological Sciences*, pp. 3.1–3.2. BNWL-1306 PT2, Battelle, Pacific Northwest Laboratories, Richland, Washington.

Coutant, C.C., C.D. Becker, and E.F. Prentice. 1969. "Passage of Downstream Migrants." In: *Pacific Northwest Laboratory Annual Report for 1968 to the USAEC Division of Biology and Medicine, Vol. 1 Life Sciences*, Part 2 Ecological Sciences, pp. 2.21–2.30. BNWL-1050 PT2, Pacific Northwest Laboratory, Richland, Washington.

Cushing, C.E. 1963. "Plankton-Water Chemistry Cycles in the Columbia River." In: *Hanford Biology Research Annual Report for 1962*, pp. 221–231. HW-76000, Hanford Atomic Products Operation, Richland, Washington.

Cushing, C.E. 1964. "Plankton-Water Chemistry Cycles in the Columbia River." In: *Hanford Biology Research Annual Report for 1963*, pp. 211–218. HW-80500, Hanford Atomic Products Operation, Richland, Washington.

Cushing, C.E. 1979. "Trace Elements in a Columbia River Food Web." *Northwest Sci.* 53:111–125.

Cushing, C.E., and L. A. Rancitelli. 1972. "Trace Element Analysis of Columbia River Water and Phytoplankton." *Northwest Sci.* 46:111–121.

Dean, J.M. 1966. "Temperature Effects on Energy Reserves in Trout." In: *Pacific Northwest Laboratory Annual Report for 1965 in the Biological Sciences*, pp. 111–115. BNWL-280, Pacific Northwest Laboratory, Richland, Washington.

Dean, J.M. 1969. "The Metabolism of Tissues of Thermally Acclimated Trout (*Salmo gairdneri*)." *Comp. Biochem. Physiol.* 29:181–196.

Dean, J.M. 1973. "The Response of Fish to a Modified Thermal Environment." In: *Responses of Fish to Environmental Changes*, ed. W. Chavin, pp. 31–60. C.C. Thomas Publishers, Springfield, Virginia.

Dean, J.M. 1976. "Temperature of Tissues in Freshwater Fishes." *Trans. Am. Fish. Soc.* 105:701–711.

Dean, J.M., and J.D. Berlin. 1969. "Alterations in Hepatocyte Function of Thermally Acclimated Rainbow Trout (*Salmo gairdneri*)." *Comp. Biochem. Physiol.* 29:301–312.

Dean, J.M., and C.C. Coutant. 1968. "Lethal Temperature Relations of Juvenile Columbia River Chinook Salmon." In: *Pacific Northwest Laboratory Annual Report for 1967 to the USAEC Division of Biology and Medicine, Vol. I Biological Sciences*, pp. 9.1–9.10. BNWL-714, Pacific Northwest Laboratory, Richland, Washington.

EPA. See U.S. Environmental Protection Agency.

Foster, R.F., R.T. Jaske, and W.L. Templeton. 1972. "The Biological Cost of Discharging Heat to Rivers." In: *Peaceful Uses of Atomic Energy, Vol. 11*, pp. 631–640. A/CONF-49/P/086, International Atomic Energy Agency, Vienna, Austria.

Fujihara, M.P., and F.P. Hungate. 1971. "*Chondrococcus columnaris* Disease of Fishes: Influence of Columbia River Fish Ladders." *J. Fish. Res. Board* Can. 28:531–536.

Fujihara, M.P., and F.P. Hungate. 1972. "Seasonal Distribution of *Chondrococcus columnaris* Infection in River Fishes as Determined by Specific Agglutinins." *J. Fish. Res. Board Can.* 29:171–178.

Fujihara, M.P., and R.E. Nakatani. 1971. "Antibody Production and Immune Responses of Rainbow Trout and Coho Salmon to *Chondrococcus columnaris." J. Fish. Res. Board Can.* 28:1251–1258.

Fujihara, M.P., and P.A. Olson. 1961. "Studies on Fish Disease *Chondrococcus columnaris." In Hanford Biology Research Annual Report for 1960*, pp. 160–165. HW-69500, Hanford Atomic Products Operation, Richland, Washington.

Fujihara, M.P., P.A. Olson, and R.F. Foster. 1960. "Mutation and Temperature Effects in *C. columnaris*." In: *Hanford Biology Research Annual Report for 1959*, pp. 181–192. HW-65500, Hanford Atomic Products Operation, Richland, Washington.

Fujihara, M.P., P.A. Olson, and R.E. Nakatani. 1971. "Some Factors in Susceptibility of Juvenile Rainbow Trout and Chinook Salmon to *Chondrococcus columnaris." J. Fish. Res. Board Can.* 28:1731–1743.

Henderson, C., and R.F. Foster. 1957. "Studies of Smallmouth Black Bass (*Micropterus dolomieui*) in the Columbia River near Richland, Washington." *Trans. Am. Fish. Soc.* 86:111–127.

Honstead, J.F., and R.G. Clark. 1953. *Temperature and Activity Density Gradients in a Columbia River Gravel Bed.* HW-29217, General Electric Company, Richland, Washington.

Jaske, R.T. 1969. *Columbia River Temperature Trends – Fact and Fallacy.* BNWL-SA-2536, Battelle, Pacific Northwest Laboratories, Richland, Washington.

Jaske, R.T., and J.B. Goebel. 1967. "Effects of Dam Construction on Temperatures of the Columbia River." *J. Am. Water Works Assoc.* 59:931–942.

Jaske, R.T., and M.O. Synoground. 1970. *Effect of Hanford Plant Operations on the Temperature of the Columbia River 1964 to Present.* BNWL-1345, Battelle, Pacific Northwest Laboratories, Richland, Washington.

Jaske, R.T., W.L. Templeton, and C.C. Coutant. 1969. "Thermal Death Models." *Ind. Water Eng.* Oct. 1969.

Jaske, R.T., W .L. Templeton, and C.C. Coutant. 1970. "Methods for Evaluating Effects of Transient Conditions in Heavily Loaded and Extensively Regulated Stream." *Chem. Eng. Prog.* 67:31–39.

Johnson, T.S., F.P. Conte, and R.E. Nakatani. 1967. "The Effects of Temperature and X-Irradiation on the Intestinal Epithelial Cells of Rainbow Trout." In: *Pacific Northwest Laboratory Annual Report for 1966 to the USAEC Division of Biology and Medicine, Vol. I Biological Sciences*, pp. 151–161. BNWL-480, Pacific Northwest Laboratory, Richland, Washington.

Nakatani, R.E. 1969. "Effects of Heated Discharges on Anadromous Fishes." In: *Biological Aspects of Thermal Pollution*, eds. P.A. Krenkel and F. L. Parker, pp. 291–317. Vanderbilt University Press, Nashville, Tennessee.

Olson, P.A. 1970. "Effects of Fluctuating Temperatures on Fall Chinook Eggs and Young." In: *Pacific Northwest Laboratory Annual Report for 1969 to the USAEC Division of Biology and Medicine, Vol. I Life Sciences, Part 2 Ecological Sciences*, pp. 3.11–3.16. BNWL-1306 PT2, Battelle, Pacific Northwest Laboratories, Richland, Washington.

Olson, P.A., and R.F. Foster. 1955. "Temperature Tolerance of Eggs and Young of Columbia River Chinook Salmon." *Trans. Am. Fish. Soc.* 85:201–207.

Olson, P.A., and R.E. Nakatani. 1969. "Effects of Chronic Variable Water Temperatures on Survival and Growth of Young Chinook Salmon." In: *Pacific Northwest Laboratory Annual Report for 1968 to the USAEC Division of Biology and Medicine, Vol. I Life Sciences*, Part 2 Ecological Sciences, pp. 2.35–2.38. BNWL-1050 PT2, Pacific Northwest Laboratory, Richland, Washington.

Olson, P.A., R.E. Nakatani, and T. Meekin. 1970. *Effects of Thermal Increments on Eggs and Young of Columbia River Fall Chinook.* BNWL-1538, Battelle, Pacific Northwest Laboratories, Richland, Washington.

Olson, P.A., E.G. Tangen, and W.L. Templeton. 1973. "Effects of Temperature Increments on Juvenile Steelhead." In: *Radionuclides in Ecosystems, Third National Symposium on Radioecology*, ed. D. J. Nelson, pp. 551–557. CONF-710501-P1, National Technical Information Service, Springfield, Virginia.

Owen, B.B., Jr. 1971. "Columbia River Periphyton Communities under Thermal Stress." In: *Pacific Northwest Laboratory Annual Report for 1970 to the USAEC Division of Biology and Medicine, Vol. I Life Sciences*, Part 2 Ecological Sciences, pp. 2.11–2.20. BNWL-1550 PT2, Battelle, Pacific Northwest Laboratories, Richland, Washington.

Owen, B.B., Jr. 1973. *The Effects of Increased Temperature on Periphyton Communities of Artificial Stream Channels.* Ph.D. Dissertation, University of Alberta, Edmundton, Alberta, Canada.

Pacha, R.E., and E.J. Ordal. 1970. "Myxobacterial Diseases of Salmonids." In a *Symposium on Diseases of Fishes and Shellfishes*, ed. F. S. Sniezsko, pp. 241–257. Spec. Publ. No. 5, American Fisheries Society, Washington, D.C.

Prentice, E.F. 1969. "Gull Predation in a Reactor Discharge Plume." In: *Pacific Northwest Laboratory Annual Report for 1968 to the USAEC Division of Biology and Medicine, Vol. I Life Sciences*, Part 2 Ecological Sciences, pp. 2.31–2.34. BNWL-1050 PT2, Battelle, Pacific Northwest Laboratories, Richland, Washington.

Schneider, M.J. 1970. "Physiology of Gas Bubble Disease." In: *Pacific Northwest Laboratory Annual Report for 1969 to the USAEC Division of Biology and Medicine, Vol. I Life Sciences*, Part 2 Ecological Sciences, pp. 3.24–3.25. BNWL-1306 PT2, Battelle, Pacific Northwest Laboratories, Richland, Washington.

Schneider, M.J. 1971. "Sonic Tracking of Adult Salmonids." In: *Pacific Northwest Laboratory Annual Report for 1970 to the USAEC Division of Biology and Medicine, Vol. I Life Sciences*, Part 2 Ecological Sciences, pp. 2.25–2.27. BNWL-1550 PT2, Battelle, Pacific Northwest Laboratories, Richland, Washington.

Schneider, M.J., and W.L. Templeton. 1973. "Effect of Thermal History on the Resistance of Columbia River Steelhead Trout (*Salmo gairdneri*) to Thermal Stress." In: *Radionuclides in Ecosystems, Third National Symposium on Radioecology*, ed. D. J. Nelson, pp. 551–563. CONF-710501-P1, National Technical Information Services, Springfield, Virginia.

Silker, W.B. 1964. "Variations in Elemental Concentrations in the Columbia River." *Limnol. Oceano*. 9:540–545.

Soldat, J.K. 1962. *A Compilation of Basic Data Relating to the Columbia River. Section 8, Dispersion of Reactor Effluent in the Columbia River*. HW-69369, Hanford Atomic Products Operation, Richland, Washington.

Templeton, W.L., and C.C. Coutant. 1971. "Studies on the Biological Effects of Thermal Discharges from Nuclear Reactors to the Columbia River at Hanford." In: *Environmental Aspects of Nuclear Power Stations*, pp. 591–612. IAEA-SM-146/33, International Atomic Energy Agency, Vienna, Austria.

Templeton, W.L., and R.J. Olson. 1973. "Predictive Models of Mortality of Young Fish in a Thermal Plume." In: *Radionuclides in Ecosystems, Third National Symposium on Radioecology*, ed. D. J. Nelson, pp. 941–949. CONF-710501-P2, National Technical Information Services, Springfield, Virginia.

U.S. Environmental Protection Agency (EPA). 1971a. *Columbia River Thermal Effects Study. Vol. 1: Biological Effects Studies*. U.S. Environmental Protection Agency, Washington, D.C.

U.S. Environmental Protection Agency (EPA). 1971b. *Columbia River Thermal Effects Study. Vol. II: Temperature Prediction Studies*. U.S. Environmental Protection Agency, Washington, D.C.

Watson, D.G. 1952. "Observations on Spawning and Migration of Chinook Salmon, *Oncorhynchus tshawytscha* (Walbaum) in the Columbia River in the Vicinity of Hanford Works." In: *Biology Research – Annual Report 1951*, pp. 14–18. HW-25021, General Electric Company, Richland, Washington.

Watson, D.G. 1970. *Fall Chinook Salmon Spawning in the Columbia River near Hanford 1941–1969*. BNWL-1515, Battelle, Pacific Northwest Laboratories, Richland, Washington.

Watson, D.G. 1973. *Estimate of Steelhead Trout Spawning in the Hanford Reach of the Columbia River*. Report to U.S. Army Corps of Engineers by Battelle, Pacific Northwest Laboratories, Richland, Washington.

Watson, D.G. 1976. *Temporal and Spatial Fall Chinook Salmon Redd Distribution near Hanford, 1961–1976*. BNWL-2163, Pacific Northwest Laboratory, Richland, Washington.

Chapter 11

GENERIC STUDIES AT HANFORD AFTER CLOSURE OF THE SINGLE-PURPOSE REACTORS, 1971 TO 1981

A new era in aquatic research at Hanford begin inconspicuously in 1971. Soon after the last single-purpose reactor was shut down, a new multidiscipline facility was completed in the 300 Area north of Richland. It was named the Life Sciences Laboratory I, or 331 Building (Figure 11.1), and it contained a wet laboratory in which aquatic experiments could be conducted. The new laboratory was supplied with unfiltered or filtered water, heated or cooled as needed, from the Columbia River. The 146-FR Building in the 100-F Area, which had served as an aquatic facility since 1952, was closed and eventually demolished.

The Life Sciences I Building was financed and owned by the federal government. In 1975, Battelle Memorial Institute, operating the U.S. Department of Energy's (DOE) Pacific Northwest Laboratory (PNL) at Hanford since 1965, constructed a private research facility nearby and named it Life Sciences Laboratory II (LSL II). The LSL II Building also contained a wet laboratory supplied with Columbia River water. Research on aquatic organisms was conducted in both facilities through the 1970s.

Studies on cooling water discharges to the Hanford Reach from the single-purpose reactors were not possible after January 1971. As a result, studies to assess effects of radioactivity, heat, and other stressors from effluents generated onsite lost momentum. The transition was gradual. First, aquatic research at Hanford shifted to providing data useful in environmental impact statements for power-generating plants under the National Environmental Policy Act (NEPA) of 1969. Second, the research focused on environmental unknowns arising from the development of energy sources throughout the United States. Aquatic research at Hanford was designed to provide generic information that could be used in environmental assessments nationwide.

Fig. 11.1. The Life Sciences I Laboratory (LSL-I) or 331 Building, 300 Area, Hanford Site. This facility was a government installation built by the AEC (now DOE). It was completed January 15, 1971, and contained a fully equipped wet laboratory.

Participants in aquatic ecology research at Hanford in the 1970s included research scientists C. Dale Becker, Colbert E. Cushing, Dennis D. Dauble, Richard M. Emery, Duane H. Fickeisen, M. Paul Fujihara, Robert H. Gray, Thomas L. Page, Duane A. Neitzel, Mark J. Schneider, John A. Strand, William L. Templeton, Donald G. Watson, and Raymond E. Wildung. Technicians playing major roles included C. Scott Abernethy, Steve A. Barraclough, Robert G. Genoway, Edward W. Lusty, Donald C. Klopfer, Jerry C. Montgomery, Alan J. Scott, and Eugene G. Tangen. For most of the decade, the staff were members of the Ecosystems Department, which was reorganized in August 1971 and directed by Burton E. Vaughan.

The entire Hanford Site was dedicated as a National Environmental Research Park (NERP) on March 18, 1977. Hanford was uniquely suited as a NERP because it contained large tracts of land in relatively pristine condition between isolated operational areas. Furthermore, onsite activities had been conducted over the years, for the most part, in such a manner that they met EPA's stringent air and water quality standards. No significant impact on the land, wildlife, and natural vegetation had been revealed by ecological monitoring at Hanford from more than 25 years of operating nuclear facilities within the site boundary (Vaughan and Rickard 1977).

Studies with radioactivity after reactor closure

Closure of the last single-purpose reactor provided an opportunity to evaluate the depletion of reactor-derived radionuclides from the river ecosystem. After 1971, only limited amounts of radioactivity generated onsite entered the Hanford Reach. This input occurred mainly through the slow movement of contaminated groundwater and seepages of intragravel water along the shoreline. Input was closely monitored and could not be detected in the Columbia River flow.

In 1974, studies on biogeochemical cycling of radionuclides shifted from the Columbia River inland to Gable Mountain Pond and U Pond. These artificial lakes were created by the discharge of low-level radioactive liquids in the 200-East and 200-West Areas at Hanford. The radioactivity in these wastes came from the chemical processing of irradiated reactor fuel, and they differed from those that had appeared in the cooling water effluents. These studies were relevant to radioactive waste disposal management rather than the Columbia River ecosystem and are not discussed here.

Laboratory experiments with tritium were initiated in 1970. Tritium was a major contributor of radioactivity in cooling water effluents from commercial, nuclear power plants sited elsewhere (Strand 1973). It was also expected in gaseous and liquid discharges from nuclear power plants of advanced design. Because tritium is a hydrogen derivative and part of the water molecule, it cannot be easily removed from effluents by conventional methods. Tritium has a half-life of more than 12 years and emits a weak beta particle.

Radioactivity in Biota Downstream from the Hanford Reach

The decline of radionuclides in biota was followed in and below the Hanford Reach from July 1971 to June 1973. Organisms from different trophic levels were collected at three sites: below the closed reactors at Hanford, in Lake Wallula (behind McNary Dam), and in Bonneville

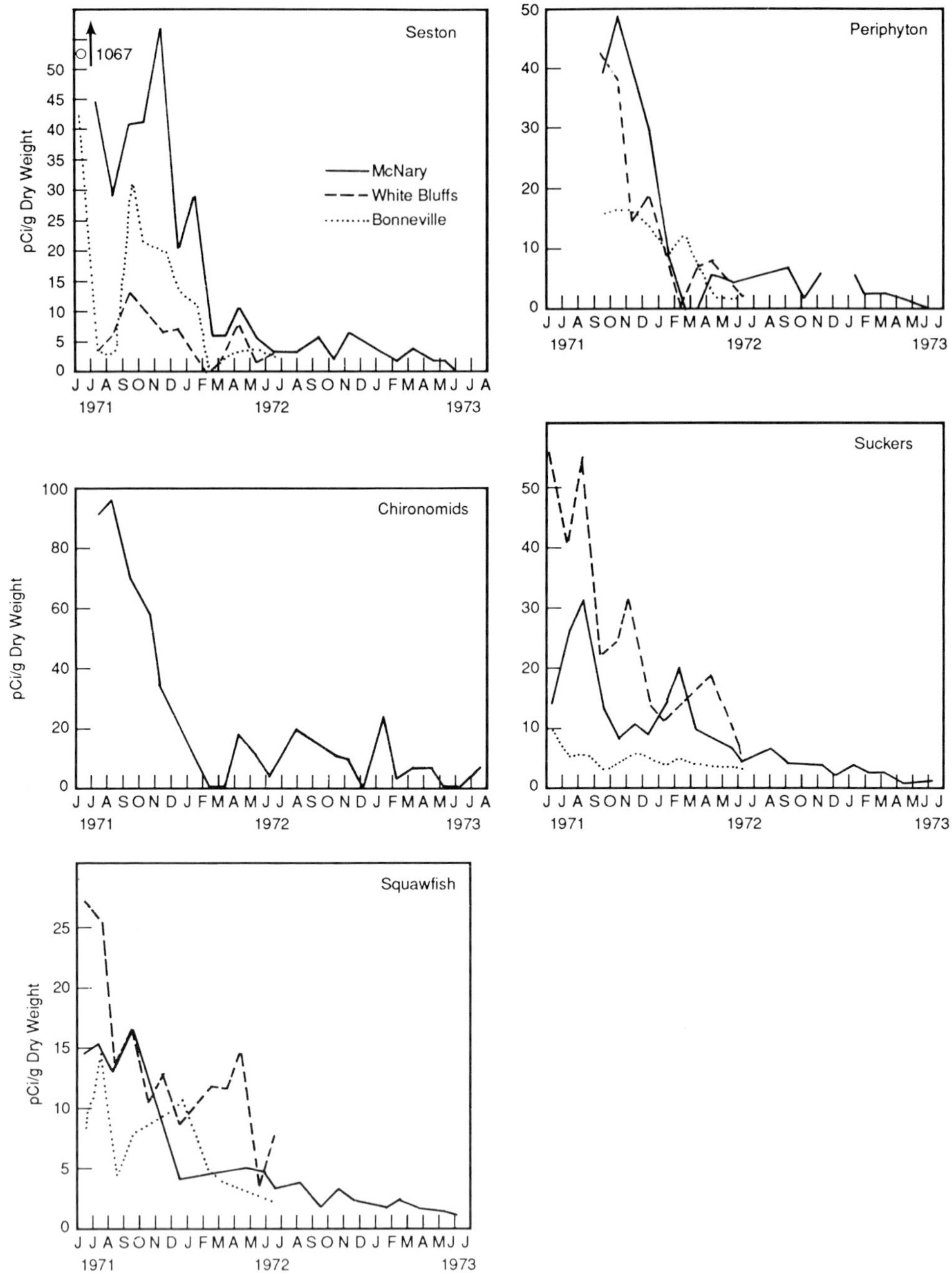

Fig. 11.2. Concentrations of Zn-65 in selected Columbia River biota after shutdown of the last single-purpose reactor in 1971 (from Cushing et al. 1980).

Reservoir 80 kilometers (50 miles) from the Pacific Ocean. Radioactivity in downstream biota decreased by three simultaneous processes: 1) physical decay of radionuclides, 2) biological turnover of radionuclides by the organisms, and 3) decreasing availability of radionuclides in water and food supply (Cushing et al. 1981).

Concentrations of Co-60 (5.24-year half-life) in seston, periphyton, and invertebrates did not decrease as did other radionuclides because, in part, some Co-60 seeped into the river from a disposal trench near the N Reactor. It decreased somewhat in fish, but there was no obvious trend. Biota retained substantial amounts of Zn-65 (245-day half-life). Amounts of Zn-65 in seston and periphyton decreased rapidly and were unmeasurable after the spring of 1973 (Figure 11.2). By February 1973, Zn-65 was unmeasurable in caddis fly larvae. Concentrations in midge larvae from Lake Wallula varied, however, because they were ingesting contaminated sediment. Zinc-65 fell to fairly low, constant levels near 1 and 3 picocuries per gram (pCi/g) dry weight in suckers and squawfish (Figure 11.2), respectively (Cushing et al. 1974, 1980, 1981).

In the river-reservoir complex of the Columbia River, concentrations of fission-produced radionuclides decreased to extremely low or unmeasurable levels within 18 to 24 months after input of reactor effluents ceased. Radioactivity declined rapidly in the Hanford Reach as contaminated seston and fine sediment were transported downstream.

Transport and Depletion of Radionuclides Downstream from Hanford

The transport and depletion of radionuclides from nonbiotic components of the Columbia River downstream from Hanford were also examined after reactor closure. Starting in April 1971, water samples were collected at McNary, John Day, The Dalles, and Bonneville dams. Sediment cores were taken periodically between Hanford and Bonneville Dam. The largest sediment deposits were in Lake Wallula, the first slackwater area below Hanford. Measurements showed radionuclides from both the closed reactors and atmospheric fallout.

All short-lived radionuclides from the closed reactors had decayed. The most abundant radionuclides were Fe-55, Zn-65, Eu-155, Co-60, Eu-152, Eu-154, Mn-54, and Sc-46. Also present in much lower concentrations were Sb-125, Cs-137, Ce-144, and Pu-229. The residual radionuclides were closely bound to bottom sediments, particularly the deposits in Lake Wallula.

More than 98% of the radionuclides that originated at Hanford occurred in sediments in the form of particulates. The main mechanism for removing the bound radionuclides from the river ecosystem, other than radioactive decay, was scouring of sediments by high spring flows, which then transported the radionuclides to the Pacific Ocean. Eventually, major deposits of radioactive sediment in Lake Wallula would be covered by new, uncontaminated layers of silt from upriver (Robertson et al. 1973).

Uptake of Tritium from an Aquatic Microcosm

Initially, the accumulation of tritium by aquatic organisms was examined in a large outdoor microcosm, a pond stocked with aquatic biota. Tritium was introduced continuously as tritiated water at 1 microcurie per liter (μCi/L) for 8 months. Subsequently, depuration of tritium from the system was followed for 8 additional months. Analysis differentiated between volatile or "water-bound" tritium and nonvolatile or "bound" tritium.

All biota rapidly took up tritium, but tissue levels did not reach equilibrium with pond water. Concentrations of volatile tritium neared

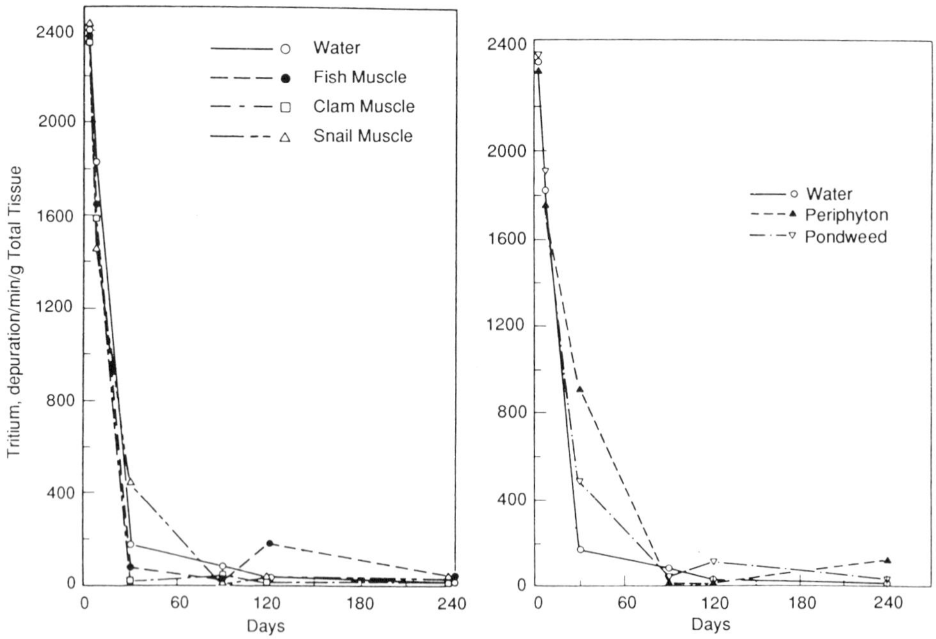

Fig. 11.3. Reduction of tritium from water and organisms in an aquatic microcosm after cessation of continuous spiking with 1 μCi/L of tritium for 8 months (from Strand et al. 1976).

90% of the introduced level in freshwater mussels, crayfish, and carp in the first 2 days. Tissue levels then remained steady for the rest of the 8-month exposure. Concentrations of tritium in filamentous algae exceeded 90% of the introduced level, and those in emergent vegetation reached 70%. Nonvolatile tritium accounted for only 4% to 21% of the total isotope acquired by animal or plant tissues, depending on the organism (Strand et al. 1976).

Tritium was not significantly bioconcentrated. After tritium input ceased, less than 10% of the initial concentration remained in the water after 1 month. Tritium was rapidly depurated from tissues of aquatic organisms. Animals eliminated tritium more rapidly than plants (Figure 11.3).

Exposure of Early Development Phases of Trout to Tritium

Rainbow trout eggs were exposed to tritium irradiation during embryogenesis to establish dose-response relationship. Eggs were reared in uncontaminated, recirculating spring water containing 0.0 (control), 0.01, 0.1, 1.0, and 10.0 μCi/mL of tritium (biological grade) for 28.5 days at 10.6°C. Direct effects were assessed by the proportion of hatched eggs and abnormal embryos. Indirect effects were assessed by select behavioral and physiological tests of young fish.

No reduction in the number of eggs hatched nor any delay in hatching time could be detected. In fact, hatching was slightly enhanced in 0.01 and 0.1 μCi/mL tritium. Exposure to 0.1 and 1.0 μCi/mL tritium, but not 10.0 μCi/mL, produced a few more abnormal embryos. Companion experiments revealed no consistent impairment in relative susceptibility of juvenile fish to predation, thermal death times, or growth rates (Strand et al. 1972).

Uptake of tritium by rainbow trout eggs was also examined. Again, fertilized eggs were reared in uncontaminated, recirculating spring water containing 0.0 (control), 0.01, 0.1, 1.0, and 10.0 μCi/mL of tritium (biological grade), this time for 25 days at 10.5°C. After 20 days, some embryos were removed to examine retention and turnover. Analysis distinguished between the volatile ("water-body") fraction and the nonvolatile ("bound") fraction.

Tritium was acquired rapidly at all concentrations for 1 to 2 days, and equilibrium levels were maintained in eggs and fry during subsequent exposure. The "bound" tritium accounted for about 20% of the total radioactivity in each egg. When transferred to flowing water, both volatile and nonvolatile fractions eluted rapidly; their half-lives were estimated to be about 1 and 2 hours, respectively. A slower part of the volatile fraction persisted for about 17 days in hatched fry. Behavioral and physiological tests disclosed no impairment from exposure to tritium at such low levels (Strand et al. 1973a).

Effect of Tritium on Immune Response of Trout

The effect of tritium irradiation on the susceptibility of fish to infectious disease was next examined in a series of experiments. Exploratory efforts indicated that juvenile and adult rainbow trout surviving exposure to tritiated water were more susceptible to *Flexibacter columnaris* than nonirradiated fish (Strand et al. 1973b). Subsequent studies delved into this possibility.

Trout eggs were exposed during embryogenesis to 0 (control), 0.04, 0.4, 4.0 and 40 roentgen (R) doses of tritium for 20 days. Synthesis of columnaris antibodies was initiated in irradiated and unirradiated 5-month-old fish by injection of a heat-killed vaccine. The development of agglutins specific to columnaris in these fish was then monitored. Agglutinin production was suppressed to 50% of control levels during the ninth week in trout exposed to 40 R, and during the eleventh week in trout exposed to as low as 4.0 R. The serum protein in the blood of irradiated and vaccinated fish contained a protein fraction that was significantly reduced by tritium irradiation. This protein was apparently active in antibody production (Strand et al. 1977; Becker and Fujihara 1978).

Additional efforts were made to determine if suppression of the immune response persisted as the fish aged. Trout from irradiated eggs were exposed a second time to inactivated columnaris cells at the age of 17 months. The suppression effect appeared again and, under experimental conditions, was assumed to be permanent. Suppression in yearling trout equaled or exceeded suppression in 5-month-old trout. Sensitivity was detected when fertilized eggs were exposed to as little as 0.1 pCi/mL of tritium. This concentration was no less than four orders of magnitude above present levels of tritium in the aquatic environment (Strand et al. 1982).

Toxicity of Lithium to Freshwater Organisms

Lithium is not radioactive; rather, it is an element unusually toxic to plants. Lithium was an element of concern because it would be used in advanced technology, fusion reactors to generate deuterium and, possibly, as a liquid-metal coolant. Thus, the effects of lithium on aquatic life became an energy-related issue. In the late 1970s, the toxicity of lithium to aquatic organisms was examined in bioassays at Hanford.

When exposed to lithium in water, an incipient or "threshold" level of toxicity in rainbow trout, aquatic insect larvae (*Chironomus spp.*), and periphyton occurred at concentrations between 0.1 and 1 milligrams per liter (mg/L). The toxicity of lithium was comparable to that of beryllium, a chemically similar component of fusion reactor cores. Natural concentrations of lithium in freshwater are usually less than 0.01 mg/L (Emery et al. 1981). The effects of provisional releases of lithium in reactor effluents have not been documented.

Thermal effect studies with aquatic biota

Potentially adverse effects from the discharge of heated water at steam-electric power plants, both nuclear- and fossil-fueled, remained an environmental issue through the 1970s. Information on the response of aquatic organisms to acute changes in temperature (in mixing zones), or

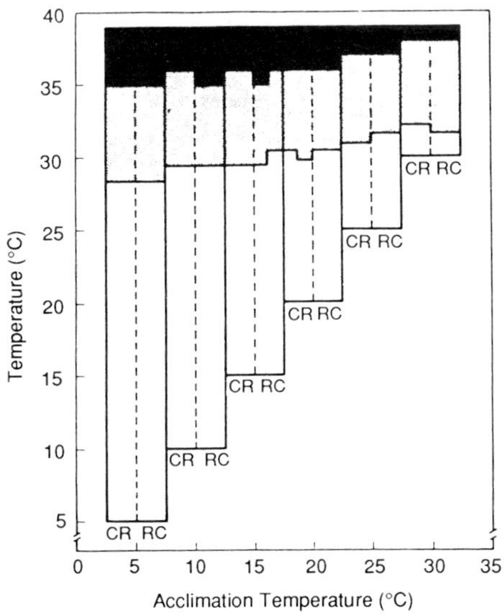

Fig. 11.4. Thermal resistance of a northwestern crayfish over 48-hour abrupt exposures, based on acclimation temperature and median tolerance (CR = Columbia River and RC = Rock Creek populations). Open space represents the zone of thermal tolerance, shaded space the zone of thermal resistance, and black space the zone of thermal death (from Becker et al. 1975).

to consistently warmer temperatures (below mixing zones), was vital to the nation's environmental assessment programs. Thermal studies at Hanford were, for the most part, conducted in the laboratory after January 1971 when heated discharges ceased.

Thermal Resistance of Crayfish

The resistance of a common crayfish in the Columbia River to abrupt thermal change was determined by bioassays. Crayfish from two populations, one in the Hanford Reach and the other in a downriver tributary, were acclimated at 5°C intervals and abruptly exposed to a higher water temperature for 48 hours.

The upper lethal limit for crayfish ranged from 28.5° to 31.5°C, depending on acclimation level. The ultimate upper lethal limit, based on the highest acclimation, was near 32° to 33°C. Thermal resistance of the two populations was similar (Figure 11.4). Molting crayfish were particularly susceptible to thermal stress. Large crayfish from the Hanford Reach were somewhat less resistant than small ones, and female crayfish were somewhat more resistant than males. The crayfish was more tolerant of higher temperatures than native salmon and trout (Becker and Genoway 1974; Becker et al. 1975).

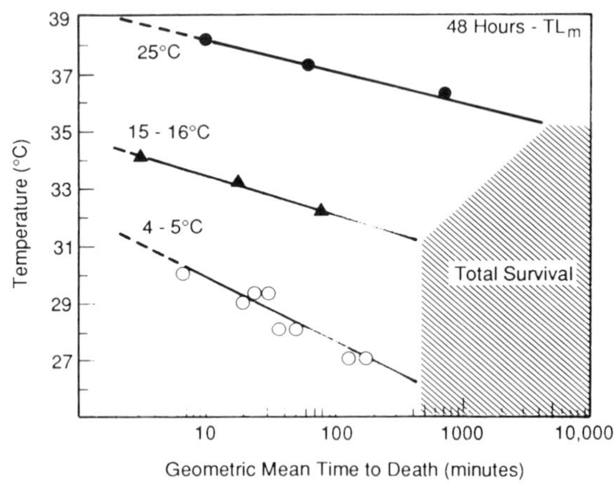

Fig. 11.5. Geometric mean times to thermal death of juvenile brown bullhead when acclimated to 4--5°, 15--16°, and 25°C and exposed to the indicated temperature level (from Becker et al. 1972).

Thermal Resistance of Brown Bullhead

The resistance of an introduced "warm-water" fish, the brown bullhead, to thermal stress was also examined in bioassays. Previous thermal studies with fish at Hanford had involved only native, "cold-water" salmonids. Juvenile brown bullhead from the Hanford Reach were acclimated to water temperatures of 4° to 5°C, 15° to 16°C, and 25°C, then abruptly exposed to a higher temperature for 48 hours.

The brown bullhead was more tolerant of elevated temperatures than native salmonids (Becker 1972). When acclimated to 25°C, the brown bullhead survived exposure to temperatures as high as 35°C (Figure 11.5). Introduced "warm-water" species of fish were in little danger from slightly elevated temperatures, as once caused by reactor discharges in the Hanford Reach, and may even have benefitted from the warmer water.

Cold Resistance in Fish and Crayfish

Most studies on thermal resistance of fish focused on upper rather than lower temperatures. Information on resistance of aquatic organisms

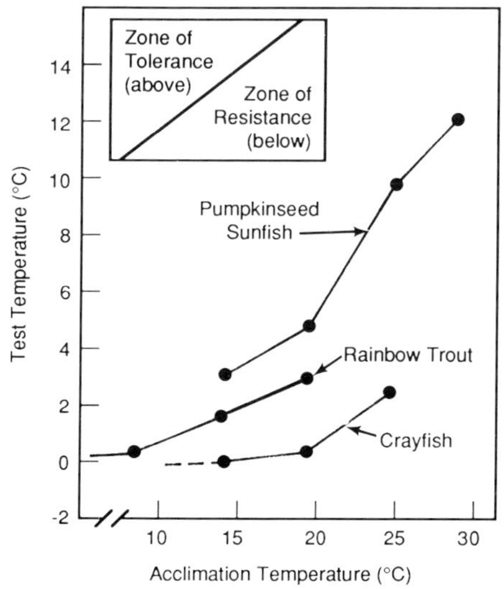

Fig. 11.6. Comparative resistance of pumpkinseed sunfish, rainbow trout, and the crayfish to abrupt cold shock. The data points are lower median tolerance limits after 96-hour exposures based on acclimation at 5°C intervals. Zones of tolerance and resistance are indicated by the inset (from Becker et al. 1977a).

to abrupt and gradual temperature declines was limited. Yet fish in the warmed discharges of power plants will be suddenly exposed to cold temperatures during shutdowns. In winter, temperature declines can be extreme, and the resulting cold is as lethal to fish as added heat.

In bioassays started in 1974, resistance to abrupt and gradual cold stress was examined in three Columbia River organisms: rainbow trout, pumpkinseed sunfish, and crayfish. Test species were acclimated to high temperatures at 5°C intervals. The criteria sought were 96-hour median tolerance limits for abrupt exposure to cold, and 50% loss of equilibrium under temperature declines of 18°, 15°, 10°, 5°, and 1°C per hour.

Resistance to abrupt and gradual cold shock varied with acclimation temperature and differed among test species, in the order of least resistance: pumpkinseed sunfish, rainbow trout, and crayfish. For abrupt exposure to cold water, medium tolerance values were 12.3°C for pumpkinseed sunfish acclimated at 30°C, 3.3°C for rainbow trout at 20°C, and 2.5°C for crayfish at 25°C (Figure 11.6). Loss of equilibrium values were slightly below medium tolerance values for both species of fish, but well above for crayfish (Becker et al. 1977a).

Physiology of Cold Shock in Channel Catfish

The physiological effects of cold shock were examined in juvenile channel catfish abruptly subjected to a cool water temperature of 10°C after acclimation to a warm 30°C. Cold shock, similar to heat shock, had both direct and indirect effects.

After 24 hours at 10°C, plasma osmolarity and chloride ions in channel catfish declined, while plasma glucose and tissue water increased. After 48 hours exposure to 10°C, glucose levels increased by a factor of 2.8, while plasma osmolarity decreased 34% and chloride ions decreased 55%. When fish were returned to acclimation temperature, plasma concentrations of glucose, chloride, and osmolarity gradually returned to control values in 4 days (Block 1974).

Response of Young Salmonids to a Simulated Thermal Plume

Some juvenile salmonids that migrated seaward through the Hanford Reach before 1971 might have encountered a reactor discharge plume, but nothing was known about their response. Outmigrants may be passive to hydraulic forces and temperature differences of a plume, or they may react in ways that increase or decrease exposure.

A special model raceway was built in 1974 to assess the reaction of juvenile chinook salmon that encounter a simulated thermal plume. Heated water was discharged upward from a point midway down the raceway. Test fish were chambered at the head of the raceway, allowed to swim for several minutes, and then released to move "downstream" with the flow under three conditions: no discharge plume, discharge plume at ambient temperature, and heated discharge plume. Responses were videotaped.

Young salmon avoided a discharge plume when it was heated 9° to 11°C above the water temperature in the raceway. Fish occasionally oriented to the discharge current, but were not attracted to a discharge plume heated less than 11°C. Further, fish did not pass to the lower end of the raceway when a plume was heated 9° to 11°C above ambient (Gray et al. 1977). At higher ambient temperatures in the raceway, the plume temperature causing fish avoidance also increased. However, on the average, the increment causing avoidance remained at 9° to 11°C (Gray 1977).

Modeling of Temperature Declines in a Thermal Plume

In field situations, temperatures in mixing zones undergo transitional decline when a thermal discharge terminates. The rate and dispersion of a decline depends largely on the physical characteristics of the water mass and its movement, which are site-specific. Temperature declines in mixing zones were modeled in 1974 on the basis of known physical laws. When applied to site-specific conditions, the model provided insight for assessing the effects of cold shock on aquatic biota.

The model was formulated for a time-temperature distribution of an established thermal plume after a hypothetical power plant shutdown. The mathematical expression was analyzed for an offshore, submerged point discharge where shear current and boundary effects were minimal. Results were valid for heated discharges in either freshwater lakes and reservoirs or for coastal regions, but not for discharges in one-directional rivers (Becker et al. 1977b).

Effect of Thermal Shock on Swimming Ability of Trout

Thermal shock, at sublethal levels, may have an adverse affect on the ability of fish to swim. In field situations, migration upstream in heated water may become an imposing stress. This possibility had not been examined experimentally.

The swimming performance of rainbow trout acclimated to water temperature of 10°C (a "normal" temperature) was determined after their exposure to temperatures of 10°, 15°, 20°, and 25°C. The trout were then induced to swim against a current at 20 centimeters per second (cm/s). The current was raised in 10-cm/s increments every 20 minutes until a fish was overcome by fatigue and swept downstream. Performance was calculated as critical swim speed on the basis of absolute values (cm/s) and relative values (body lengths/s).

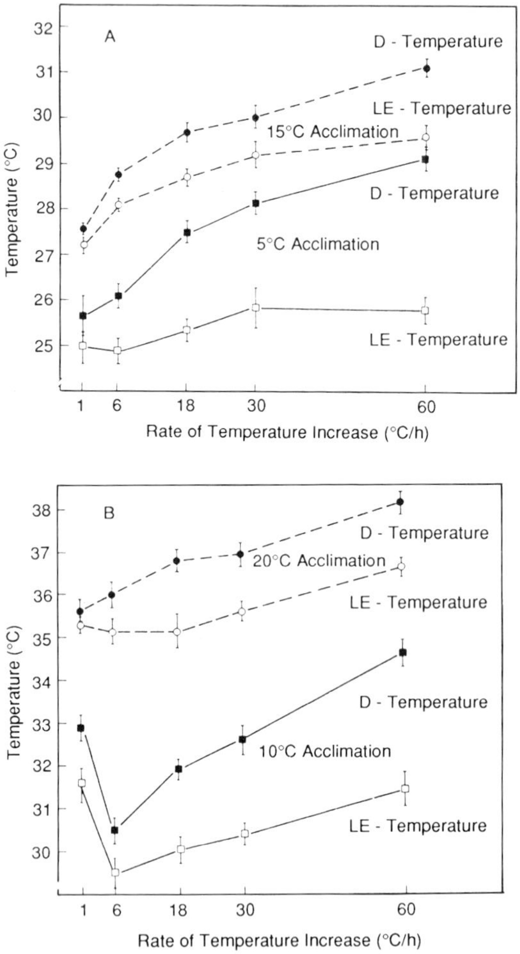

Fig. 11.7. Time and temperature relationships in CTM tests with coho salmon fingerlings (a) and pumpkinseed sunfish (b) acclimated to the indicated temperatures (LE = Loss of Equilibrium and D = Death). The data points are means of several tests (from Becker and Genoway 1979).

Critical swim speeds were similar at 10°, 15°, and 20°C thermal increments, but declined at 25°C. Thus, the ability of trout to swim was impaired only at high incremental temperatures (Schneider and Connors 1982).

Evaluation of the Critical Thermal Maximum

The highest temperature tolerated by fish is often called the critical thermal maximum (CTM). The CTM is essentially a line separating survival and death of a cold-blooded animal subject to a rise in environmental temperature. It varies with ambient water temperature, to which an animal will be acclimated, and other factors. The CTM is usually determined experimentally by gradually increasing the water temperature. In fish, the usual responses are loss of equilibrium (LE) followed by thermal death (D).

The CTM was used to experimentally evaluate the effects of increasing water temperatures at different rates on the upper lethal limits of fish. Tests were conducted with two species, pumpkinseed sunfish (warm-water species) and juvenile coho salmon (cold-water species). Groups of each fish were acclimated to selected low and high temperatures and exposed to five rates of thermal increase. Endpoints were LE and death D.

CTM values for pumpkinseed sunfish were higher than those for coho salmon. Generally, higher acclimations and more rapid temperature increases resulted in higher CTM values (Figure 11.7). Rates of temperature increase also controlled the range of time and temperature between LE and D. Results indicate that, procedurally, the methods used to determine the CTM should be standardized. Only pumpkinseed sunfish acclimated at 10°C and exposed to the slowest rate of increase, 1°C per hour, demonstrated the ability to acclimate upward while the water warmed. Ways to standardize CTM determinations were identified, and the CTM description was amended (Becker and Genoway 1979).

Combined effect studies involving temperature

In most cases, organisms are affected by several interacting factors in aquatic habitats. Examining the effects of one factor is relatively simple. But an experimental design becomes more complex with each additional factor, or test variable. Two-factor or multiple-factor studies gained importance for assessing potential impacts of energy technologies simply because the effects of combined factors could not be ignored.

Uptake of Mercury at Two Temperatures

Mercury, a heavy metal, is a widespread aquatic pollutant and occurs in the flesh of many fish throughout the world. By use of its radioactive form, Hg-203, the uptake and depuration of mercury by fish could be examined in relation to water temperature.

Juvenile carp and brown bullhead were exposed to Hg-203 at 50 pCi/mL in flow-through systems maintained at 7° and 14°C. After 4 weeks of exposure, both species of fish held at 14°C contained about twice as much mercury as those held at 7°C. Carp acquired greater than 50% more of the metal than bullhead. Rates of uptake and loss were also greater at 14°C, reflecting the influence of water temperature on fish metabolic activity (Figure 11.8). The effective half-life of the radionuclide was about 17 and 26 days in carp, and 25 and 31 days in bullhead at 7° and 14°C, respectively (Watson 1972).

Interaction of Mercury and Temperature

The metabolism of a cold-blooded aquatic organism increases as the water warms. Therefore, fish may take up a potential toxicant more

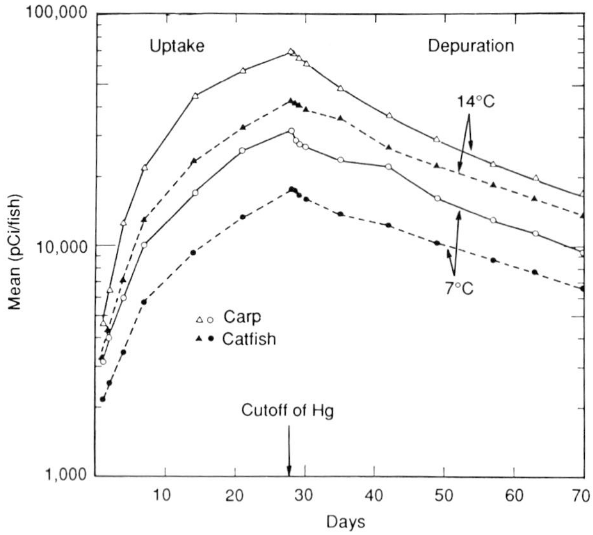

g. 11.8. Uptake and depuration of mercury (as Hg-203) by juvenile carp and brown bullhead at different temperatures (from Watson 1972).

rapidly in warmer water. At the same time, a toxicant may interact with temperature to affect fish response.

Bioassays lasting 96 hours were used to assess the combined effect of mercury and an altered temperature on rainbow trout. The resistance of trout to mercury varied by factors of two to three depending on acclimation and exposure temperatures. Trout acclimated to 10°C were more resistant when exposed to mercury at 15°C than when exposed at 10° or 20°C. Temperatures near 15°C were most favorable for trout metabolism and seemed to elicit the greatest resistance (Thatcher 1974). No data were available to determine if the chemical nature of mercury and/or its availability to trout were altered by a shift to 15°C.

Fatigue and Thermal Resistance of Trout

The thermal resistance of fish in water receiving heated discharges might be modified by level of fatigue. Accordingly, the effects induced by forced swimming on temperature stress were examined in rainbow trout. Levels of lactic acid and glucose in circulating blood plasma were used to indicate fatigue because of their role in energy metabolism.

Initial studies in 1973 examined only the effect of thermal shock on the indicators. Unexercised rainbow trout acclimated to water temperatures of 4.3°C were exposed abruptly to 27°C water for 3 minutes. After return to 4.3°C, lactic acid and glucose were measured in shocked and control fish. Both indices rose abruptly during thermal shock, then fluctuated for at least 2 hours (Schneider et al. 1974, 1975). A special trap was devised to minimize stress when trout were selected as physiological samples (Schneider et al. 1977).

Normal levels of glucose and lactate in the plasma of rainbow trout at different temperatures were then determined. The greatest amounts of glucose occurred in fish acclimated to 8° and 12°C (average 76 and 67 mg/dL, respectively), while amounts of lactate did not differ significantly among fish acclimated to 8°, 12°, and 16°C (average 4.2 to 5.3 mg/dL) (Connors et al. 1978).

The simultaneous effects of elevated temperature and exercise were examined last. Rainbow trout acclimated to 12°C were exposed to 12°C (control), 17°C (delta T 5°C), 22°C (delta T 10°C), and 27° C (delta T 15°C) while either resting or swimming at cruising speed. Glucose and lactate levels in blood plasma were measured after rest or after 30 minutes of exercise. A temperature effect on plasma glucose and lactate was apparent in rested and exercised fish only at 27°C (Figure 11.9). Both glucose and lactate were significantly higher in exercised and rested fish

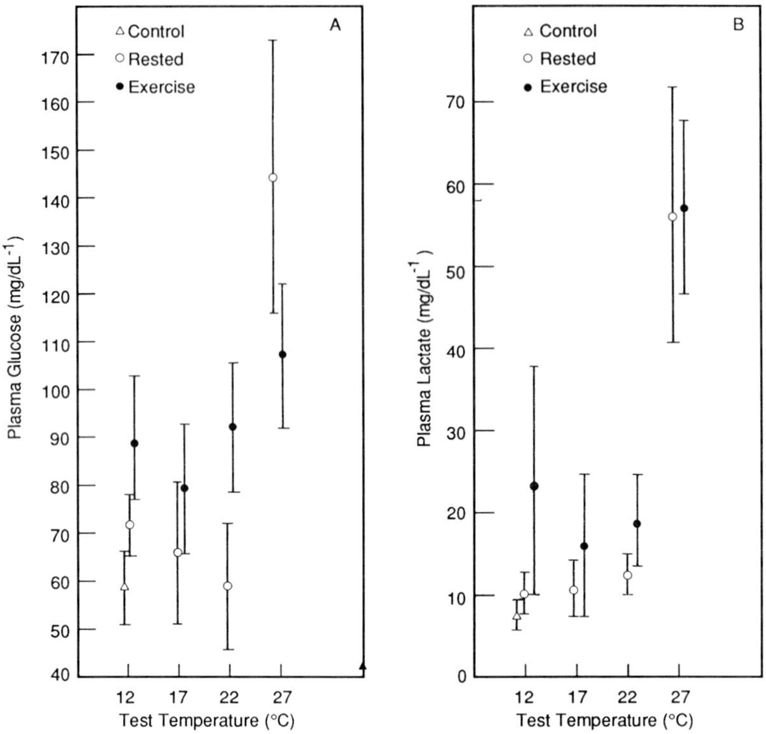

Fig. 11.9. Levels of glucose (a) and lactate (b) in the blood plasma of control, rested, and exercised rainbow trout at 12°, 17°, 22°, and 27°C. The means and 95% confidence intervals are shown by dots and vertical lines (from Schneider et al. 1981).

at 12° and 22°C, but not at 17° or 27°C. This study established a quantitative difference between trout subjected to a temperature increase and trout forced to swim at a higher temperature. A measurable stress from interaction occurred when the fish were forced to swim at temperatures above 22°C (10°C) that was not produced by either factor acting alone (Schneider et al. 1981).

Interaction of Temperature and Acute Radiation

The effect of acute radiation on fish may be influenced by water temperature. This possibility was examined with juvenile rainbow trout acclimated to a high temperature (for this species) of 19° or 22°C and exposed to 250, 500, 1000, and 2000 R for 2.2 to 12.5 minutes from a Co-60 source. In 1973 tests, only fish receiving doses of 2000 R had consistently higher mortalities. Yet losses were somewhat higher among all irradiated groups, even at low exposures, than among controls. Total losses were slightly higher at 22°C than at 19°C among both exposed and control fish (Duever and Abernethy 1974).

Subsequent tests involved rainbow trout acclimated to a wider range of water temperatures: 13°, 16°, and 22°. Mortalities were higher in all groups at higher temperatures. Fish exposed at 250 and 500 R survived as well as controls in all tests. Losses among fish exposed at 1000 R were slightly higher than among controls held at 13° and 22°C. Fish exposed at 2000 R had heavy losses at 22°C, and moderate losses at 13° and 16°C. Losses attributed to irradiation began to occur regularly 10 days after exposure but, at higher temperatures, were partially due to pathogens (Abernethy and Watson 1975).

The last test involved rainbow trout acclimated to 7° and 10°C. The trend of reduced mortality at low temperatures continued, and only fish exposed to 2000 R had high losses from irradiation. After 8 weeks, all irradiated fish were gradually acclimated to 20°C, a level where columnaris disease broke out. In these tests, temperatures above and below optimum (13° to 16°C) clearly affected fish growth. Further, ability to produce antibodies (an immune response) was apparently suppressed at 500 and 1000 R, and destroyed at 2000 R (Abernethy and Watson 1976).

Interaction of Temperature and Chlorine

Chlorine is commonly used to prevent growth of microorganisms in cooling water piping at commercial power plants, from which the water is discharged to aquatic ecosystems. Bioassays were initiated in 1974 to examine the combined effects of chlorine and warm water on rainbow trout and brook trout. Basically, fish acclimated to several temperature levels were abruptly exposed to warmer water containing chlorine for 96 hours in flow-through systems.

In initial tests, both species of salmonid died when exposed to chlorine beyond a certain threshold. Chlorine was a more toxic agent than heat. Combined effects appeared after rainbow trout acclimated to water temperatures of 5°, 10°, 15°, and 20°C were exposed to chlorine in water warmed above 10°C. Small brook trout were more vulnerable to chlorine-temperature interaction than large ones. Most fish appeared to recover when returned to fresh water after first showing chlorine stress (Wolf 1976).

Further tests with brook trout showed no temperature-induced effects at 10° and 15°C regardless of acclimation level. Fish tested at 20°C were more susceptible to chlorine, indicating a synergistic response, but the

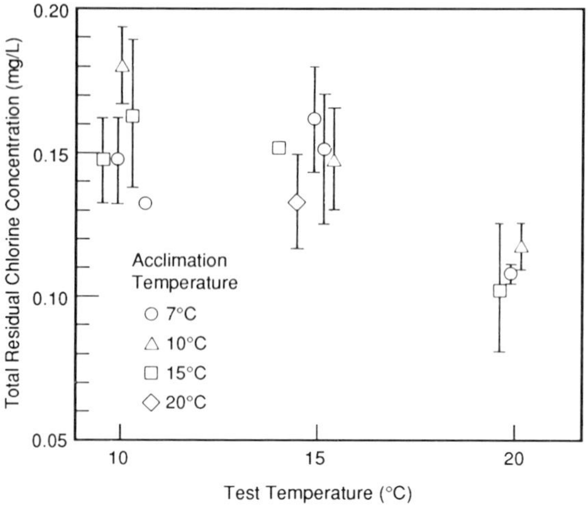

Fig. 11.10. Medium lethal concentration values for juvenile brook trout exposed to chlorine and acclimated to different temperatures (from Thatcher et al. 1976).

effect was again independent of acclimation level (Figure 11.10). In field situations where cooling effluents warm the water to near 20°C, the amount of total residual chlorine (in particular, chloramines) causing acute mortality would be about 30% less than in areas where temperatures were cooler (Thatcher et al. 1976).

Combined Effect of Temperature, Chlorine, and Nickel

Both chlorine (a biocide) and nickel (an alloy in condenser piping) may appear in heated effluents from commercial power plants. Bioassays were initiated in 1976 to examine the interactions and mechanisms of the two toxicants and temperature on the physiology of young rainbow trout. The factorial test design included three levels each of chlorine and nickel. Fish were acclimated and exposed at different temperatures.

Initially, a synergistic toxic effect was identified. Concentrations of 0.05 parts per million (ppm) total residual chlorine (TRC) alone, caused 5% mortality of rainbow trout. Nickel alone, at concentrations of 4.0 to 8.5 ppm, caused no mortality in 96 hours. When chlorine and nickel were combined at these levels, 95% to 100% of the fish died in 96 hours (Anderson and Schneider 1979). Nickel, as the radionuclide Ni-31, accumulated in the blood and eyes of trout and was excreted by the kidneys and lower gastrointestinal tract.

Final results indicated that combinations of chlorine and nickel produced a synergistic toxic interaction in rainbow trout, and that chlorine was the key factor causing mortality. Individually, the 96-hour median lethal concentrations were 0.90 ppm TRC for chlorine and 17.1 ppm for nickel. Small changes in concentrations of chlorine, which alone did not cause mortality, significantly increased losses of fish in chlorine-nickel exposures. Further, the uptake of nickel (as Ni-63) was higher in fish jointly exposed to chlorine and nickel than in fish exposed only to nickel. Chlorine apparently increased the permeability of gill tissues to nickel. As a result, bioaccumulation factors for radionuclides may have been underestimated if the water contained chlorine (Anderson 1981, 1983).

Effect of Nickel on Thermal Tolerance of Fish

Bioassays were conducted in 1976 and 1977 to determine if sublethal exposure to nickel would alter the thermal resistance of young salmonids. First, coho salmon and rainbow trout were exposed to sublethal amounts of nickel. Their thermal resistance was then determined by the CTM method. Exposure to nickel (acclimation) and initial test temperatures were 15°C for coho salmon and 8°C for rainbow trout. Temperatures were increased at a rate of 6°C per hour until LE and D occurred.

Sublethal exposure of juvenile rainbow trout to 1.5 mg/L nickel for 7 to 21 days significantly lowered their resistance to elevated temperatures, but exposure to 0.9 mg/L nickel for 28 days did not. The suppressed thermal resistance corresponded to a twofold increase of nickel in gill and liver tissues. Thermal resistance also seemed lower in juvenile coho salmon surviving exposure to higher levels of nickel for 14 days (Becker and Wolford 1980). Recognizing that the resistance of fish to higher temperatures may be lowered when a toxicant is present at sublethal levels could be an important factor to consider in assessment programs.

Effects of hydroelectric generation

Part of the nation's electricity is generated at hydroelectric dams. This method of generating electric power, while relatively benign in its environmental impact, is not without risk to aquatic biota. Some environmental effects of hydrogeneration were investigated at Hanford as part of a national effort to evaluate further development of nonnuclear energy resources.

Issues investigated at Hanford included supersaturation of river water with air, fluctuation of water-levels in the Hanford Reach, movement and reproduction of smallmouth bass in Hanford sloughs, and dewatering of salmonid redds in exposed gravel. Expertise acquired at Hanford led to a review of research on aquatic effects of hydroelectric generation, and an assessment of research needs for the DOE (Fickeisen et al. 1981).

Air Supersaturation of River Water

Adverse effects on fish from supersaturation of water with air appeared below hydroelectric dams on the mainstem Columbia River in the late 1960s. The primary cause was the passage of excess flows during the spring over spillways, where hydrostatic pressures in the plunge basin forced entrained air to dissolve. As river water flowed downstream and pressure lessened, it became supersaturated with dissolved atmospheric gases, predominantly nitrogen. Supersaturation may also occur in heated effluent zones because air becomes less soluble in warmer water.

When fish in the Columbia River were exposed to supersaturated water, their tissues filled with air and gas-phase emboli formed at nucleation sites, resulting in gas-bubble disease. Exposure to highly supersaturated water caused mortality.

In 1972, air supersaturation in the Columbia and Snake rivers exceeded 120% as early as March, and it eventually reached 141% (Fickeisen and Schneider 1973). In 1973, spring flows in the Columbia and Snake rivers were low, there was little spill, and saturation levels reached only 95% to 105%. The Yakima River at its outlet was usually saturated between 97% and 105%, but saturation peaked at between 163% and 180% in mid-May of 1973 (Fickeisen 1974a). In an effort to improve precision, the accuracy of the Weiss saturometer for measuring supersaturation was evaluated (Fickeisen et al. 1975). This effort led to further studies of gas supersaturation at Hanford. A workshop was held October 8 and 9, 1974, to address gas-bubble disease. The workshop provided a forum for an interagency exchange of information on existing problems and state-of-the-art relating to air supersaturation (Fickeisen and Schneider 1976). Subsequently, some aquatic biologists at Hanford participated in the Interagency Nitrogen Task Force, a group formed to examine supersaturation problems in the Columbia River. Supersaturation levels below most dams on the Columbia River were reduced by modification of spillways in the late 1970s.

Supersaturation Effects Among River Fish

The histopathology of gas-bubble disease was first described at Hanford in chinook salmon fingerlings from an aquarium at Rocky Reach

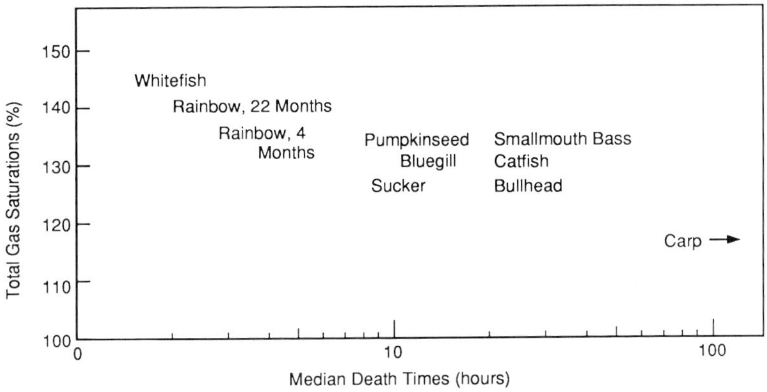

Fig. 11.11. Relative survival times of ten species of resident fish common to the Hanford Reach when exposed to supersaturated air in water (from Fickeisen 1974a).

Dam. Tissues in the heart and stomach of afflicted fish appeared normal, but all other tissues showed cellular damage. Necrosis probably resulted when capillaries providing oxygen to tissues were blocked by gas emboli. Pathological changes were most striking in the roof of the mouth (Pauley and Nakatani 1967).

In 1975 and 1976, 179 smallmouth bass and 85 northern squawfish were collected by angling from the lower Snake and middle Columbia rivers. Emboli were present in 72% of the bass and 84% of the squawfish. The prevalence of gas-bubble symptoms corresponded to the spring runoff when total dissolved supersaturations in river water exceeded 115% (Montgomery and Becker 1980).

Bioassays were used to examine the relative tolerance to supersaturated water of ten species of fish from the Hanford Reach. Fish were acclimated and tested at 20°C (68°C). Tolerance varied among species (Figure 11.11). Fish preferring cold water (whitefish, rainbow trout) were less tolerant of supersaturation than those preferring warm water. Most exposed fish developed skin blisters, an external symptom of supersaturation. Mortalities were usually caused by emboli in the heart (85%) or gills (10%), or from internal hemorrhage when gas bubbles ruptured (5%). Many fish showing symptoms recovered when transferred from highly supersaturated water (Fickeisen 1974a).

Temperature and Tolerance of Fish to Supersaturation

Bioassays began in 1974 to quantify the tolerance of rainbow trout, black bullhead, and pumpkinseed sunfish to supersaturated water at different temperatures. The fish were acclimated at 4°C intervals from 8° to 32°C water temperature, and exposed to supersaturation for 96 hours in hatchery troughs.

Rainbow trout were most susceptible to supersaturation at 20°C, the highest point they could be acclimated, and all lower temperatures. At 8°C, median tolerance limits (in percent total gas supersaturation) were 120.5 for trout, 126.9 for bullhead, and 126.6 for sunfish. At 20°C, median tolerance limits were 117.8 for trout, 124.4 for bullhead, and 123.8 for sunfish. At 32°C, median tolerance limits were 120.2 for bullhead and 115.1 for sunfish. Tolerance of all three species decreased as temperature increased. Therefore, if the water was supersaturated, fish would be more susceptible to incremental temperatures in a heated discharge plume (Fickeisen 1974b, Fickeisen et al. 1976, Fickeisen and Montgomery 1977).

Depth and Tolerance of Fish to Supersaturation

Additional bioassays in 1976 and 1977 examined the phenomenon of hydrostatic pressure compensation by fish moving to deeper water. Four species of Columbia River fish were exposed to water supersaturated with air in a tank 3.2 meters (10.5 feet) deep. The fish were caged, placed at various depths, and held under supersaturated conditions for 10 days at 10°C. Saturation levels ranged from 132% at the surface of the tank to 100% at the bottom.

Based on medium times to death, tolerance to supersaturation increased among test species in the following order: mountain whitefish, cutthroat trout, largescale sucker, and torrent sculpin. Hydrostatic pressure increased with water depth, thereby reducing supersaturation levels and increasing the survival of exposed fish. Typical symptoms of gas-bubble disease accompanied mortality. Torrent sculpin, the most resistant species, usually died of exhaustion from struggling when large bubbles caused them to float. Hydrostatic pressure compensation would not be available to fish in shallow rivers (Fickeisen and Montgomery 1978).

Migration of Adult Salmon in a Supersaturated River

The depth inhabited by fish in a supersaturated river may be crucial for their survival. As depth increases, the gap between dissolved gas tension and the partial pressure in water decreases, which helps maintain dissolved gases in solution in fish blood. Thus, fish can tolerate greater supersaturation at a depth of 1 meter (3.3 feet) than at the surface. In 1975, a field study was made to determine if adult salmon migrating upstream in the Columbia River would regulate swimming depths in response to air supersaturation, thus avoiding critical zones.

An initial step was to attach radio transmitters to returning adult salmon to record depth and temperature. Suitable pressure-sensitive transmitter tags were developed in cooperation with the National Marine Fisheries Service. The tags were tested on adult chinook salmon returning to the lower Snake River (Gray and Haynes 1976). External tags were more suitable than internal tags because they did not affect upstream travel times (Gray and Haynes 1979).

The next step was to tag and monitor the upstream movement of adult chinook salmon in the Snake River under prevailing field conditions. In spring 1976, air supersaturations ranged from 120% to 130%, and returning salmon migrated at a depth near 6.4 meters (21 feet), on the average. In fall 1976 and spring 1977, when saturations were less than 108%, average migration depths were near 3.0 and 4.0 meters (9.8 and 13.1 feet), respectively. Therefore, chinook salmon passing upriver usually chose a depth below a zone of 110% saturation, which served to decrease exposure and aid survival (Gray and Haynes 1977; Haynes 1978). Delays in upstream migration at Little Goose Dam, particularly in 1976, were due to fishway design rather than to supersaturation levels (Haynes and Gray 1980).

Water-Level Fluctuations in Hanford Reach

During the 1970s, water levels in the Hanford Reach fluctuated daily and weekly because hydroelectric generation at Priest Rapids Dam was regulated by power demand. However, frequent changes in water level impacted ecological functions along downriver shorelines. Obvious impacts included entrapment, stranding, desiccation, and exposure to predators of fish and aquatic invertebrates. Field assessment studies began in 1975 (Fickeisen and Montgomery 1977).

One phase of the studies examined the effects of water-level changes on the movement and spawning of smallmouth bass. Adult bass were marked either with standard tags or radiotransmitters to monitor their activity (Figure 11.12). Adult bass spawned in backwater areas and sloughs warmed by insolation from April through June, then moved back to the mainstem Columbia River. Fluctuations in water level impacted spawning, and some fish were lost by entrapment and dewatering of shoreline areas (Montgomery and Fickeisen 1978). Further field work revealed that bass nests were flooded with cool water from the main channel as river flows increased each spring, impairing the development of fertilized eggs.

Fig. 11.12. Dart tags and radiofrequency transmitters were used to monitor the movement of smallmouth bass in the Hanford Reach during 1975 and 1976 (from Montgomery et al. 1980).

Spawning success was poor in 1976, a high-water year, in comparison with 1977, a low-water year. Young bass fry in inshore areas were often stranded by water-level changes (Montgomery et al. 1980; Becker et al. 1981).

Another phase of the studies monitored water-level changes in sloughs along the Hanford Reach, and required development of remote, pressure-sensitive radiotelemetry systems. Water levels fluctuated as much as 2.0 meters (6.5 feet) in 24 hours in some sloughs. Impacts on aquatic communities in sloughs, including spawning smallmouth bass, varied with the stage of river discharge and with air and water temperatures. In some sloughs, temperatures changed as much as 14°C in 24 hours (Neitzel et al. 1982).

River conditions were compared in 1976 (high river flows) and 1977 (low river flows) to assess effects related to water-level fluctuations in the Hanford Reach. Fluctuations were more extreme in 1976 than in 1977.

Fig. 11.13. The physical features of the F Area slough in the Hanford Reach led to entrapment and standing of juvenile fish as water levels fluctuate, primarily from hydroelectric power generation at Priest Rapids Dam (from Montgomery et al. 1980).

River temperatures were higher during the lower flows of 1977. Observations in three major sloughs, Hanford, F-Area (Figure 11.13), and White Bluffs, revealed losses of fish and benthic invertebrates from stranding, entrapment (with or without dewatering), warming of water, and predation. Juvenile fish, especially chinook salmon and smallmouth bass fry in the spring, were more susceptible to water-level fluctuations than adult fish. Accurate estimates of losses could not be made, but they probably had little effect on ecosystem dynamics (Becker et al. 1981, 1983a).

A mathematical model was developed to enable prediction of water level changes in the Hanford Reach in support of field work. The model covered flow releases from 36,000 to nearly 200,000 cubic feet per second (ft^3/s) at Priest Rapids Dam (RKm 639) above Hanford. The model was calibrated and verified for resulting water levels at a site downstream (RKm 582). Accuracy allowed close correlation with impacts from water-level changes, and could be adapted to other river systems (Sneider and Skaggs 1983).

Dewatering of Salmonid Redds in Gravel

Mature salmonids deposit fertilized eggs in redds excavated in streambed gravel. Redds can be exposed by low stream flows when, for example, upstream storage reservoirs are filled or too much irrigation water is withdrawn. Starting in 1979, the tolerance of salmonid embryos

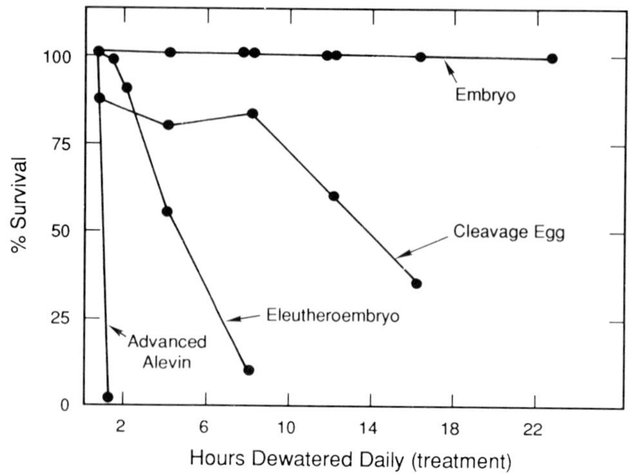

Fig. 11.14. Tolerance of four intergravel development phases of chinook salmon to daily dewatering for 20 consecutive days. The embryo (fertilized egg) was the phase most tolerant of dewatering while the alevin (egg-sac fry) was the least tolerant phase (modified from Becker et al. 1982).

to dewatering in simulated redds was investigated at Hanford. The redds consisted of aquaria filled with a gravel mix, planted with one of four intergravel development phases, and supplied with 4 liters per minute of river water at 10°C. The development phases were cleavage eggs and embryos (egg phases) and eleutheroembryos and pre-emergent alevins (alevin phases). Four series of tests were completed.

First, effects of daily dewaterings were examined, representing a peaking mode of power generation at an upstream dam. Early phases of fall chinook salmon were used, and water was drained from the aquaria gravel at consistent intervals each day. Results showed egg phases were more tolerant of daily dewatering than alevin phases (Figure 11.14). Some cleavage eggs were killed by 12- and 16-hour daily dewaterings. Embryos survived up to 22-hour daily dewaterings and, in some tests, continuous dewatering for several days. In contrast, about half the eleutheroembryos were killed by 4-hour daily dewatering, and nearly all pre-emergent alevins were killed by 1-hour daily dewatering (Becker et al. 1982).

Second, effects were examined of one-time extended dewatering, representing prolonged low water conditions or stream drawdown. Again, early phases of fall chinook salmon were used, but this time water was drained from the aquaria gravel for extended intervals. Results again showed egg phases were more tolerant than alevin phases. Embryos and cleavage eggs tolerated continuous dewatering for up to 12 days. In fact, embryos tolerated dewatering up to 16 to 20 days (through cleavage egg phase), when premature hatching initiated mortalities. Eleutheroembryos tolerated 6-hour dewatering, an exposure that killed almost all pre-emergent alevins (Becker et al. 1983b).

Third, effects were examined of temperature and humidity changes that might occur in dewatered redds. Early development phases of fall chinook salmon were exposed abruptly to a different temperature. Embryos tolerated 25°C for 8 hours and 26.5°C for 2 hours. Eleutheroembryos and alevins tolerated 23.5°C for 4 hours and 25°C for 1 hour. Abrupt exposure to temperatures just above freezing did not reduce survival. However, relatively minor reductions in relative humidity, such as from 100% to 90% for 8-hours, resulted in higher mortalities (Neitzel and Becker 1985).

Fourth, effects of one-time dewatering were examined on intergravel phases of rainbow trout so that results could be compared with those from fall chinook salmon. Again, egg phases were considerably more tolerant than alevin phases, and mortalities increased when hatching began. However, the time trout eggs tolerated dewatering was less than that of salmon eggs because, at 10°C, trout eggs developed faster. Survival times of trout alevins in redds when water ceased flowing was limited by oxygen deficits (Becker et al. 1986).

These tests suggested several intergravel conditions that influenced survival of salmonid eggs and alevins during dewatering. The major factors appeared to be residual intragravel flow, retention of moisture, intergravel temperature, gravel composition, and dissolved oxygen. Alevin behavior and species differences would also influence survival (Becker and Neitzel 1985).

Site characterization studies

During the 1970s, site characterization studies for new projects were initiated at Hanford. This effort reflected DOE's commitment to examine potential environmental effects from onsite activities under NEPA. Of equal importance was a realization that available baseline data needed to be expanded to enable the prediction of effects from future power-generating plants and other industrial facilities built along the Columbia River. Water quality data continued to be collected from the Hanford Reach as one phase of the expanding environmental monitoring program at Hanford.

Mercury in the Columbia River

The amount of mercury in the Columbia River and its biota, possibly from past agricultural and industrial activities, had never been determined. In late 1971 and early 1972, the presence of mercury was evaluated by sampling water and suspended particulate matter ($< 0.1\ \mu$) at selected stations below Hanford. Sample sites included the outlets of the Yakima, Snake, and Walla Walla rivers, and irrigation water wasteways (Wildung et al. 1972).

Mercury was below detectable limits [0.1 microgram per liter (μg/L)] in all river and wastewater samples from which particulates were removed. However, small amounts of mercury were detected in sediments behind Priest Rapids Dam [0.115 microgram per gram (μg/g)] of sediment, dry weight), McNary Dam (0.331 μg/g), and Bonneville Dam (0.096 μg/g). Thus, mercury was closely bound to particulates in the Columbia River ecosystem.

Particulate concentrations suspended in river water ranged from 1.2 to 100 mg/L, and were highest during the spring at runoff times of surrounding areas. Suspended particulates in the Columbia River contained from 1.5 to 120 mg/kg mercury. Concentrations in suspended particulates were markedly higher in December at all locations on the Columbia River and in river and wastewater tributaries. Apparently, considerable amounts of mercury were transported during winter periods of relatively low water and minimum runoff. Amounts of mercury in sediments behind Columbia River dams were low in relation to amounts in suspended matter at the same locations (Wildung et al. 1973).

From 1973 through 1974, water and sediment samples were analyzed seasonally for mercury. Concentrations peaked in the summer and fall when water temperatures were highest, a factor not considered previously. During the spring freshet, mercury concentration was higher in the suspended matter from two major tributaries, the Snake and Yakima rivers, than in that from the Columbia River at Priest Rapids Dam. Comparison of the relative amounts in suspended matter during the freshet (0.11 μg/g) and during the high-temperature period (52 μg/g) suggested that runoff from land and resuspension of sediments during the spring freshet did not contribute directly to maximum observed mercury levels. Perhaps mercury was released from unspecified "sinks" during high-temperature periods. Quantities of the metal increased with depth in sediments behind McNary Dam (Garland et al. 1974).

Levels of methylmercury were examined in suckers (a bottom-feeding fish) taken from behind dams below Hanford to help identify conditions

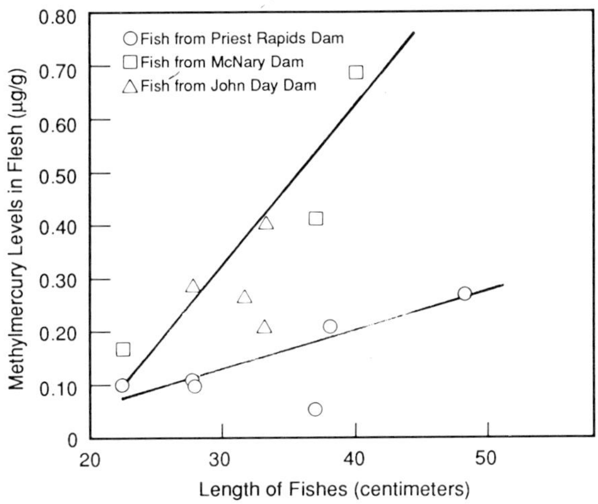

Fig. 11.15. Analytical amounts of methylmercury in the flesh of suckers (Catostomidae) behind McNary, Priest Rapids, and John Day dams in April 1974 (from Garland et al. 1974).

influencing the availability of mercury to aquatic biota (Figure 11.15). Methylmercury was lower in suckers at Priest Rapids Dam than at McNary Dam, suggesting different mechanisms of bioavailability at the two sites. Levels of methylmercury were greater in the skin and bone than in the gut (Garland et al. 1974).

Conceptual Model for Biogeochemical Cycling

Years of research at Hanford had, by 1973, provided considerable data on the transfer of radionuclides and nutrients in the Columbia River ecosystem. A conceptual model for biogeochemical cycling was developed from these data. One purpose was to examine questions necessary to comply with NEPA.

The conceptual model (Figure 11.16) identified likely major pathways for transfer of elements and energy among river components. While the components were real, some features needed further study under controlled conditions. Relationships between dissolved and suspended matter and sediments were, at that time, the least understood portions of the model (Thomas et al. 1974).

The validity of a model for cycling of minerals by periphyton was examined later in a simulated stream ecosystem. Evidence indicated that simple uptake and retention models were not adequate. Most unresolved questions dealt with adsorption versus absorption, and the varied sizes or

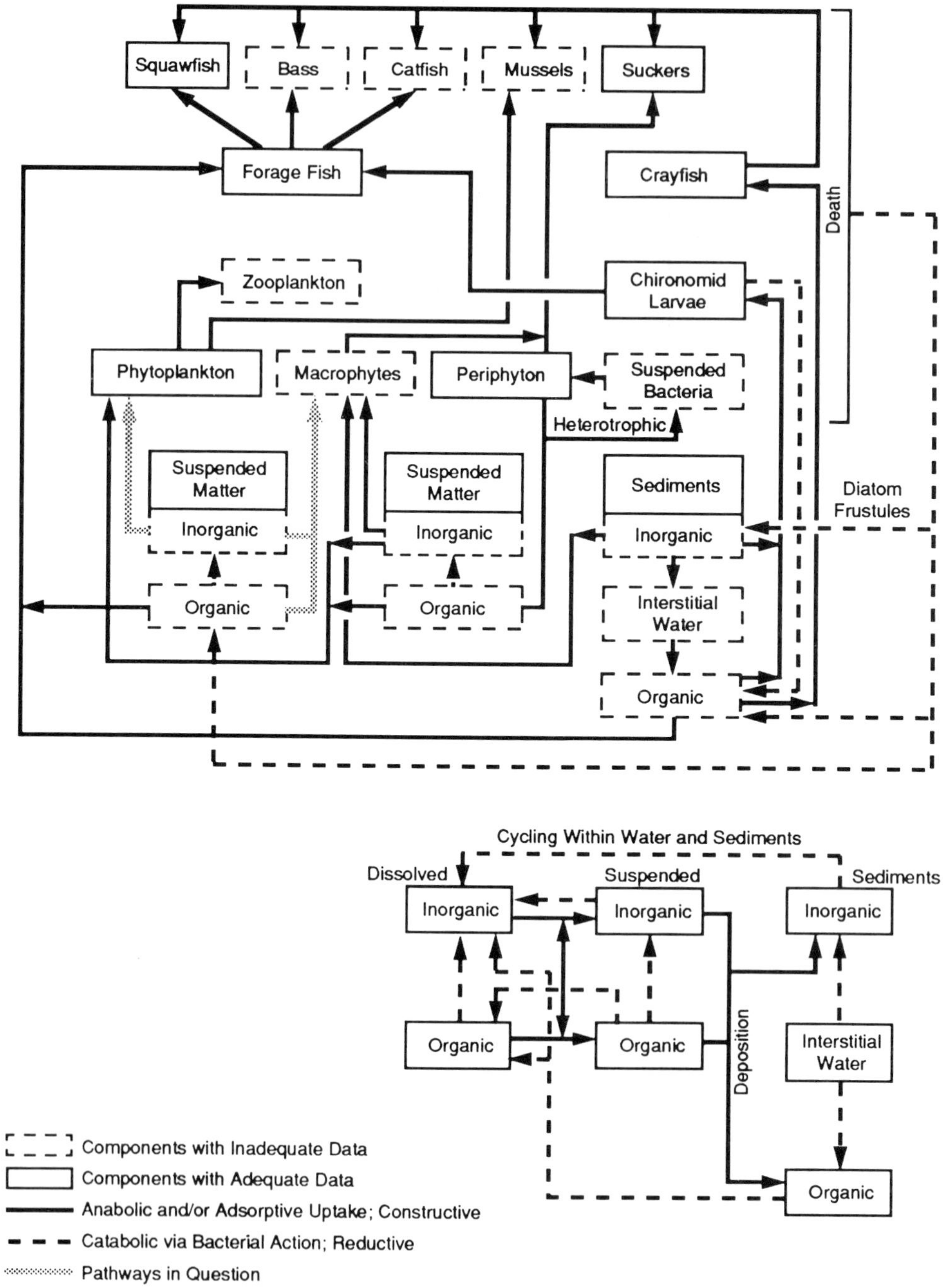

Fig. 11.16. Probable pathways for movement of elements and energy among components of the Columbia River ecosystem, based on assessment and modeling of available data (from Thomas et al. 1974).

masses of community components. The model, while primarily descriptive, did provide a basis for hypotheses to test in large-scale experiments or field situations (Cushing et al. 1975).

Nonaquatic Resources of the Hanford Reach

Many forms of wildlife use the Hanford Reach other than those that live underwater. By the 1970s, these resources were recognized as having unique ecological value. The Hanford Reach was the last segment of the mainstem Columbia River below the international border that still remained flowing. Furthermore, public use of the Hanford Site had always been restricted.

The Hanford Reach, closed to public use in 1943, was opened to upstream boating to the old Hanford townsite during the late 1960s. It

Fig. 11.17. Large numbers of Canada geese use islands in the Hanford Reach for nesting, resting, and other activities each year. Today, islands and riparian areas in the Hanford Reach provide essential habitat for many endemic but nonaquatic animals.

was then completely opened to boating in 1978. Public activities along the Hanford Site shoreline remained prohibited. Because of this land-use pattern, Hanford was a refuge for wildlife for more than 35 years.

Riparian areas along the Hanford Reach were essential habitats for wildlife, particularly the bald eagle, mule deer, coyote, and resident Great Basin Canada goose. Islands in the Hanford Reach were nesting sites for geese (Figure 11.17) and fawning sites for mule deer, where they offered protection from coyotes. The American osprey occasionally fished the Hanford Reach but did not pair and reproduce, possibly because there were no suitable nest trees (Rickard et al. 1982). Eagles fed on salmon carcasses during late fall and winter.

Fig. 11.18. White sturgeon were measured before being tagged and released in the Hanford Reach to enable the monitoring of seasonal feeding and spawning movements.

The most striking feature of Hanford's shoreline was the paucity of tree corridors, willows and cottonwoods, which border most streams and rivers in the semiarid northwest. Instead, vegetation along the shoreline consisted of a narrow zone of scrub-willows and mixes of various rushes, grasses, and forbs. These plants could establish in a rooting substrate consisting largely of water-worn pebbles and assorted gravels (Rickard et al. 1982). Only plants that could tolerate inundation from daily or periodic changes in water level grew near the water line.

Movement of White Sturgeon in the Hanford Reach

Populations of the white sturgeon, a large and primitive fish, reside in the Hanford Reach. Their presence is due, in part, to the unimpounded, flowing nature of the middle Columbia and lower Snake rivers. Information on white sturgeon in the area was needed for effective management of future changes.

From 1975 through 1977, the activity of 29 adult white sturgeon was monitored with attached radiotransmitters (Figure 11.18). Movements of these slow-growing fish were apparently in response to sexual maturity and feeding requirements. Migratory movement was largely seasonal and corresponded with water temperatures above 13°C (55°F). Sturgeon became active in June when the water warmed and inactive in the fall when the water cooled. Some sturgeon moved inshore at night and fed in shallow areas that were warmed during the day. After midnight, when temperatures inshore declined, these fish returned to the deeper, main channel (Haynes 1978; Haynes et al. 1978). Sturgeon in the Hanford Reach were dimorphic, characterized by having either short snouts or long snouts (Crass and Gray 1982).

Water Quality in the Hanford Reach

Data on water quality features other than radioactivity were obtained as part of Hanford's environmental surveillance program. During the 1970s, selected measurements from the Columbia River were used for two main purposes: 1) to detect any impact of onsite waste disposal practices on water quality in the river, and 2) to demonstrate compliance with applicable state and federal standards for water quality. These data were published in annual reports prepared by Pacific Northwest Laboratory.

Measurements of water quality were also obtained for the Hanford Reach by the U.S. Geological Survey (USGS). Because the two efforts were independent, results from one provided backup data for the other. The USGS also monitored river temperature and flow.

Nonradiological monitoring remained important during the 1970s because effluent from onsite activities was discharged at eight points along the Hanford Reach. The discharges included the backwash from water intake screens, cooling water, river bank springs, water storage tank overflow, and waste water from research wet laboratories. Each discharge was identified in National Pollutant Discharge Elimination System permits issued by the U.S. Environmental Protection Agency. Effluent from each outfall was monitored and reported by onsite contractors, as required by each permit (Sula et al. 1982).

Little impact on water quality was detected from either current or past operations on the Hanford Site from 1971 (Bramson and Corley 1972) to 1984 (Price et al. 1985). Nonradiological values for water quality were usually within standards established by the state of Washington for the Hanford Reach. In most isolated cases when state standards appeared to be exceeded, there was no apparent association with Hanford operations.

Significance of generic studies at Hanford, 1971 to 1981

Environmental studies to assess the effects of energy technologies, including operation of power plants throughout the United States, gained increased importance in the early 1970s, following passage of the NEPA and the creation of the Presidential Council on Environmental Protection. The NEPA required the U.S. Atomic Energy Commission (AEC), and later the U.S. Nuclear Regulatory Commission, to prepare environmental impact statements for all nuclear power plants.

From 1971 to 1973, the research staff at PNL responded. They assisted the AEC's Directorate of Licensing in evaluating proposed and operating nuclear power plants throughout the nation. Each assessment required detailed examination of construction and operating specifications and site-specific conditions in adjacent terrestrial and aquatic habitats. Furthermore, each assessment had to be done quickly to keep construction and licensing activities on schedule. There was little time for onsite research. Experience gained in the past with the single-purpose reactors and the Hanford Reach was invaluable in performing these assessments.

A national effort to develop new sources of energy and extend the use of old sources soon followed, with the goal of loosening the growing dependence of the United States on imported oil. At Hanford, aquatic research shifted direction to examine potential effects associated with hydroelectric generation and pumped storage, and with development of new energy technologies such as advanced reactor systems and solvent-refined coal. Aquatic research that delved into generic issues related to energy development supported a nationwide effort to draft knowledgeable and scientifically sound environmental impact assessments. This focus was not unique to Hanford. Other national laboratories were involved in related investigations through the 1970s.

Aquatic issues investigated at Hanford in the 1970s, all in energy-related fields, included

- fate of residual radionuclides in the Columbia River downstream of Hanford (from past operation of single-purpose reactors)
- effects of tritium and tritiated water (effluent from nuclear power plants, both fission and fusion designs)
- heat and cold effects on aquatic organisms (effluent of commercial power plants)
- combined (synergistic) effects of heat, chemicals, and radioactivity (effluent of commercial power plants)
- effect of air supersaturation in river water (hydroelectric generation)
- effect of water-level fluctuations, including dewatering of intergravel salmon redds (hydroelectric generation)
- environmental characterization of Hanford Reach (Hanford Site operational monitoring).

Results from thermal effects studies at Hanford in the 1960s (Chapter 10), although illuminating, could be extrapolated to other power plant sites only by considering the uniqueness of each facility and the complexity of the ecosystem receiving waste heat. Thermal studies conducted at Hanford after 1971 were planned on the basis of their potential application to other power plant sites (generic application) under the NEPA of 1969.

Information from this research effort helped clarify the likelihood of ecological impact on aquatic habitats from use and development of energy resources in the United States.

References

Abernethy, C.S., and D.G. Watson. 1974. "Effect of Temperature and Acute External Radiation on Trout." In: *Pacific Northwest Laboratory Annual Report for 1974 to the USAEC Division of Biomedical and Environmental Research, Part 2 Ecological Sciences*, pp. 90–91. BNWL-1950 PT2, Battelle, Pacific Northwest Laboratories, Richland, Washington.

Abernethy, C.S., and D.G. Watson. 1976. "Effect of Temperature and Acute Irradiation on Trout." In: *Pacific Northwest Laboratory Annual Report for 1975 to the USERDA Division of Biomedical and Environmental Research, Part 2 Ecological Sciences*, pp. 81–86. BNWL-2000 PT2, Battelle, Pacific Northwest Laboratories, Richland, Washington.

Anderson, D.R. 1981. *The Combined Effects of Nickel, Chlorine, and Temperature on the Mortality of Rainbow Trout, Salmo gairdneri*. PhD Dissertation, University of Washington, Seattle, Washington.

Anderson, D.R. 1983. "Chlorine-Heavy Metals Interaction on Toxicity and Metal Accumulation." In: *Water Chlorination. Environmental Impact and Health Effects, Vol. 4, Environment, Health, and Risk*, eds. R.L. Jolley, W.A. Brungs, et al. pp. 811–826. Ann Arbor Science, Ann Arbor, Michigan.

Anderson, D.R., and M.J. Schneider. 1979. "Ecological Effects of Combined Aquatic Stressors." In: *Pacific Northwest Laboratory Annual Report for 1978 to the Department of Energy Assistant Secretary for Environment, Part 2 Ecological Sciences*, pp. 9.1–9.9. PNL-2850 PT2, Pacific Northwest Laboratory, Richland, Washington.

Becker, C.D. 1972. Thermal Resistance of Aquatic Invertebrates." In: *Pacific Northwest Laboratory Annual Report for 1971 to the USAEC Division of Biology and Medicine, Vol. I Life Sciences*, Part 2, Ecological Sciences, pp. 1.1–1.4. BNWL-1650 PT2, Battelle, Pacific Northwest Laboratories, Richland, Washington.

Becker, C.D., and M.P. Fujihara. 1978. *The Bacterial Pathogen Flexibacter columnaris and Its Epizootiology Among Columbia River Fish. A Review and Synthesis*. Monograph No. 2, American Fisheries Society, Washington, D.C.

Becker, C.D., and R.G. Genoway. 1974. "Resistance of Crayfish to Acute Thermal Shock: Preliminary Studies." In: *Thermal Ecology*, eds. J. W. Gibbons and R. R. Sharitz, pp. 141–150. CONF-730505, Technical Information Center, U.S. Atomic Energy Commission, Washington, D.C.

Becker, C.D., and R.G. Genoway. 1979. "Evaluation of the Critical Thermal Maximum for Determining Thermal Tolerance of Freshwater Fish." *Environ. Biol. Fish.* 4:241–256.

Becker, C.D., and D.A. Neitzel. 1985. "Assessment of Intergravel Conditions Influencing Egg and Alevin Survival During Salmonid Redd Dewatering." *Environ. Biol. Fish.* 12:31–46.

Becker, C.D., and M.G. Wolford. 1980. "Thermal Resistance of Juvenile Salmonids Sublethally Exposed to Nickel, Determined by the Critical Thermal Maximum Method." *Environ. Pollut.* (Series A) 21:181–189.

Becker, C.D., R.G. Genoway, and J. A. Merrill. 1975. "Resistance of a Northwestern Crayfish, *Pacifasticus leniusculus* (Dana), to Elevated Temperatures." *Trans. Am. Fish. Soc.* 104:371–387.

Becker, C.D., R.G. Genoway, and M.J. Schneider. 1977a. "Comparative Cold Resistance of Three Columbia River Organisms." *Trans. Am. Fish. Soc.* 106:171–184.

Becker, C.D., D.S. Trent, and M .J. Schneider. 1977b. *Predicting Effects of Cold Shock: Modeling the Decline of a Thermal Plume*. PNL-2411, Pacific Northwest Laboratory, Richland, Washington.

Becker, C.D., D.H. Fickeisen, and J. C. Montgomery. 1981. *Assessment of Impacts from Water Level Fluctuations on Fish in the Hanford Reach, Columbia River*. PNL-3813, Pacific Northwest Laboratory, Richland, Washington.

Becker, C.D., D.A. Neitzel, and D.H. Fickeisen. 1982. "Effects of Dewatering on Chinook Salmon Redds: Tolerance of Four Developmental Phases to Daily Dewaterings." *Trans. Am. Fish. Soc.* 111:621–637.

Becker, C.D., D.H. Fickeisen, and D .A. Neitzel. 1983a. "Some Effects of Power Peaking on Fish in the Hanford Reach of the Columbia River." In: *Western Proceedings, 62nd Annual Conference of the Western Association of Fish and Wildlife Agencies*, pp. 501–507. Las Vegas, Nevada (1982).

Becker, C.D., D.A. Neitzel, and C.S. Abernethy. 1983b. "Effects of Dewatering on Chinook Salmon Redds: Tolerance of Four Developmental Phases to One-Time Dewatering." *J. North Am. Fish. Manage.* 3:371–382.

Becker, C.D., D.A. Neitzel, and D. W. Carlile. 1986. "Survival Data for Dewatered Rainbow Trout (*Salmo gairdneri* Rich.) Eggs and Alevins." *J. Appl. Ichthyol.* 3:1–2-110.

Block, R.M. 1974. "Effects of Acute Cold Shock on the Channel Catfish." In: *Thermal Ecology*, eds. J. W. Gibbons and R. R. Sharitz, pp. 109–118. CONF-730505, Technical Information Center, U.S. Atomic Energy Commission, Washington, D.C.

Bramson, P.A., and J.P. Corley. 1972. *Environmental Surveillance at Hanford for CY-1971*. BNWL-1683, Battelle, Pacific Northwest Laboratories, Richland, Washington

Connors, T.J., M.J. Schneider, R.G. Genoway, and S. A. Barraclough. 1978. "Effect of Acclimation Temperature on Plasma Levels of Glucose and Lactate in Rainbow Trout, *Salmo gairdneri." J. Exp. Zool.* 206:441–449.

Crass, D.W., and R.H. Gray. 1982. "Snout Dimorphism in White Sturgeon, *Acipenser transmontanus*, from the Columbia River at Hanford, Washington." *Fish. Bull.* 80:151–160.

Cushing, C.E., D.G. Watson, D. E. Robertson, and W. B. Silker. 1974. "Decline of Radioactivity in the Columbia River – McNary Reservoir Ecosystems Following Shutdown of Hanford Reactors." In: *Pacific Northwest Laboratory Annual Report for 1973 to the USAEC Division of Biomedical and Environmental Research, Part 2 Ecological Sciences*, pp. 81–89. BNWL-1850 PT2, Battelle, Pacific Northwest Laboratories, Richland, Washington.

Cushing, C.E., J.M. Thomas, and L. L. Eberhardt. 1975. "Modeling Mineral Cycling by Periphyton in a Simulated Stream System." *Verh. Internat. Limnol.* Bd. 19:1591–1598.

Cushing, C.E., D.G. Watson, A.J. Scott, and J.M. Gurtisen. 1980. *Decline of Radionuclides in Columbia River Biota*. PNL-3269, Pacific Northwest Laboratory, Richland, Washington.

Cushing, C.E., D.G. Watson, A.J. Scott, and J.M. Gurtisen. 1981. "Decrease of Radionuclides in Columbia River Biota Following Closure of Hanford Reactors." *Health Phys.* 7(41):51–67.

Duever, M.J., and C.S. Abernethy. 1974. "Synergistic Effects of Temperature and Acute Radiation. In: *Pacific Northwest Laboratory Annual Report for 1973 to the USAEC Division of Biomedical and Environmental Research, Part 2 Ecological Sciences*, pp. 71–78. BNWL-1850 PT2, Battelle, Pacific Northwest Laboratories, Richland, Washington.

Emery, R.M., D.C. Klopfer, and J. R. Skalski. 1981. *The Incipient Toxicity of Lithium to Freshwater Organisms Representing a Salmonid Habitat*. PNL-3640, Pacific Northwest Laboratory, Richland, Washington.

Fickeisen, D.H. 1974a. "Dissolved Gas Studies." In: *Pacific Northwest Laboratory Annual Report for 1973 to the USAEC Division of Biomedical and Environmental Research, Part 2, Ecological Sciences*, BNWL-1850 PT2, Battelle, Pacific Northwest Laboratories, Richland, Washington.

Fickeisen, D.H. 1974b. "Gas Bubble Disease." In: *Pacific Northwest Laboratory Annual Report for 1974 to the USAEC Division of Biomedical and Environmental Research, Part 2 Ecological Sciences*, pp. 61–69. BNWL-1950 PT2, Battelle, Pacific Northwest Laboratories, Richland, Washington.

Fickeisen, D.H., and J.C. Montgomery. 1977. "Effects of Hydroelectric Generation on Riverine Ecology." In: *Pacific Northwest Laboratory Annual Report for 1976 to the ERDA Assistant Administrator for Environment and Safety, Part 2 Ecological Sciences*, pp. 6.1–6.4. BNWL-2100 PT2, Battelle, Pacific Northwest Laboratories, Richland, Washington.

Fickeisen, D.H., and J.C. Montgomery. 1978. "Tolerance of Fishes to Dissolved Gas Supersaturation in Deep Tank Bioassays." *Trans. Am. Fish.* Soc. 107:371–381.

Fickeisen, D.H., and M.J. Schneider. 1973. "Dissolved Gas Supersaturation." In: *Pacific Northwest Laboratory Annual Report for 1972 to the USAEC Division of Biomedical and Environmental Research, Vol. I Life Sciences, Part 2 Ecological Sciences*, pp. 6.21–6.23. BNWL-1750 PT2, Pacific Northwest Laboratories, Richland, Washington.

Fickeisen, D.H., and M.J. Schneider, eds. 1976. *Gas Bubble Disease.* CONF-741033, National Technical Information Service, Springfield, Virginia.

Fickeisen, D.H., C.D. Becker, and D.A. Neitzel. 1981. *Review of Pacific Northwest Laboratory Research on Aquatic Effects of Hydroelectric Generation and Assessment of Research Needs*. PNL-3816, Pacific Northwest Laboratory, Richland, Washington.

Fickeisen, D.H., J.C. Montgomery, and R.W. Hanf, Jr. 1976. "Effect of Temperature on Tolerance to Dissolved Gas Supersaturation of Black Bullhead, *Ictalurus melas*." In *Gas Bubble Disease*, eds. D.H. Fickeisen and M.J. Schneider, pp. 71–74. CONF-741033, National Technical Information Service, Springfield, Virginia.

Fickeisen, D.H., J.C. Montgomery, and M.J. Schneider. 1975. "A Comparative Evaluation of the Weiss Saturometer." *Trans. Am. Fish. Soc.* 104:816–820.

Garland, T.R., R.E. Wildung, A.J. Scott, and A.V. Robinson. 1974. "Distribution of Mercury in Water, Suspended Matter, and Sediments of the Lower Columbia River Watershed." In: *Pacific Northwest Laboratory Annual Report for 1974 to the USAEC Division of Biomedical and Environmental Research, Part 2 Ecological Sciences*, pp. 41–49. BNWL-1950 PT2, Battelle, Pacific Northwest Laboratories, Richland, Washington.

Gray, R.H. 1977. "Avoidance of Thermal Effluent by Juvenile Chinook Salmon (*Oncorhynchus tshawytscha*) and Its Implications in Waste Heat Management." In: *Proceedings of the Conference on Waste Heat Management and Utilization*, Miami Beach, Florida.

Gray, R.H., and J.M. Haynes. 1976. "Upstream Movement of Adult Salmonids in Relation to Gas Supersaturated Water." In: *Pacific Northwest Laboratory Annual Report for 1975 to the USERDA Division of Biomedical and Environmental Research, Part 2 Ecological Sciences*, pp. 71–76. BNWL-2000 PT2, Pacific Northwest Laboratory, Richland, Washington.

Gray, R.H., and J.M. Haynes. 1977. "Depth Distribution of Adult Chinook Salmon (*Oncorhynchus tshawytscha*) in Relation to Season and Gas-Supersaturated Water." *Trans. Am. Fish. Soc.* 106:611–620.

Gray, R.H., R.G. Genoway, and S.A. Barraclough. 1977. "Behavior of Juvenile Chinook Salmon (*Oncorhynchus tshawytscha*) in Relation to Simulated Thermal Effluent." *Trans. Am. Fish. Soc.* 106:361–370.

Gray, R.H., and J.M. Haynes. 1979. "Spawning Migration of Adult Chinook Salmon (*Oncorhynchus tshawytscha*) Carrying External and Internal Transmitters." *J. Fish. Res. Board Can.* 36:1060–1064.

Haynes, J.M. 1978. *Movement and Habitat Studies of Chinook Salmon and White Sturgeon*. PNL-2471, Pacific Northwest Laboratory, Richland, Washington.

Haynes, J.M., and R.H. Gray. 1980. "Influence of Little Goose Dam on Upstream Movements of Adult Chinook Salmon, *Oncorhynchus tshawytscha*." *Fish. Bull.* 78:181–190.

Haynes, J.M., R.H. Gray, and J.C. Montgomery. 1978. "Seasonal Movements of White Sturgeon (*Acipenser transmontanus*) in the Mid-Columbia River." *Trans. Am. Fish. Soc.* 107:271–280.

Montgomery, J.C., and C.D. Becker. 1980. "Gas Bubble Disease in Smallmouth Bass and Northern Squawfish from the Snake and Columbia Rivers." *Trans. Am. Fish. Soc.* 109:731–736.

Montgomery, J.C., and D.H. Fickeisen. 1978. *Spawning and Movements of Smallmouth Bass (Micropterus dolomieui) in the Mid-Columbia River. PNL-2785, Pacific Northwest Laboratory, Richland, Washington.*

Montgomery, J.C., D.H. Fickeisen, and C.D. Becker. 1980. "Factors Influencing Smallmouth Bass Production in the Hanford Area, Columbia River." Northwest Sci. 54:291–305.

Neitzel, D.A., and C.D. Becker. 1985. "Tolerance of Eggs, Embryos, and Alevins of Chinook Salmon to Temperature Changes and Reduced Humidity in Dewatered Redds." *Trans. Am. Fish. Soc.* 114:261–273.

Neitzel, D.A., C.D. Becker, and D.H. Fickeisen. 1982. "System for Monitoring of Changes in Aquatic Habitat Resulting from Water-Level Fluctuations." In: *Acquisition and Utilization of Aquatic Habitat Inventory Information*, pp. 131–137. Western Division of the American Fisheries Society, Bethesda, Maryland.

Pauley, G.B., and R.E. Nakatani. 1967. "Histopathology of "Gas-Bubble" Disease in Salmon Fingerlings." *J. Fish. Res. Board Can.* 24:861–871.

Price, K.R., J.M.V. Carlile, R.L. Dirkes, R.E. Jaquish, M.S. Trevathan, and R.K. Woodruff. 1985. *Environmental Monitoring at Hanford for 1984*. PNL-5407, Pacific Northwest Laboratory, Richland, Washington.

Rickard, W.H., W.C. Hanson, and R.E. Fitzner. 1982. "The Non-Fisheries Biological Resources of the Hanford Reach of the Columbia River." *Northwest Sci.* 56:61–76.

Robertson, D.E., W.B. Silker, J.C. Langford, M.R. Petersen, and R.W. Perkins. 1973. "Transport and Depletion of Radionuclides in the Columbia River." In: *Radioactive Contamination in the Marine Environment*, pp. 141–157. IAEA-SM-158/9, International Atomic Energy Agency, Vienna, Austria.

Schneider, M.J. and T.J. Connors. 1982. "Effects of Elevated Water Temperature on Critical Swim Speeds of Yearling Rainbow Trout, *Salmo gairdneri*." *J. Therm. Biol.* 7:221–229.

Schneider, M.J., R.G. Genoway, and S.A. Barraclough. 1974. "Preliminary Studies on the Effect of Fatigue on Thermal Tolerance of Rainbow Trout." In: *Thermal Ecology*, eds. J. W. Gibbons and R. R. Sharitz, pp. 171–185. CONF-730505, Technical Information Center, U.S. Atomic Energy Commission, Washington, D.C.

Schneider, M.J., C.D. Becker, D.H. Fickeisen, T.O. Thatcher, E.G. Wolf, and E. L. Hunt. 1975. "Aquatic Physiology of Thermal and Chemical Discharges." In: *Environmental Effects of Cooling Systems at Nuclear Power Plants*, pp. 541–561. IAEA-SM-187/15, International Atomic Energy Agency, Vienna, Austria.

Schneider, M.J., R.G. Genoway, T.J. Connors, and S.A. Barraclough. 1977. "Tank Trap for Nontraumatic Serial Sampling of Fish Stocks in Psychological Studies." *Prog. Fish-Cult.* 39:131–137.

Schneider, M.J., T.J. Connors, R.G. Genoway, and S.A. Barraclough. 1981. "Some Effects of Simultaneous Thermal and Exercise Stress in Rainbow Trout (*Salmo gairdneri*)." *Northwest Sci.* 55:61–69.

Sneider, S.C., and R.L. Skaggs. 1983. *Unsteady Flow Model of Priest Rapids Dam Releases at Hanford Reach, Columbia River, Washington*. PNL-4527, Pacific Northwest Laboratory, Richland, Washington.

Strand, J.A. 1973. *Suppression of Primary Immune Response in Rainbow Trout, Salmo gairdneri, Sublethally Exposed to Tritiated Water During Embryogenesis*. PhD Dissertation, University of Washington, Seattle, Washington.

Strand, J.A., M.P. Fujihara, W.L. Templeton, and R.G. Genoway. 1972. "Effect of Short-Range Particle Irradiation of Embryogenesis of Teleost Fish." In: *Pacific Northwest Laboratory Annual Report for 1971 to the USAEC Division of Biology and Medicine, Vol. I Life Sciences, Part 2 Ecological Sciences*, pp. 1.1–1.13. BNWL-1650 PT2, Battelle, Pacific Northwest Laboratories, Richland, Washington.

Strand, J.A., W.L. Templeton, and E.G. Tangen. 1973a. "Accumulation and Retention of Tritium (Tritiated Water) in Embryonic and Larval Fish, and Radiation Effect." In: *Radionuclides in Ecosystems, Third National Symposium on Radioecology*, ed. D.J. Nelson, pp. 441–451. CONF-710501-P1, National Technical Information Service, Springfield, Virginia.

Strand, J.A., M.P. Fujihara, W.L. Templeton, and E.G. Tangen. 1973b. Suppression of *Chondroccocus columnaris* Immune Response in Rainbow Trout Sub-Lethally Exposed to Tritiated Water During Embryogenesis." In: *Radioactive Contamination of the Marine Environment*, pp. 541–549. International Atomic Energy Agency, Vienna, Austria.

Strand, J.A., W.L. Templeton, and P.A. Olson. 1976. "Fixation and Long-Term Accumulation of Tritium from Tritiated Water in an Experimental Aquatic Environment" In: *Proceedings of the International Conference on Radiation Effects and Tritium Technology for Fusion Reactors*, pp. 71–95. CONF-7509089, Gatlinburg, Tennessee.

Strand, J.A., M.P. Fujihara, R.D. Burdett, and T.M. Poston. 1977. "Suppression of the Primary Immune Response in Rainbow Trout, *Salmo gairdneri*, Sublethally Exposed to Tritiated Water During Embryogenesis." *J. Fish. Res. Board Can.* 34:1291–1304.

Strand, J.A., M.P. Fujihara, T.M. Poston, and C.S. Abernethy. 1982. "Permanence of Suppression of the Primary Immune Response in Rainbow Trout, *Salmo gairdneri*, Sublethally Exposed to Tritiated Water During Embryogenesis." *Radiat. Res.* 91:531–541.

Sula, M.J., W.D. McCormack, R.L. Dirkes, K.R. Price, and P.A. Eddy. 1982. *Environmental Surveillance at Hanford for CY-1981*. PNL-4211, Pacific Northwest Laboratory, Richland, Washington.

Thatcher, T.O. 1974. "Combined Effects of Mercury and Temperature on the Mortality of Rainbow Trout." In: *Thermal Ecology*, eds. J. W. Gibbons and R. R. Schwartz, pp. 51–58. CONF-730505, Technical Information Center, U.S. Atomic Energy Commission, Washington, D.C.

Thatcher, T.O., M.J. Schneider, and E.G. Wolf. 1976. "Bioassays on the Combined Effects of Chlorine, Heavy Metals and Temperature on Fishes and Fish Food Organisms Part I. Effects of Chlorine and Temperature on Juvenile Brook Trout (*Salvelinus fontinalis*)." *Bull. Environ. Contam. Toxicol.* 15:40–48.

Thomas, J.M., C.E. Cushing, and L.L. Eberhardt. 1974. "A Conceptual Model of Radionuclide Transfer in Columbia River Biota." In: *Pacific Northwest Laboratory Annual Report for 1973 to the USAEC Division of Biomedical and Environmental Research, Part 2 Ecological Sciences* ll, pp. 81–91. BNWL-1850 PT2, Battelle, Pacific Northwest Laboratories, Richland, Washington.

Vaughan, B.E., and W.H. Rickard. 1977. *Hanford National Environmental Research Park (NERP). A Descriptive Summary of the Site and Site-Related Research Programs, 1951–1977*. PNL-2299, Pacific Northwest Laboratory, Richland, Washington.

Watson, D.G. 1972. "Uptake of Mercury-203 by Fish." In: *Pacific Northwest Laboratory Annual Report for 1971 to the USAEC Division of Biomedical and Environmental Research, Vol. I Life Sciences, Part 2 Ecological Sciences*, pp. 1.11–1.17. BNWL-1650 PT2, Battelle, Pacific Northwest Laboratories, Richland, Washington.

Wildung, R.E., T.R. Garland, and R.L. Schmidt. 1973. "Mercury Levels in Particulate Matter Suspended in Waters of the Lower Columbia River Watershed." In: *Pacific Northwest Laboratory Annual Report for 1972 to the USAEC Division of Biomedical and Environmental Research, Vol. 1 Life Sciences, Part 2 Ecological Sciences*, pp. 5.1–5.5. BNWL-1750 PT2, Pacific Northwest Laboratory, Richland, Washington.

Wildung, R.E., R.L. Schmidt, D.G. Watson, and W.L. Templeton. 1972. "Mercury Levels in the Lower Columbia River Watershed." In: *Pacific Northwest Laboratory Annual Report for 1971 to the USAEC Division of Biomedical and Environmental Research, Vol. I Life Sciences, Part 2 Ecological Sciences*, pp. 1.17–1.18. BNWL-1650 PT2, Battelle, Pacific Northwest Laboratories, Richland, Washington.

Wolf, E.G. 1976. "Combined Effects of Waste Heat and Chlorine on Juvenile Rainbow and Eastern Brook Trout." In: *Pacific Northwest Laboratory Annual Report for 1975 to the ERDA Division of Biomedical and Environmental Research, Part 2 Ecological Sciences*, pp. 61–68. BNWL-2000 PT2, Battelle, Pacific Northwest Laboratories, Richland, Washington.

Chapter 12

FACILITY-SPECIFIC STUDIES IN HANFORD REACH AFTER CLOSURE OF SINGLE-PURPOSE REACTORS, 1971 TO 1984

In the 1970s and early 1980s, three energy-producing facilities on the Hanford Site used Columbia River water for cooling: Hanford Generating Project (HGP), N Reactor, and Washington Public Power Supply Nuclear Project No. 2 (WNP-2).

HGP and WNP-2, operated by the Washington Public Power Supply System (Supply System), produced electrical power. N Reactor, operated by private contractors for the U.S. Department of Energy (DOE), produced special nuclear materials for national defense. N Reactor also provided the steam that ran HGP. Together, N Reactor and HGP formed a unique, dual-purpose unit.

Aquatic specialists at Battelle conducted studies in the Hanford Reach to support the start up and operation of these facilities. Some preoperational studies were also conducted in support of WNP-1 and -4, two commercial nuclear plants that were never completed. For the most part, the studies were site-specific. They gathered ecological baseline data and examined potential operational impacts under requirements of the National Environmental Policy Act (NEPA) of 1969, the Clean Water Act of 1977 (CWA), other applicable federal and state regulations, and siting agreements with the state of Washington. [a]

In most cases, potential impacts on aquatic biota by energy-producing facilities are related to the intake and discharge of cooling water. Phyto-

[a] During the 1970s, river flow regulation became very important to offsite management agencies to integrate hydropower production, irrigation withdrawal, and downstream passage of juvenile salmonids. River flows in the Hanford Reach were altered as computerized methods were developed to closely coordinate releases at upriver dams. Because any change in flow regulation affects facilities at Hanford that use river water for cooling, the existing data base on flows and temperatures in the Hanford Reach was updated (Whelan and Newbill 1983).

plankton, zooplankton, and larval fish may be withdrawn (entrained) in water pumped into the cooling system intakes. Entrained organisms experienced considerable stress from, first, injury and rapid heating in the condenser system and, second, from rapid cooling at re-entry to the river. Larger fish can be stuck (impinged) against the intake screens by velocity of the incoming water. In the river proper, young fish and small organisms carried downstream in the current can enter the discharge plume (entrained) and experience thermal or chemical shock. Further, the life cycles of benthic organisms downstream of discharge zones can be modified by warmer temperatures. Evaluation of potential impacts in the Hanford Reach focused on these site-specific unknowns.

Four Battelle staff members were involved in preoperational and operational assessment studies in the Hanford Reach through the 1970s. They were Dennis D. Dauble, Robert H. Gray, Duane A. Neitzel, and Thomas L. Page. Periodic contributors included C. Scott Abernethy, Robert W. Hanf, Jr., Donald C. Klopfer, E. William Lusty, Mark J. Schneider, and Alan J. Scott. Ecological data to support HGP studies were collected from March 1973 to June 1974. Effort shifted to WNP-2 in September 1974 and continued through 1984. This work was conducted by Battelle from 1974 to 1978; Beak Consultants, Inc., from 1979 to 1980; and the Supply System after 1980.

Hanford Generating Project

The Hanford Generating Project (HGP), a 860-megawatt-electric steam-electric plant (enough power for a city with a population of 500,000), is sited in the 100 N Area at RKm 611. HGP is just upriver from N Reactor, which provides low-pressure steam to operate HGP's power generators through large, aboveground pipes. River water is used at HGP only as on-line cooling to condense the steam after part of its heat has been extracted for power generation. Cooling water is returned to the Hanford Reach immediately downstream of HGP.

HGP was initially called "Hanford No. 1." Construction began in September 1963. The plant started up for testing in April 1966, and produced electricity in November 1966. This event was important for two reasons. First, HGP initiated a transition from all-hydro to hydro-thermal electrical power production in the Pacific Northwest. Second, HGP extended the use of N Reactor by capturing the steam created during plutonium production to generate electrical power, making N Reactor the first dual-purpose reactor in the United States.

HGP generated electricity only when N Reactor operates in a dual-purpose mode (Supply System 1977a). Because only expended steam was cooled, its cooling water effluent contained less heat than the effluent of most power-generating reactors in the United States. Furthermore, the temperature of the effluent was much lower than that of the effluent discharged by the eight single-purpose production reactors at Hanford from 1943 to 1971.

The HGP Cooling System – Intake and Discharge

The cooling water system at HGP consists of a shoreline intake to withdraw raw river water, condensers to cool steam from N Reactor, and a conduit to return heated effluent to the Hanford Reach (Figure 12.1).

The intake is barred by trash racks consisting of metal bars spaced 8.3 centimeters (3.25 inches) apart. A curtain wall extends within 0.9 to 1.5 meters (3 to 5 feet) of the water level behind the trash racks. Behind the wall are two pump bays, each faced with three rows of vertical traveling

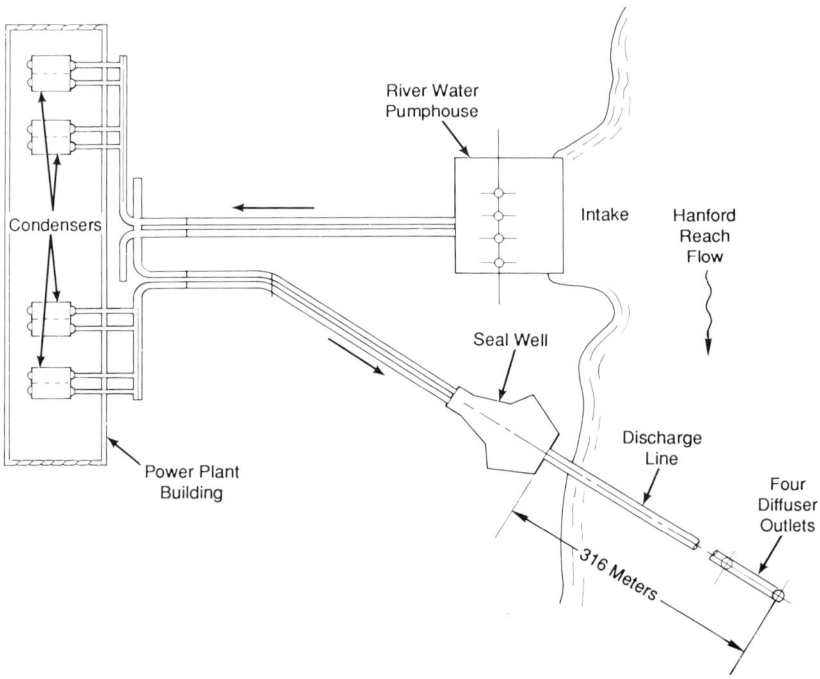

Fig. 12.1. The cooling water system at the Hanford Generating Plant (HGP). Main design features include an inshoreline intake, pumphouse, condenser system, and effluent discharge line to Hanford Reach (from RKE 1982).

Fig. 12.2. The cooling water intake of HGP. Juvenile fish impinged on the traveling screens are carried to the top, washed off, and returned to the Hanford Reach below the intake through a 16-inch diameter pipe (from Gray et al. 1979).

screens. Four cooling-water pumps are installed behind the screens, two in each pump bay. When all four pumps are working, total pumping capacity is 35.6 cubic meters per second (m^3/s) [(1257 cubic feet per second (ft^3/s)]. Only three pumps with a capacity of 26.7 m^3/s (942 ft^3/s) are used when river temperatures fall below 7.2°C (Gray et al. 1975; Page et al. 1978).

The traveling screens consist of connected, horizontal panels with small mesh openings. Fish and small debris were caught against the screens, carried by rotation to the top of the structure, and then washed off with a spray. Until a few years ago, the screens were rotated and washed three times a day (once per shift) with a high-pressure spray, and all washed material was passed to a sump. Today, the screens are rotated from March to June, and impinged materials (including fish) are washed from the screens with a low-pressure spray and returned to the Hanford Reach via a large pipe (Figure 12.2).

At HGP's intake, the flow through the screen mesh is about 0.72 meter per second (2.36 feet per second) when all four pumps are working and the river is at minimum low flow. This calculated impingement velocity is similar to that at the intake of N Reactor (Page et al. 1978). The intake structure of HGP (and of N Reactor) has fish escape ports in its exterior, downstream wall.

The heat load from HGP to the Hanford Reach is about 2700 megawatts-thermal. Heated effluent leaving the condenser first enters a seal well on the river bank, then flows through a 3.35-meter (11-foot) -diameter pipe for 319 meters (1050 feet) back to the Hanford Reach. The effluent emerges through four diffuser ports near midriver about 183 meters (600 feet) upstream of the N Reactor discharge.

From 1966 to 1983, the outlets in HGP's discharge line consisted of four low-velocity capped diffusers. Forty to 60% of the effluent emerged from the last port. The discharge line contained six vents for air relief within 61 meters (200 feet) of the shoreline. Each diffuser opened upward

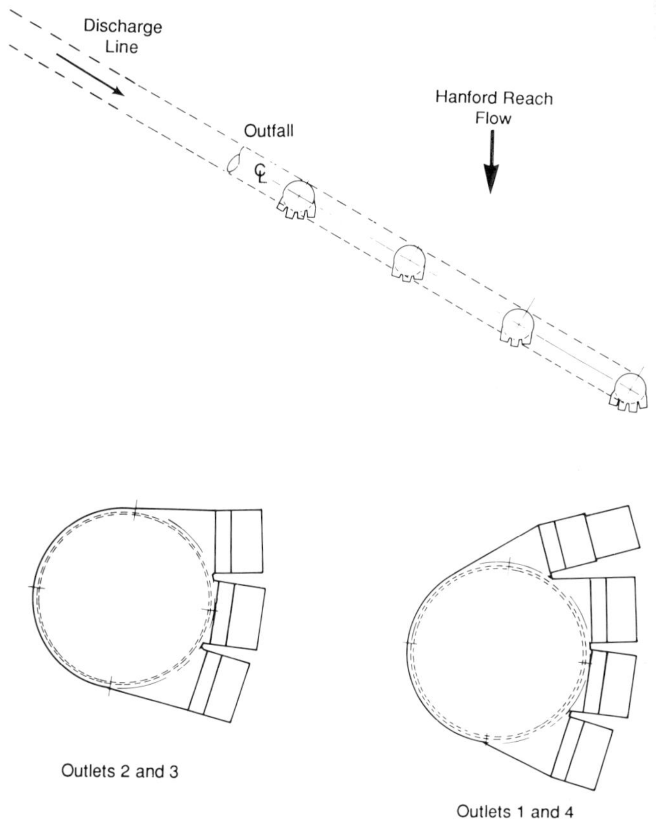

Fig. 12.3. Four multiport diffusers were installed on the effluent discharge line of HGP in 1983 to enhance mixing near the release points. Outlets 1 and 4 have four nozzles facing downstream; outlets 2 and 3 have three nozzles.

and had a stainless steel cap that diverted the effluent horizontally into river flow. In the summer of 1983, the existing capped diffusers were modified into multiport diffusers to enhance mixing in the "near field." The redesigned diffusers (Figure 12.3) consisted of three to four nozzles directed 5 degrees below horizontal and sized to an average discharge velocity of 4.9 meters per second (16 feet per second) (RKE 1982; Singarella and Brocard 1982).

Ecological Baseline Data

Field studies at HGP were initiated from March 1973 to June 1974. Objectives were to characterize aquatic biota near the plant, and to assess interactions between the plant's cooling water system and biota in the Hanford Reach.

Baseline information, obtained seasonally, included 1) abundance and primary productivity of phytoplankton, 2) abundance of zooplankton, 3) abundance of benthic organisms, and 4) abundance of different species of fish, adult and young.

Assuming an average flow of 3398 m^3/s (120,000 ft^3/s) in the Hanford Reach and pumping at the maximum volume of about 36 m^3/s (560,000 gallons per minute) the entire year, less than 1% of the river's flow would enter HGP's condensers. During periods of minimum regulated flows of 1,019 m^3/s (36,000 gallons per minute), a condition rare in the Hanford Reach, about 3.4% of the river volume would enter HGP's condensers. Under these scenarios, only 1% of the phytoplankton and zooplankton drifting downriver past HGP each year might be entrained, on the average, which increases to 3.4% maximum on rare intervals. Thus, any effect of HGP's discharge on planktonic organisms was judged to be insignificant. Few ichthyoplankton were entrained because fish larvae were uncommon in the river drift (Battelle 1976). Confirmation that drift organisms in the Hanford Reach included few fish larvae was obtained in repeated field studies.

Entrainment and Impingement at Cooling Water Intake

Screen passage (entrainment) and impingement of fish was studied at the cooling water intake of HGP from March 1973 through April 1976. Ways to reduce mortalities by preventing entrainment and mitigating impingement were evaluated, implemented, and found successful.

Initial efforts in 1973 and 1974 showed that zero-age fry of fall chinook salmon were more susceptible at the intake of HGP than any other species or life stage of fish in the Hanford Reach. Affected fry were small, measuring less than 50 millimeters (2 inches) fork length. Losses occurred only during a brief period each spring, in late March, April, and May, soon after fry emerged from streambed gravel. Furthermore, only fry emerging upstream between HGP and Priest Rapids Dam were impacted. Although outmigration of juvenile salmonids through the Hanford Reach continued through June, larger fish were not vulnerable. Far fewer numbers of other species of small fish were impacted than chinook salmon fry (Gray et al. 1975; Page et al. 1975).

Initially, more chinook salmon fry passed through or around the traveling screens than were impinged. Only the smallest fry could pass through the 0.64-centimeter (0.25-inch) mesh used in HGP's screens. Therefore, the other fry apparently passed through openings between the screen panels and at the sides and between screen panels (Gray et al. 1975; Page et al. 1976, 1977a). To reduce screen passage, which allowed fry to enter the pump bay and be entrained, the major suspected entry

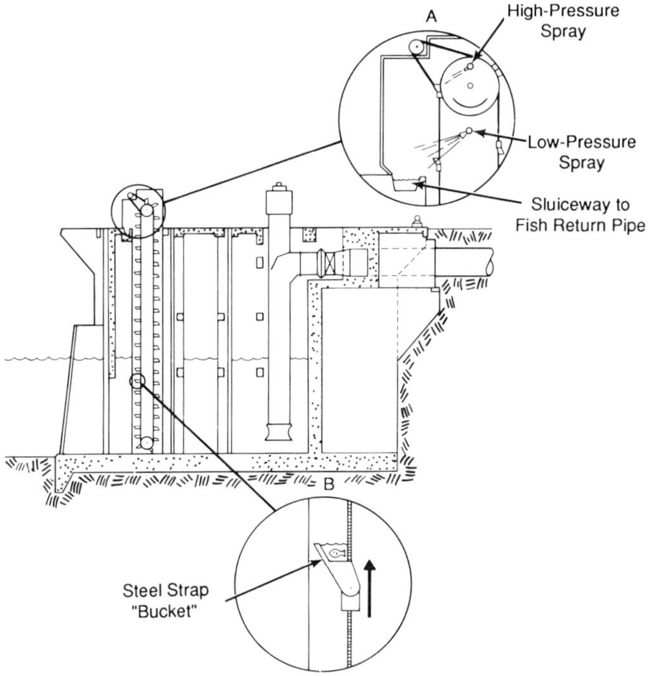

Fig. 12.4. Cooling water intake at HGP after modification to eliminate loss of juvenile fish from impingement in 1976. Steel buckets added to the traveling screens collected impinged fish and carried them up to the spray removal system (insert B). A low-pressure spray removed fish, then a high-pressure spray removed debris (insert A).

points were sealed. Furthermore, the screens were replaced by screens with 0.318-centimeter (0.125-inch) mesh. As a result, chinook salmon fry no longer bypassed the traveling screens to be entrained through the cooling system (Page et al. 1978; Gray et al. 1979).

At this point, impingement replaced entrainment as the major problem at HGP's intake. A solution was sought by changing the structure and operation of the traveling screens in late 1975 and 1976. First, steel straps were installed at the bottom of each panel that made up the traveling screens (Figure 12.4). Small impinged fish fell into the filled "buckets" and were lifted as the screens rotated. At the top, they were dumped and washed into a trough. Second, a low-pressure spray (15 psi) was installed below the high-pressure, trash-removal spray (90 psi) to wash fry from the screens (Figure 12.4). They were then flushed through a sluiceway and to the river below the intake through a 40.6-centimeter (16-inch) -diameter pipe. Third, the screens were continuously rotated and washed to prevent impinged fish from suffocating while out of water. Last, experimental studies confirmed the safe return of fry from screens back to the river via the sluiceway and pipe (Page et al. 1978; Gray et al. 1979).

This effort increased survival of impinged chinook salmon fry at HGP's intake to more than 93%. All other species of small fish from the Hanford Reach subject to screen impingement benefited.

Features of Discharge Plume

At the site of HGP and N Reactor, the Columbia River varies from about 421 to 488 meters (1400 to 1600 feet) wide and 7 to 11 meters (25 to 35 feet) deep, depending on river flow. Surface velocities range from 0.3 to 3.3 meters per second (3 to 11 feet per second) (Ecker et al. 1983a). River flows change daily, weekly, and seasonally from releases at upriver dams.

HGP's contribution of heat to the Hanford Reach is normally 2700 megawatts-thermal. If river temperatures are low, three pumps circulate water at 26.7 m^3/s (423,000 gallons per minute) and discharge it about 23.9°C above ambient. If river temperatures are high, requiring greater cooling capacity, four pumps circulate water at 35.6 m^3/s (564,000 gallons per minute) and discharge it about 19.4°C above ambient. The size of HGP's discharge plume, compared to N Reactor's, is relatively large. However, HGP's plume enters the river farther out than N Reactor's plume, and there is little interaction (Ecker et al. 1983a).

Temperatures below HGP's outfalls were surveyed in October 1978. At this time, the original low-velocity capped diffusers were still in use. Plume distributions were examined at four stages of river flow between

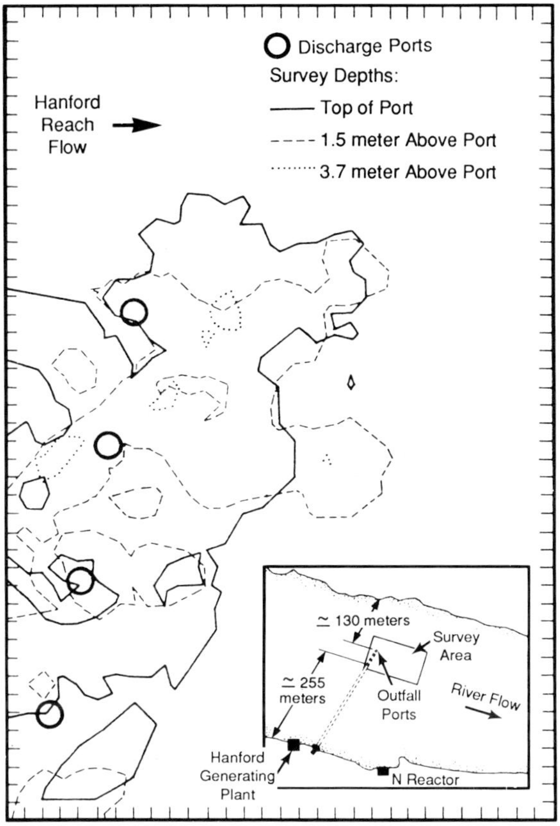

Fig. 12.5. Estimated configuration of the 3°C elevated temperature isotherm at the four discharge ports of HGP during low flows in October 1978. Effluent was emerging from the original capped-port diffusers in use until 1983 (from Neitzel et al. 1982c).

1416 m^3/s and 2974 m^3/s. The plume was dynamic, and surged from the diffusers to mix rapidly with downstream currents. Temperatures more than 3°C above ambient were confined to small areas within 20 to 30 meters (66 to 98 feet) of the diffusers and did not reach the surface (Figure 12.5). No area warmed 1°C (1.8°F) or more above ambient intersected the shoreline or an island. Temperatures 2°C (3.6°F) or more above ambient reached the river's bottom at a point about 80 m downstream before disseminating (Supply System 1978; Neitzel et al. 1982c). Furthermore, when fully mixed downstream, HGP's discharge raised the ambient temperature of the Columbia River less than 0.7°C. Temperatures in the Hanford Reach varied more each day from insolation and other natural phenomena.

The diffuser design was changed at HGP in 1983 as a response to concerns of regulatory agencies about the near-field mixing zone, which might have led to a modified, more restrictive National Pollution Discharge Elimination System (NPDES) permit. The new diffusers were intended to uniformly distribute the heated effluent and increase mixing so that temperatures in the upper two thirds of the river would not exceed 25°C at the infrequent times both the ambient river temperature was 20°C and the river was at the regulated 1019 m^3/s minimum flow (RKE 1982). A resurvey in the summer of 1984 indicated that thermal increments in the near field of HGP's discharge plume were now limited (Larsen 1985) as intended.

Evaluation of Thermal Effects from HGP

HGP must be operated within specific federal and state regulations for release of cooling water effluent to the Hanford Reach. Through the early 1970s, HGP was exempt from backfitting to closed-cycle cooling under the Federal Water Pollution Control Act (FWPCA), as amended by the CWA of 1977, because of its age. However, the thermal effluent of HGP did not always comply with water quality standards set in the Supply System's NPDES permit. To meet these standards, HGP would be required to convert to an alternate cooling system. Plant operators believed that the standards were more stringent than necessary to protect aquatic resources in the Hanford Reach under the intent of law.

In October 1977, the Supply System applied for a NPDES permit for cooling water discharge at HGP. Application required assessment of potential impacts from HGP's heated effluent as specified in Section 316(a) of the amended FWPCA.

In September 1978, HGP operators submitted information to the state's Department of Ecology indicating that the site was a "low potential impact area" for phytoplankton, zooplankton and meroplankton, habitat formers, shellfish and macroinvertebrates, and wildlife. The information was accepted by the department in November 1978. Further assessment focused on eight representative important species (RIS) of fish in the Hanford Reach, including four salmonids and four resident fish. Ecological data related to plume dispersion and possible adverse effects of the intake and discharge were assembled and addressed.

The Section 316(a) Demonstration showed that on-line cooling at HGP was environmentally acceptable. There was little potential for direct or indirect mortality to adult or juvenile salmonids, or to their eggs, from thermal increments of the discharge plume. Sublethal exposures to the dispersing plume would not increase the susceptibility of RIS of fish to predation, gas-bubble disease, or infectious disease. Further, the plume would not affect reproduction, growth, or exclusion of RIS. Once the screens were modified to reduce loss of fish in 1976, operation of the intake would not appreciably alter populations of RIS in the Hanford Reach (Supply System 1978; Neitzel et al. 1982).

N Reactor

The N Reactor was built in the 100 N Area at RKm 611 downriver from HGP (Figure 12.6). The facility was authorized by Congress in 1958 as a "New Production Reactor" to provide special nuclear materials, primarily plutonium, for the nation's defense arsenal. Construction began in May 1959, and the reactor reached full operation in 1964. It was rated to operate at 4000 megawatts-thermal (DOE 1988).

Fig. 12.6. N Reactor, a dual-purpose facility producing special nuclear materials and steam for HGP: (a) cooling water intake, (b) inshore part of effluent discharge line, (c) main reactor building, (d) line transferring steam to HGP.

N Reactor is cooled by purified water under pressure in a closed, primary recirculating system. Heat from fission in the reactor core is transported by the primary coolant to steam generators that provide electricity for in-plant use and steam to operate HGP. Nearly all radioactivity is retained in the primary cooling loop. Large amounts of steam are transferred to HGP through large, above-ground pipes. Raw river water is withdrawn from the Hanford Reach, used to cool excess steam in "dump condensers" at N Reactor, and returned in an on-line, open-cycle system. Water in the secondary cooling system is untreated, except for small amounts withdrawn for in-plant use ($< 0.1\%$ of cooling water discharge).

N Reactor can operate in two modes, production of special nuclear materials (single-purpose mode) or production of nuclear materials plus steam for HGP (dual-purpose mode). Thus, N Reactor can operate when HGP is closed. But when N Reactor is shut down for refueling, maintenance, or safety improvement, HGP must also shut down. Temperatures in the cooling effluent of N Reactor are higher when HGP is shut down, because little thermal energy from the reactor's core is extracted as steam. N Reactor almost always operated in a dual-purpose mode until it was shut down for safety upgrading in January 1987.

N Reactor is exempt from backfitting to closed-cycle cooling under the FWPCA because of its age. Eventually, effects from the discharge of warmed water from N Reactor while in a normal, dual-purpose mode of operation and an experimental, single-purpose mode were assessed in 316(a) Demonstrations (DOE 1982, 1988).

The N Reactor Cooling System – Intake and Discharge

The cooling water system at N Reactor consists of a shoreline intake to withdraw raw river water, piping to pass the water through the steam condenser, and a conduit to carry heated effluent back to the Hanford Reach.

The cooling water intake of N Reactor (Figure 12.7) is similar to the intake at HGP. It is located 276 meters (905 feet) downriver. The intake at N Reactor is also barred by trash racks, behind which is a curtain wall descending 2 meters (6.5 feet) below the water surface at average river flow. Two pump bays, each faced with three rows of vertical traveling screens, are located behind the curtain wall. Each bay contains two large pumps to provide water for the reactor's secondary cooling loop, and one small pump to provide water during emergencies. Capacity of each of the four cooling-water pumps is 6.6 m^3/s (233.9 ft^3/s), and total pumping

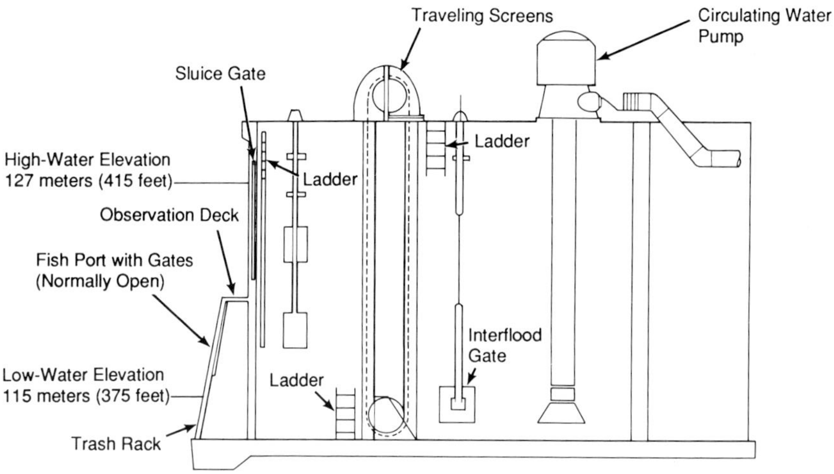

Fig. 12.7. Cross section from upstream side of the cooling water intake at N Reactor. The essential features are similar to those at HGP (see Figure 12.5), located just upriver.

capacity is 26.4 m^3/s (935.6 ft^3/s). Only three cooling-water pumps with combined capacity of 19.8 m^3/s (701.8 ft^3/s) are needed for normal operation.

Calculated impingement velocity, the flow through the screen mesh, is 0.89 m/sec (2.86 ft/s), assuming one traveling screen for each circulating water pump and minimum river level. If the incoming water passes through the six screens equally, the calculated intake velocity is 0.58 meters per second (1.9 feet per second). Actual intake velocity probably falls between these two values (Page et al. 1978). During operation, the traveling screens at N Reactor are rotated and washed continually from April through June to limit impingement of small chinook salmon fry. A pipeline returns screenwash water, impinged fish, and other material from the top of the screens to a point downstream.

When N Reactor is in a dual-purpose mode with three pumps operating, the maximum temperature of the effluent is about 27.2°C, and the heat load to the Hanford Reach is about 550 megawatts-thermal (DOE 1988). When in a single-purpose mode with four pumps operating (total discharge of 390,000 gpm), the effluent is heated about 21.1°C above ambient river temperature (DOE 1982). In this configuration, no thermal energy goes to HGP as steam.

Heated effluent from N Reactor passes through two seal wells and a 2.6-meter (8.5-foot) -diameter pipe extending below the river bed for 263 meters (862 feet) into the Hanford Reach. It then discharges upward through one port (HGP has four ports) capped with a circular plate that diverts the effluent horizontally (DOE 1988). Heating slightly raises saturations of dissolved gas in the effluent, but these soon dissipate as the effluent mixes and cools (Neitzel and Page 1979).

Entrainment and Impingement at Cooling Water Intake

N Reactor and HGP, although representing a dual-purpose facility, have separate cooling water systems. The type of traveling screens and the calculated water velocity through the screens at the two adjacent intakes were similar. Yet features of fish impingement and entrainment at the two intakes differed.

Entrainment and impingement of fish were examined at the cooling water intake of N Reactor from late April to August 1977. Results were then compared with data obtained earlier from HGP's intake. Both chinook salmon and yellow perch fry were commonly impinged on N Reactor's intake screens (Figure 12.8). But chinook salmon fry were impinged far less frequently at N Reactor than at HGP. While 95% of the

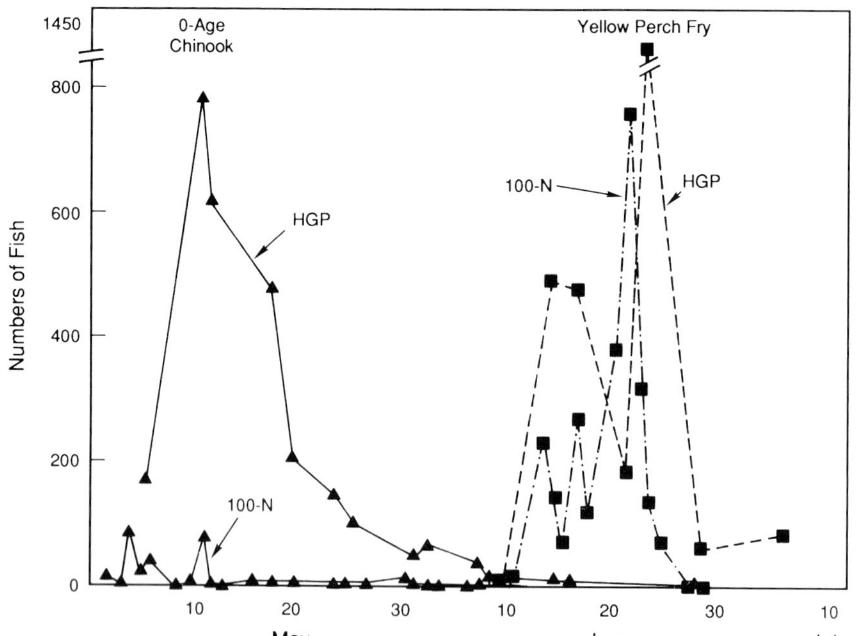

Fig. 12.8. Comparative impingement of zero-age chinook salmon in May and yellow perch fry in June at the cooling water intakes of N Reactor and HGP in 1977 (modified from Page et al. 1978).

fish impinged at HGP survived, all fish impinged at N Reactor were killed (Page et al. 1977b, 1978). Rainbow trout fry, when released in front of N Reactor's intake, were less likely to enter than fry released at HGP.

At N Reactor, as well as at HGP, only chinook salmon fry emerging from redds in the Hanford Reach upstream to Priest Rapids Dam were likely to be impacted. Because the two intakes were adjacent, equal numbers of chinook salmon fry should have been affected at each site. Yet releases of live and dead rainbow trout fry in front of traveling screens at both intakes showed that HGP impinged six times more live fish than N Reactor, and N Reactor impinged about 1.3 times more dead fish than HGP. Environmental stimuli somehow induced salmon fry to avoid impingement at the intake of N Reactor (Page et al. 1978). Screen passage (entrainment) appeared to be less of a problem at N Reactor than it formerly was at HGP.

Discharge Plume During Dual-Purpose Mode

As is the case for facilities releasing effluent to the Hanford Reach, N Reactor must operate within federal and state requirements. When contractors operating N Reactor applied for a NPDES permit in 1982, one requirement was that the distribution of the discharge plume from N Reactor must be documented during seasonally low flows. The study was conducted when N Reactor was in its normal, dual-purpose mode of operation and providing steam for HGP.

Heat is transferred from N Reactor to the Hanford Reach at 550 megawatts-thermal, and the maximum discharge temperature is 27.2°C. The effluent exits upward from the discharge pipe from a single outlet near midriver at about 1.5 meters per second (5 feet per second). A flat circular plate (velocity cap) diverts the effluent 360 degrees horizontally. The N Reactor's plume is relatively small, compared to HGP's, and releases closer inshore.

The distribution of N Reactor's plume was surveyed during September 1982 at three stages of low river flows: 1501, 2209, and 2690 m^3/s (53,000, 78,000, and 95,000 ft^3/s). Effluent temperatures varied from 20° to 24.4°C, and ambient river temperatures from 17.3° to 17.6°C. The low flows and high temperatures during late August and early September represented "worst case' conditions for the NPDES assessment.

The discharge plume was generally narrow, and could be traced as far as 610 meters (2000 feet) below the discharge outlet. Plume temperatures at the surface of the river were highest within 8 meters (25 feet) of the outlet, then dropped rapidly over the first 46 meters (150 feet) below that point. Temperatures were maximum near the outlet at a depth of 3 meters (10 feet). Computer modeling indicated the temperatures along the centerline of N Reactor's plume would, as expected, not only be higher but extend farther downriver at the regulated minimum low flow (Ecker et al. 1983a; DOE 1988).

HGP's thermal plume, extending downriver past N Reactor, did not merge with N Reactor's plume in the confines of its NPDES mixing zone in the September 1982 survey (Ecker et al. 1983a). After HGP's outlet diffusers were modified in the fall 1983, the two plumes merged slightly near the upstream edge of N Reactor's plume (Larsen 1985).

Discharge Plume During Single-Purpose Mode

In 1982, plant operators applied for a special NPDES permit for the period December 1982 through March 1983, when N Reactor would run temporarily in a single-purpose mode. A special test of reactor systems would be conducted at this time, and the discharge would not comply with stringent state standards.

When not providing steam for HGP, N Reactor's effluent was heated about 21.1°C above ambient river water temperature, and the maximum discharge would increase to 390,000 gallons per minute. The heat dissipation system would require use of all four intake pumps, and increase the velocity of water entering the traveling screens about 24%, to between 45.7 and 61.0 centimeters per second (1.5 and 2.0 feet per second). Heat would be extracted from the reactor's steam generators by 14 dump condensers (DOE 1982). The distribution of N Reactor's effluent plume would change somewhat.

N Reactor's plume was modeled for a single-purpose mode. Plume dispersions were computed at 1862 and 3398 m^3/s (66,000 and 120,000 f^3/s) river flow rates. Depending on flow, the 2.2°C (4°F) elevated isotherm involved 3.6% to 8.9% of the river cross section, and covered 0.3 to 1.2 acres. Based on the model, a passively drifting organism would be exposed to temperatures above ambient of 2.2°C or more for not more than 3.6 to 6.7 minutes (Ecker et al. 1983b).

Ecological data were assembled to form a 316(a) Demonstration. Even when N Reactor was operating in an atypical, single-purpose mode, its discharge was considered to have low impact on all groups of aquatic invertebrates in the river drift or on the river bottom. Most importantly, no appreciable harm would occur to RIS of fish, including anadromous salmonids, under normal flow levels from December to March, when the special conditions would prevail (DOE 1982). No assessment was made for operation of N Reactor in a single-purpose mode the entire year.

Outmigration of Juvenile Salmon Past N Reactor

Information on distribution of juvenile salmon during their annual seaward migration was needed to evaluate possible interactions of outmigrants with N Reactor's discharge plume. To quantify passage in terms of time and space, fyke nets were fished from five barges anchored across the Hanford Reach at the N Reactor site. Other types of collecting gear were fished inshore.

Distribution of juvenile salmonids near the N Reactor was expected to be influenced by currents in the river cross section. Water velocity varied greatly at the site. During high-flow periods, surface velocities in midriver approached 1.8 meters per second (6 feet per second) (Dauble and Page 1984). Currents slowed towards the shoreline and were, at times, imperceptible at some inshore locations. The bottom sloped gradually from the reactor towards a deep channel off the opposite shoreline.

In late summer of 1983, older, yearling-sized sockeye salmon, chinook salmon, and steelhead originating above Priest Rapids Dam passed N Reactor primarily in the depths at midriver where current velocities were high. Principal downstream movement was at night. Apparently, passage was direct and few of these fish lingered in the Hanford Reach. Juvenile resident fish (nonmigratory) were restricted primarily to inshore areas and depths less than 5 meters (16 feet) (Dauble and Page 1984).

In the spring of 1984, recently emerged, zero-age chinook salmon from spawning areas in the upper Hanford Reach and hatchery releases near Priest Rapids Dam (90% of the salmonids collected) appeared throughout the river's cross-section (Figure 12.9). Naturally produced chinook salmon fry seemed to prefer inshore areas, while most hatchery releases were caught near midriver. Again, most older sockeye salmon, chinook salmon, and steelhead originating above Priest Rapids Dam passed in the depths near midriver, and the main outmigration took place at night. Juvenile resident fish were uncommon in the river drift (Dauble et al. 1984).

Thermal Shock from Simulated Plume Conditions

Outmigration studies indicated that some juvenile salmonids may encounter N Reactor's discharge plume and experience thermal shock. While the plume rapidly mixed with river water because of outfall design and downstream currents, potentially lethal temperatures existed near

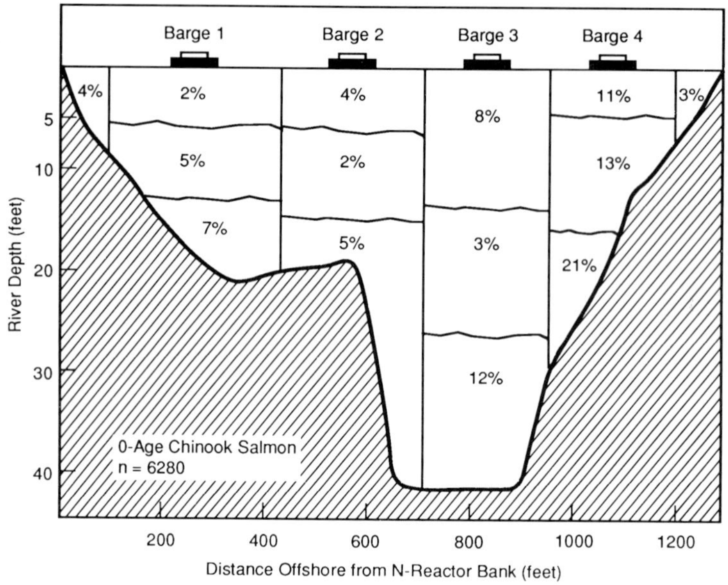

Fig. 12.9. Catch of zero-age chinook salmon in a cross section of the Columbia River at the N Reactor site during spring 1984. The catch reflects the early outmigration of fry produced, for the most part, upriver in the Hanford Reach (from Dauble et al. 1984).

the outfall. The exact conditions that could kill fish needed to be quantified in laboratory experiments to assess possible impacts when N Reactor was in a single-purpose mode.

Potentially lethal combinations for juvenile chinook salmon in the N Reactor plume were investigated in 1983. Tests were planned from computer simulations, calibrated from field measurements of the plume (Ecker et al. 1983b), of exposure times and temperatures that outmigrants might encounter during plume entrainment. Tests were run under conditions representing the warm temperatures (17.2°C), low flows of late summer (Figure 12.10) – a seasonal "worst-case" situation.

Exposures were lethal only under some later summer conditions, a time when only young salmonids from upriver still passed through the Hanford Reach. Results indicated that passage was safe whenever river discharges reached 4078 m^3/s (144,000 ft^3/s) or more, as occurs each spring. At flows less than 4078 m^3/s, the size of the area with potentially lethal temperatures depended on river flow. At 1020 m^3/s (36,000 ft^3/s), the regulatory minimum flow in the Hanford Reach, potentially lethal temperatures occurred to 18.5 meters (60 feet) to either side of the discharge outfall. At 3398 m^3/s (120,000 ft^3/s), potentially lethal temperatures occurred only along the plume centerline. The tests showed that

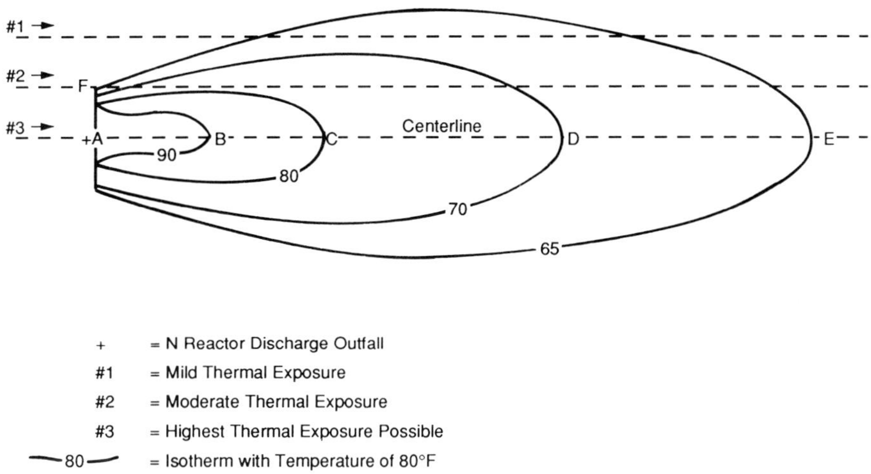

Fig. 12.10. Computer-simulated time-temperature distribution in the discharge plume of N Reactor (single-purpose mode). The three dashed lines represent possible exposure of juvenile salmonids during downstream passage through the plume. Exposure is related to position of outmigrants in relation to the plume centerline (from Neitzel et al. 1984a).

indirect mortality from disease or predation after sublethal exposures to the plume were unlikely (Neitzel et al. 1984a, 1986).

Additional tests were conducted in 1984 with four representative important resident and anadromous fish from the Hanford Reach under conditions representing cool water (10°C) and high flow conditions of spring. Tests were planned from computer simulations of exposure times and temperatures that the fish might encounter in the plume during the season the annual outmigration of salmonids normally takes place.

Passage through N Reactor's plume was less apt to be lethal to juvenile fish in the spring than during late summer (Figure 12.11). Potentially lethal combinations of time and temperature during spring occurred only at unseasonably low discharges of less than 1529 m^3/s (54,000 ft^3/s). Further, the maximum width of a plume with potentially lethal temperatures was 18 meters (60 feet) at the minimum low flow of 1020 m^3/s. Juvenile chinook salmon were less tolerant to test exposures than juvenile coho salmon, steelhead, and northern squawfish (Neitzel et al. 1984b, 1986).

Evaluation of Thermal Effects from N Reactor

Normally, the thermal discharge from N Reactor in a dual-purpose mode complied with limits specified in NPDES permits. But water

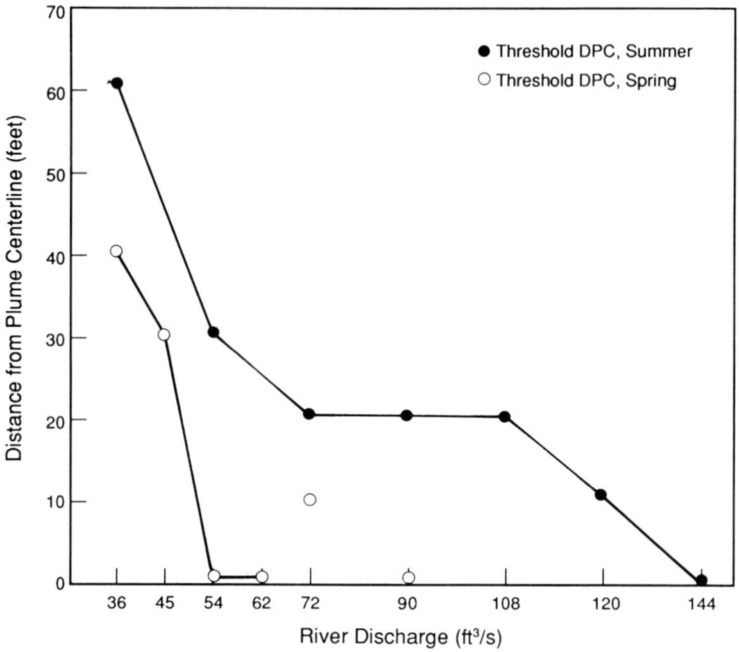

Fig. 12.11. Comparison of conditions potentially lethal to juvenile fish passing through the discharge plume of N Reactor (single-purpose mode) during high spring and low summer flows. The highest temperature encountered by fish, in relation to the discharge plume centerline (DPC) is crucial for their survival (from Neitzel et al. 1984b).

quality standards were exceeded at certain times of the year outside of the plume's mixing zone. In 1988, a Section 316(a) Demonstration addressed this problem (DOE 1988). Ecological data related to plume dispersion and possible adverse effects were assembled and assessed. The specific studies reported above added credence to the study.

Assessment focused on potential impacts to RIS, 10 fish, 5 shellfish and macroinvertebrates in the Hanford Reach, as required by Section 316(a) of the 1977 amended FWPCA. Previously, the Hanford Reach near N Reactor had been declared a low-potential impact area for all aquatic species except fish by the Washington State Department of Ecology and Environmental Protection Agency.

The Section 316(a) Demonstration showed that RIS of fish would not be adversely affected by the thermal plume from N Reactor in a dual-purpose operation. One consideration was that, ignoring other variables, temperatures causing direct mortalities to adult anadromous fish range from 20.9° to 22.0°C. The exposure time causing fish mortality ranges downward from 10,000 minutes at 22.5°C to 12 minutes at 30°C. During dual-purpose operation, the maximum increase in river temperature from N Reactor's plume would be about 6.6° to 7.7°C. Temperatures in the mixing plume would exceed 22.2°C only during the period that water temperatures in the Hanford Reach exceeded 15°C and the effluent was heated 7.7°C above ambient. Even so, any possible exposure to lethal conditions would be brief. The plume becomes diluted within a few feet of the outfall, and any temperatures greater than 22.2°C would occur in less than 1% of the river's cross section (DOE 1988).

Washington Public Power Supply System Nuclear Plant No. 2 (WNP-2)

WNP-2 was built about 19 kilometers (12 miles) north of Richland near RKm 566. Site preparation began in March 1973, fuel was loaded in December 1983, and commercial operation began in December 1984. The plant was designed to produce up to 1100 megawatts-electric from fission of nuclear fuels.

A Site Certification Agreement for WNP-2 was executed between the State of Washington and the Supply System on May 17, 1972. An environmental statement for WNP-2 was published by the Directorate of Licensing, U.S. Atomic Energy Commission in December 1972 (AEC 1972). The Environmental Report for the operating license stage was released in March 1977 (Supply System 1977b).

The Supply System retained Battelle in 1974 to conduct preoperational studies in the Hanford Reach at the future site of WNP-2. (Three similar power plants using nuclear fuel were planned, licensed, and started but only WNP-2 was completed.) Five "baseline" reports describing aquatic biota in the Hanford Reach near the construction site were issued (Battelle 1977a,b, 1978, 1979a,b). From August 1978 through March 1980, an offsite group conducted additional studies in the Hanford Reach to support the startup of WNP-2 (Beak Consultants, Inc. 1980).

The Supply System analyzed all preoperational data on aquatic biota from 1974 to 1980 and presented, in 1982, an operational monitoring program to Washington State's Energy Facility Site Evaluation Council (Supply System 1982). The Council approved a modified program, and operational monitoring begin in March 1983. Release of effluent from WNP-2's cooling towers was intermittent until the plant reached full production in fall 1984. No significant effects on the Hanford Reach were detected during the initial years of plant operation (Supply System 1985, 1986, 1987).

Cooling System Operation – Intake and Discharge

The cooling system at WNP-2 uses river water, but is "off-line" or limited cycle. The plant has six mechanical draft cooling towers and, for emergency cooling, two spray ponds. Raw river water, primarily to replace evaporative losses in the towers, comes from the Hanford Reach via offshore, perforated-pipe intakes. WNP-2 has no cooling-water discharge in the typical sense. Limited amounts of tower blowdown are diluted with river water before release to eliminate measurable chemical and thermal changes in water quality downstream.

Because of environmental concerns, the cooling water intake for WNP-2 was designed to minimize impingement and entrainment of juvenile fish. Preliminary studies pointed to the advantages of an intake consisting of perforated pipes. This type of intake would greatly reduce velocities of the water pulled inward through the perforations by pumps, thus preventing impingement. Entrainment would be restricted to microscopic river organisms in the river drift. Of the designs considered, the perforated intake was also the least costly to construct, operate, and maintain (Schreiber et al. 1973, 1974).

When installed, the cooling water intake at WNP-2 consisted of two large, stainless steel pipes placed well offshore, parallel to the current, and just above the river bed (Figure 12.12). Water entered at opposite ends of each structure through two perforated sections, each 2 meters (6.5 feet) long and containing a network of 1-centimeter (0.375-inch) round holes. Velocities of incoming water are equalized by movable, internal, perforated sleeves. Under normal conditions, 47,000 liters per minute (12,500 gallons per minute) of water enter both structures at an estimated intake velocity of 0.05 meter per second (0.15 feet per second). The river usually flows past the intake pipes at 1.2 to 1.5 meters per second (4 to 5 feet per second) (Supply System 1977b, 1985). Thus, water enters the intake at a lower velocity than the water passing downstream, greatly reducing any chance of impinging fish.

The six cooling towers at WNP-2 remove heat from recirculating river water. As it cools, about 10,000 gallons per minute are released to the atmosphere by evapotranspiration. More river water is added to make up this loss. After the coolant has cycled about 10 to 12 times though the reactor's system, minerals are concentrated and cooling efficiency is lost. Used coolant is mixed with river water and discharged as blowdown to the Hanford Reach through a buried pipe.

Blowdown emerges from a submerged port below the intakes and closer inshore. The exit is a rectangular slot angled upward and surrounded by

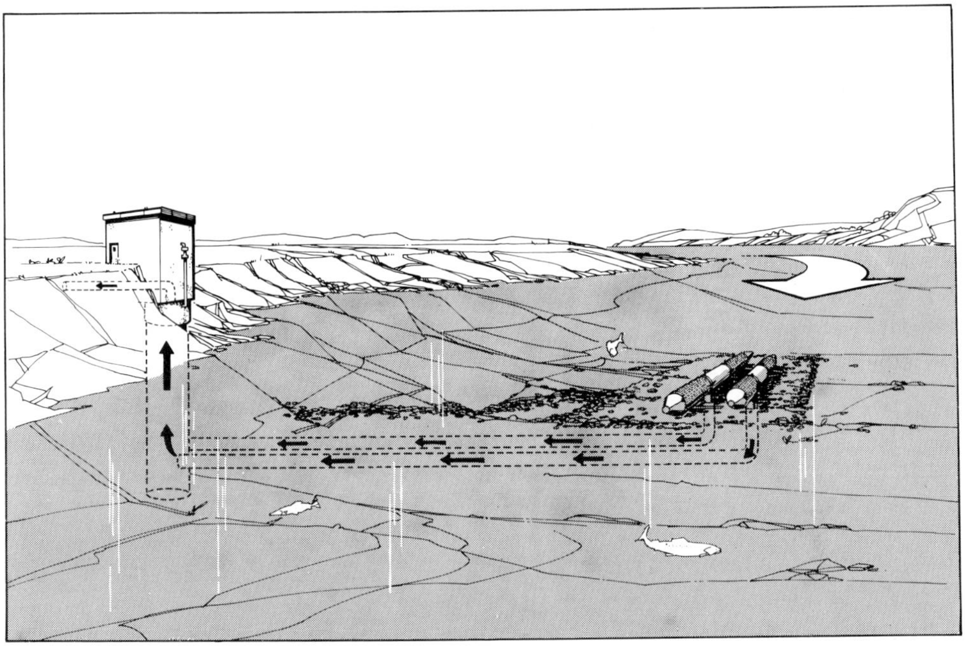

Fig. 12.12. Conceptual rendition of the perforated-pipe, water intake at Washington Nuclear Project-2, which was designed to eliminate impingement and entrainment of juvenile fish (from Gray et al. 1979).

riprap to prevent erosion. The mixing zone allowed by WNP-2's NPDES permit is limited to within 91 meters (300 feet) of the release point. Water samples from downstream are analyzed, as required by the permit, to document any affect on water quality from blowdown.

The lowest regulated flow in the Hanford Reach is 1019 m^3/s (36,000 ft^3/s). At minimum flow, the most river water that could be withdrawn at HGP's intake is less than 0.15% of the water passing downstream.

Design Features of Discharge Plume

According to design, cooling tower blowdown could discharge to the Hanford Reach at a temperature about 9.5° to 14°C higher from January to June, and about 4.5°C to 6.7°C higher from July through December. The maximum rate of blowdown would be 0.4 m^3/s (14.5 ft^3/s), and the maximum temperature of the blowdown would be 28°C. The calculations indicated that, under minimum regulated flows, the Hanford Reach would be warmed less than 0.11°C at a point 229 meters (750 feet) below the discharge, and less than 0.0006°C after the effluent was completely mixed (Supply System 1982). Computer models (Kannberg 1980) before plant startup defined the expected dispersal of the effluent plume.

The NPDES permit allowed the boundaries of the mixing zone to extend from a point 15 meters (50 feet) upstream to 91 meters (300 feet) downstream, and from surface to riverbed, with a width of 30 meters (100 feet). Class A water quality standards for the state of Washington required that water temperatures outside the mixing zone could not be raised more than 0.3°C whenever ambient temperatures reached 20°C.

Since start up in 1984, discharges at WNP-2 have complied with regulatory requirements and exerted no or little effect on water quality. Further, young chinook salmon that were drifted through the plume survived even when blowdown temperatures were greater than 23.9°C (Supply System 1985). This supported the NRC's contention that the temperature of cooling effluent from WNP-2 would have no significant environmental impacts (NRC 1981).

Preoperational Quantification of Biota

Preoperational studies at the proposed WNP-2 site defined aquatic communities and provided baseline data on related ecological features. The data were also used to evaluate the accuracy of information assembled for the site's environmental impact statement of 1972. More than 5 years of field work were completed before the Supply System issued its preoperational summary report.

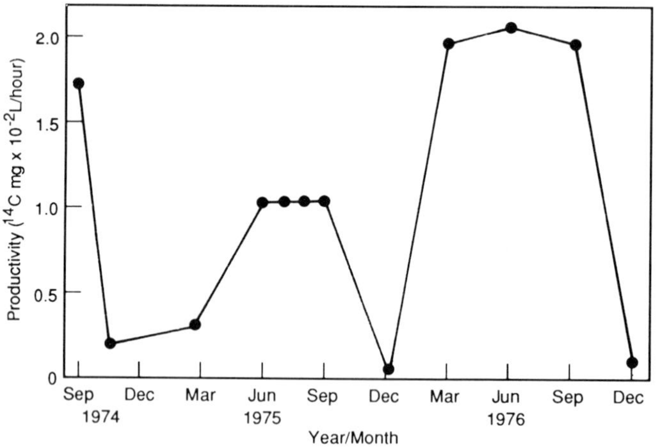

Fig. 12.13. Rates of primary production by drift phytoplankton in the Hanford Reach near Washington Public Power Supply System Nuclear Project-2, 1974 through 1976 (from Supply System 1982).

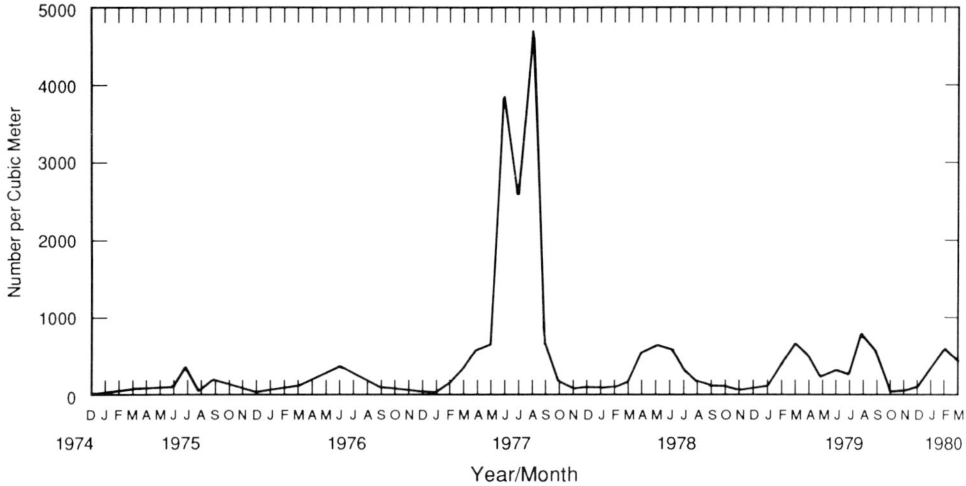

Fig. 12.14. Densities of drift zooplankton in the Hanford Reach near Washington Public Power Supply System Nuclear Project-2, 1975 to 1980 (from Supply System 1982).

Biological data for the WNP-2 site were characterized by seasonal peaks and by year-to-year variability in abundance of aquatic organisms. Primary production by drift phytoplankton underwent one main peak

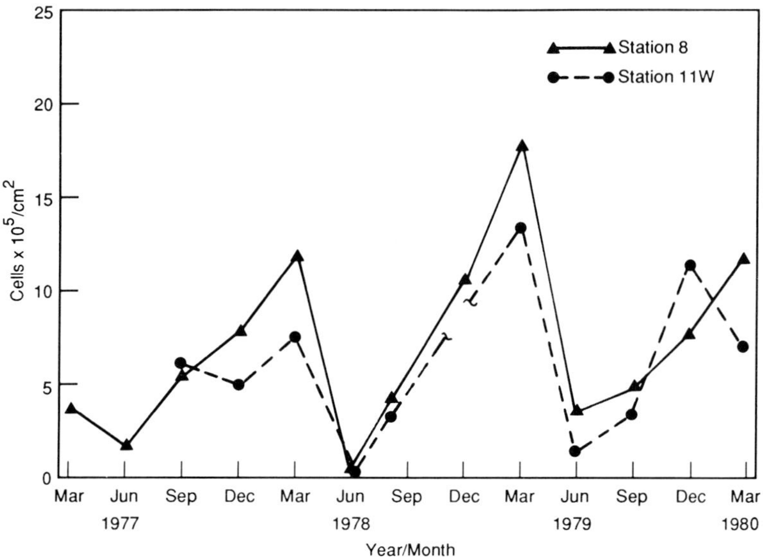

Fig. 12.15. Biomass of attached periphyton in the Hanford Reach at two stations near Washington Public Power Supply System Nuclear Project-2, 1977 to 1980 (from Supply System 1982).

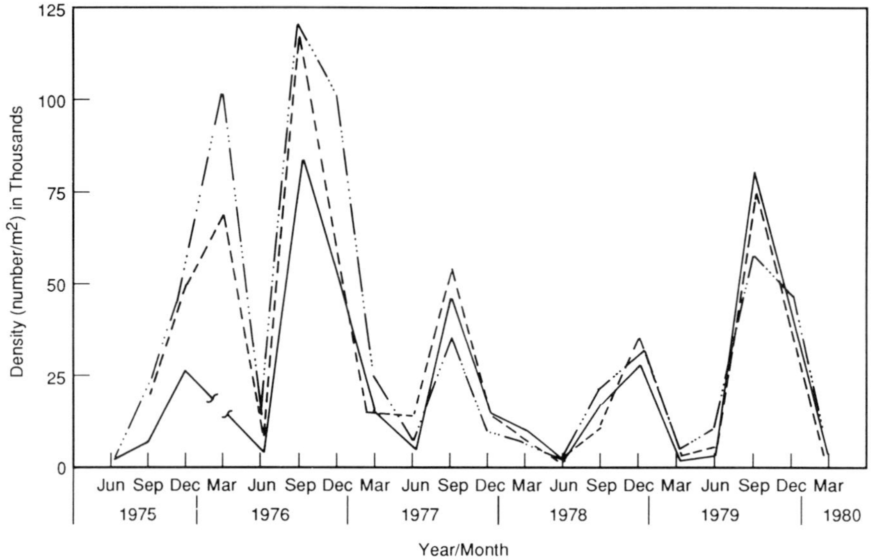

Fig. 12.16. Densities of benthic invertebrates at three stations in the Hanford Reach near Washington Public Power Supply System Nuclear Project-2, 1975 to 1980 (from Supply System 1982).

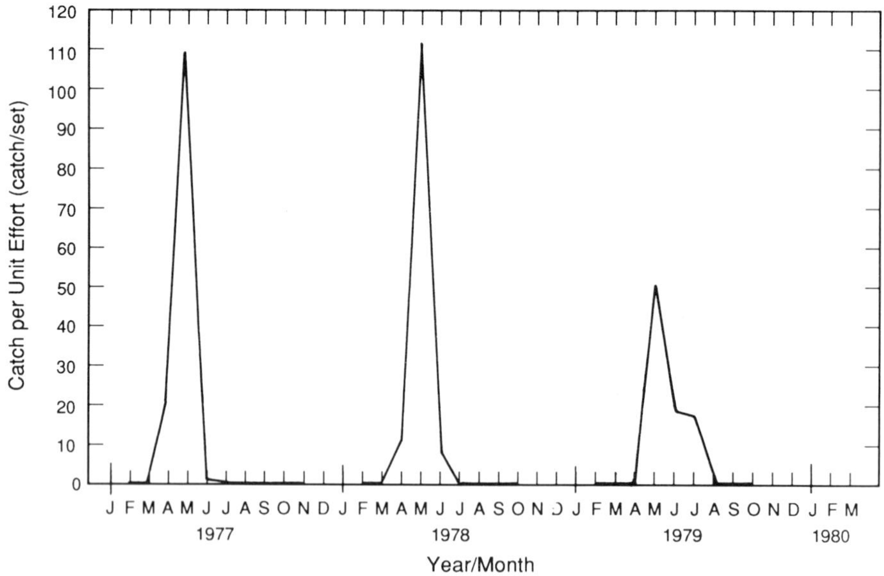

Fig. 12.17. Catch per unit of effort for juvenile chinook salmon in beach seine collections near Washington Public Power Supply System Nuclear Project-2, January 1977 to March 1980 (from Supply System 1982).

each summer (Figure 12.13). Numbers of drift zooplankton also peaked during the summer, and zooplankton densities in 1977 were high compared to other years (Figure 12.14). Attached periphyton reached greatest biomass in early spring (Figure 12.15). Benthic invertebrates were most abundant in the fall, resulting largely from production during summer of aquatic insects (Figure 12.16). Juvenile chinook salmon, emerging from the streambed in the Hanford Reach above the WNP-2 site, were present inshore from April through June (Figure 12.17).

Operational Ecological Monitoring Program

After reviewing its 1974 to 1980 data base, the Supply System concluded that certain phases of its preoperational program were more apt to detect operational impacts, should they occur, than others. The Supply System made three main recommendations for its postoperational monitoring program: 1) that monitoring of periphyton, benthos, fish, and water quality conditions be continued; 2) that studies of the thermal plume, effluent toxicity, and river water intake be added, as required by regulatory agencies; and 3) that monitoring of phytoplankton and zooplankton be deleted (Supply System 1982).

The operational monitoring program for WNP-2 was specific for the Hanford Reach. Phytoplankton, potentially impacted forms, were only a minor link in the food chain supporting indigenous populations of fish and wildlife. At this location, the Columbia River was a low impact area for zooplankton. Periphyton formed a vital link in the aquatic food chain, and were immobile and relatively sensitive to thermal and chemical discharges. Therefore, periphyton might reflect any impacts in relation to exposure (distance from release point and time) to the blowdown effluent. Benthic invertebrates needed to be monitored for essentially the same reasons.

The NPDES permit required that impacts on juvenile chinook salmon and other fish should be evaluated seasonally by beach seine collections, experimental drifts through the discharge plume, and entrainment sampling. In addition, juvenile salmonids must be exposed to blowdown effluent in bioassays (Supply System 1982).

Initial Evaluation of Operational Effects from WNP-2

Ecological monitoring at WNP-2 was intended to identify not only immediate operational impacts but also long-term trends. Monitoring during the first 3 years of complete operation (1984, 1985, and 1986) revealed no significant effects on aquatic organisms below the site or on water quality in the Hanford Reach.

Specifically, no measurable changes in periphyton or benthic communities were revealed when data were compared from, first, pre- and post-operations and, second, upstream and downstream stations. No fish, fish eggs, or fish larvae were entrained. There was no impingement of fish or excessive fouling (unless they were attacked by birds) of the intake perforations. Water quality was usually similar at stations above and below the effluent discharge, although a few differences were detected within the allowable mixing zone. Fish held in tower blowdown, even at 100% concentrations, survived for 96 hours (Supply System 1985, 1986).

Biological data from 1970s assessments

Baseline studies in the upper Hanford Reach at HGP and N Reactor (RKm 611), and downriver at WNP-2 (RKm 566) from July 1973 through March 1980 generated considerable information about biological communities and their interactions in the Hanford Reach. Information was initially presented in internal reports covering one study phase, or one-time studies related to a specific site. In a few cases, information on aquatic communities was later published in scientific journals and became part of public knowledge offsite.

For site assessment work, aquatic organisms in the Hanford Reach were generally divided into microflora, zooplankton, benthos, fish, and riparian vegetation. This effort was important because today, the flowing Hanford Reach retains physical characteristics that differ from those in impounded areas of the Columbia River upstream and downstream.

Microflora

Microflora in the Hanford Reach consisted of phytoplankton (drifting algae) and periphyton (attached algae), primary producers that convert energy from sunlight by photosynthesis. Data included estimates for relative abundance and density of common species, uptake of carbon, and production of chlorophyll a and organic matter.

Both communities consisted largely of diatoms (a group of algae), which usually consisted of five genera, *Asterionella, Fragilaria, Melosira, Synedra*, and *Tabellaria*. Taxa changed in relative predominance during the year, but one or two species were usually most abundant at any one time. The phytoplankton community included over 150 species; densities peaked in summer, followed by a second, lower peak in fall. Generally, primary productivity increased in summer and decreased in winter. Overall, the periphyton community was probably more productive than the phytoplankton community. Densities of phytoplankton passing through the Hanford Reach apparently have increased since 1949, indicating enhanced production in upriver reservoirs (Neitzel et al. 1982a).

Zooplankton

Zooplankton in the Hanford Reach are largely microscopic crustaceans suspended in the river drift, although some development occurs in slackwater areas inshore. Data emphasized the relative abundance and seasonal distribution of common species.

The zooplankton community consisted of 58 species, predominated by *Bosmina longirostrus, Diaptomus ashlandi*, and *Cyclops bicuspidatus*. Densities of zooplankton were lowest in winter (usually less than 50 per cubic meter) and highest in summer (up to 4500 per cubic meter). Populations consisted largely of *Bosmina* in summer, *Diaptomus* in winter, and *Cyclops* in spring (Neitzel et al. 1982b). Also common in the river drift along with zooplankton were *Daphnia* spp., rotifers, and insects.

Benthos

Benthos included all aquatic macroinvertebrates associated with the river bottom and submerged substrates in the Hanford Reach. Data identified indigenous species, and determined their relative abundance and seasonal distribution.

Aquatic larvae of the midge fly and caddis fly predominated in all benthic samples at 90% total density. Other taxa comprised less than 20% to the benthos, and included black flies, oligochaetes, mollusks, sponges, and water mites. Population densities changed seasonally, largely in response to the life cycles of the most abundant organisms. Generally, total biomass was lowest in summer, due largely to the seasonal maturation and emergence of insect larvae, and highest in fall and winter as insect eggs hatched. Most aquatic insects emerge as adults and deposit eggs from one to three times during spring and summer, depending on species, but overwinter as larvae underwater (Battelle 1977a,b, 1978, 1979a,b).

No benthic invertebrate in the Hanford Reach is on the endangered species list maintained under the Endangered Species Act of 1973 by the U.S. Fish and Wildlife Service. However, two are candidates for the threatened species list, the Giant Columbia River spire snail and the Great Columbia River limpet.

Fish Populations

Fish are the largest and best known of all aquatic organisms. Such species as chinook salmon, steelhead trout, and white sturgeon are important to the economy of the Pacific Northwest. But a number of other species of fish, native and introduced, exist in the Columbia River. All hold different roles in ecosystem functioning, and depend on other river organisms as a food supply. Since 1943, 43 species of fish representing 13 families have been identified in the Hanford Reach (Gray and Dauble 1977).

Species of fish differ in relative abundance, sensitivity to habitat change, and vulnerability to use of water from the Hanford Reach for condenser cooling and effluent disposal. Fish populations at Hanford are dominated each spring by emergence from the streambed gravel of chinook salmon fry, which feed largely on insects and zooplankton (Dauble et al. 1980) before moving seaward. Native species of suckers are the most abundant, year-around resident fish of the Hanford Reach. The life history and ecology of the largescale sucker and the bridgelip sucker were closely examined in relation to site-assessment work (Dauble 1980, 1986); the two species often hybridize (Dauble and Buschbom 1981). The life cycle of the sandroller, a species endemic to the lower and middle reaches of the Columbia River, was also clarified (Gray and Dauble 1979).

No fish in the Hanford Reach is listed as an endangered or threatened species by the U.S. Fish and Wildlife Service. Fish larvae are relatively uncommon in the river drift, but some species have temporarily planktonic life phases.

Riparian Vegetation

Plants along the shoreline and islands of the Hanford Reach vary according to distance from water, frequency of inundation, current velocity, and substrate. Changes in water level, including daily fluctuations from releases at upriver dams, strongly affect riparian structure.

The river margin through most of the Hanford Reach is a narrow zone of cobble wetted almost daily during the summer and fall growing seasons by changes in water level. Very gradual slopes consist of large areas of cobblestone with sparse or no vegetation, depending on season, of perennial forbs and grasses. In growing season, areas frequently inundated by daily fluctuations in water level are inhabited by water smartweed and water speedwell, while areas periodically wetted are inhabited by reed canarygrass, common witchgrass, barnyard grass, and summer-blooming forbs. Willows and trees are usually absent (Fickeisen et al. 1980).

One rare plant occurs in the Hanford Reach at the water's edge in the zone of fluctuating water level, the local variety of yellowcress (*Rorippa calycina* var. *columbiae*) (Sauer and Leder 1985).

Significance of facility-specific studies at Hanford, 1971 to 1984

After closure of the last single-purpose reactor at Hanford in early 1971, environmental studies in the Hanford Reach changed direction. They were conducted to comply with new environmental regulations, particularly the NEPA of 1969 and the amended FWPCA of 1972. However, the regulatory requirements differed for each of the three major facilities that used river water for cooling: HGP, N Reactor, and WNP-2.

HGP and N Reactor were both constructed before passage of the federal environmental acts that would have applied to their operation. However, both facilities came under jurisdiction of the state's water quality standards, which included well-defined thermal limits. And the amended FWPCA retained water quality standards established in earlier laws, but set forth new requirements dealing with the effects of cooling water on aquatic organisms. Specifically, Section 316(a) of the amendment was written to protect organisms entrained in the cooling water and in the discharge plume. Section 316(b) of the amendment was written to protect organisms impinged by the water intake structures. This portion of the FWPCA required that location, design, construction, and capacity of cooling systems reflect the best technology available to minimize adverse impacts.

Aquatic research near HGP in 1978 provided the Supply System with information that 1) demonstrated the absence of appreciable harm to fish communities, and 2) justified the designation of benthic and plankton communities as low potential impact areas in the Hanford Reach. On March 7, 1980, the State of Washington (Department of Ecology), with agreement of the Region X EPA, granted the Supply System a waiver of the state thermal discharge limits for HGP. There was one exception: after 1983, no heated water could be discharged from July 1 to September 7, the annual period of lowest river flow and highest river temperature.

Similar research in the Hanford Reach enabled the U.S. Department of Energy to apply for NPDES permits that allowed continued operation of N Reactor. A 316(a) Demonstration was issued for a temporary single mode of operation in 1982 and for a normal dual-purpose mode of operation in 1988.

When planned and constructed, WNP-2 was a new facility that fell under the scope of the NEPA. The Site Certification Agreement for WNP-2 was executed on May 17, 1972, between the State of Washington and the Supply System. It required that ecological monitoring be conducted during the preoperational and operational phases of site development and use. As a result, studies that characterized aquatic communities near the WNP-2 site were conducted from September 1974 through March 1980. Before plant startup, the Washington State Energy Facility Site Evaluation Council reviewed preoperational aquatic monitoring data and approved an operational monitoring program. The first comprehensive operational environmental report, prepared by Supply System scientists in 1984, revealed no significant effect on water quality in the Hanford Reach or its aquatic biota. The environmental issues raised during the design of WNP-2 contributed to making the facility environmentally benign.

HGP, N Reactor, and WNP-2 could not have started up or maintained operations, as required by today's exacting environmental regulations, without the support of aquatic studies in the Hanford Reach. Most important, however, these studies revealed no significant environmental effects from operation of these facilities.

More than any other factor, research performed for regulatory compliance established a broad biological and ecological data base for the Hanford Reach. This information is sufficient to allow detection of major changes in water quality or biota that might take place from continued use of the river ecosystem.

References

AEC. See U.S. Atomic Energy Commission.

Battelle. 1976. *Final Report on Aquatic Ecological Studies Conducted at the Hanford Generating Project, 1971–1974. WPPSS Columbia River Ecological Studies Vol. 1.* Report to United Engineers and Constructors for Washington Public Power Supply System, Richland, Washington.

Battelle. 1977a. *Aquatic Ecological Studies Conducted Near WNP 1, 2 and 4, September 1974 Through September 1975. WPPSS Columbia River Ecological Studies Vol. 2.* Report to United Engineers and Constructors for Washington Public Power Supply System, Richland, Washington.

Battelle. 1977b. *Aquatic Ecological Studies Conducted Near WNP 1, 2, and 4, October 1975 Through February 1976. WPPSS Columbia River Ecological Studies Vol. 3.* Report to United Engineers and Constructors for Washington Public Power Supply System, Richland, Washington.

Battelle. 1978. *Aquatic Ecological Studies Near WNP-1, -2 and -4 March Through December 1976. WPPSS Columbia River Ecological Studies Vol. 4.* Report to United Engineers and Constructors for Washington Public Power Supply System, Richland, Washington.

Battelle. 1979a. *Aquatic Ecological Studies Near WNP 1, 2 and 4 January Through December 1977. WPPSS Columbia River Ecological Studies Vol. 5.* Report for Washington Public Power Supply System, Richland, Washington.

Battelle. 1979b. *Aquatic Ecological Studies Near WNP 1, 2, and 4 January Through August 1978. WPPSS Columbia River Ecology Studies Vol. 6.* Report for Washington Public Power Supply System, Richland, Washington.

Beak Consultants, Inc. 1980. *Preoperational Environment Monitoring Studies Near WNP-1, 2, and 4, August 1978 through March 1980. WPPSS Columbia River Ecological Studies Vol. 7.* Report for Washington Public Power Supply System by Beak Consultants, Inc., Portland, Oregon.

Dauble, D.D. 1980. "Life History of the Bridgelip Sucker in the Central Columbia River." *Trans. Am. Fish. Soc.* 109:91–98.

Dauble, D.D. 1986. "Life History and Ecology of the Largescale Sucker (*Catostomus macrocheilus*) in the Columbia River." *Am. Midl. Nat.* 16:351–367.

Dauble, D.D., and, R.L. Buschbom. 1981. "Estimates of Hybridization Between Two Species of Catostomids in the Columbia River." *Copeia* 1981:802–810.

Dauble, D.D., and, T.L. Page. 1984. *Fish Distribution Studies Near N Reactor, Summer 1983.* UNI-2754, Prepared by Pacific Northwest Laboratory for UNC Nuclear Industries, Inc., Richland, Washington.

Dauble, D.D., R.H. Gray, and T.L. Page. 1980. "Importance of Insects and Zooplankton in the Diet of 0-age Chinook Salmon (*Oncorhynchus tshawytscha*) in the Central Columbia River." *Northwest Sci.* 54:251–258.

Dauble, D.D., T.L. Page, and R.W. Hanf, Jr. 1984. *Distribution of Juvenile Salmonids in the Columbia River Near N Reactor, Spring 1984.* WHC-EP-0175, prepared by Pacific Northwest Laboratory for UNC Nuclear Industries, Inc., Richland, Washington.

Ecker, R.M., W.H. Walters, and F.L. Thompson. 1983a. *N Reactor Thermal Plume Characterization Study During Dual-Purpose Mode of Operation. Part I – Field Investigation.* UNI-2620, prepared by Pacific Northwest Laboratory for UNC Nuclear Industries, Richland, Washington.

Ecker, R.M., R.G. Parkhurst, F.L. Thompson, and W.W. Walters. 1983b. *N Reactor Thermal Plume Characterization During Pu-Only Mode of Operation.* UNI-2618, prepared by Pacific Northwest Laboratory for UNC Nuclear Industries, Inc., Richland, Washington.

Fickeisen, D.H., D.D. Dauble, D.A. Neitzel, W.H. Rickard, R.L. Skaggs, and J.L. Warren. 1980. *Aquatic and Riparian Resource Study of the Hanford Reach, Columbia River, Washington.* Prepared by Pacific Northwest Laboratory for the U.S. Army Corps of Engineers, Seattle District, Seattle, Washington.

Gray, R.H. and D.D. Dauble. 1977. "Checklist and Relative Abundance of Fish Species from the Hanford Reach of the Columbia River." *Northwest Sci.* 51:201–215.

Gray, R.H., and D.D. Dauble. 1979. "Biology of the Sandroller in the Central Columbia River." *Trans. Am. Fish. Soc.* 108:641–649.

Gray, R.H., T.L. Page, E.G. Wolf, and M.J. Schneider. 1975. *A Study of Fish Impingement and Screen Passage at Hanford Generating Project – A Progress Report.* Prepared by Battelle, Pacific Northwest Laboratories for the Washington Public Power Supply System, Richland, Washington.

Gray, R.H., D.A. Neitzel, and T.L. Page. 1979. "Water Intake Structures: Engineering Solutions to Biological Problems." *North. Eng.* 10:21–33.

Kannberg, L. D. 1980. *Mathematical Modeling of the WNP-1, 2 and 4 Cooling Tower Blowdown Plumes.* Prepared by Battelle, Pacific Northwest Laboratories for Washington Public Supply System, Richland, Washington.

Larsen, J. 1985. *Hydrothermal Field Study of Cooling Water Diffuser Outfall Hanford Generating Station.* Prepared by Alden Research Laboratory for the Washington Public Power Supply System, Richland, Washington.

Neitzel, D.A., and T.L. Page. 1979. *Total Dissolved Gas Saturation in the Columbia River Near 100-N Reactor Outfall: March Through September 1979.* Prepared by Pacific Northwest Laboratory for UNC Nuclear Industries, Richland, Washington.

Neitzel, D.A., T.L. Page, and R.W. Hanf, Jr. 1982a. "Mid-Columbia River Microflora." *J. Freshwater Ecol.* 1:491–505.

Neitzel, D.A., T.L. Page, and R.W. Hanf, Jr. 1982b. "Mid-Columbia River Zooplankton." *Northwest Sci.* 57:111–118.

Neitzel, D.A., T. Page, R.H. Gray, and D. Dauble. 1982c. "Once-Through Cooling on the Columbia River – the Best Available Technology?" *Environ. Impact Assess. Rev.* 3:41–58.

Neitzel, D.A., T.M. Poston, C.S. Abernethy, T.L. Page, and D.W. Carlile. 1984a. *Laboratory Simulation of Late-Summer Juvenile Chinook Salmon Passage Through N Reactor Thermal Plume During Single-Purpose Mode of Operation.* UNI-2755, prepared by Pacific Northwest Laboratory for UNC Nuclear Industries, Inc., Richland, Washington.

Neitzel, D.A., T.M. Poston, C.S. Abernethy, M.T. McLane, T.L. Page, and D.W. Carlile. 1984b. *Laboratory Simulated Passage of Four Species of Fish Through the N Reactor Thermal Plume in Spring During Single-Purpose Mode of Operation.* UNI-3220, prepared by Pacific Northwest Laboratory for UNC Nuclear Industries, Inc., Richland, Washington.

Neitzel, D.A., T.M. Poston, T.L. Page, and C.S. Abernethy. 1986. "Laboratory Simulation of Fish Passage Through a Heated-Water Discharge." In: *Special Technical Publication No. 921*, pp. 121–134. American Society for Testing and Materials, Philadelphia, Pennsylvania.

NRC. See U.S. Nuclear Regulatory Commission.

Page, T.L., R.H. Gray, and E.G. Wolf. 1975. *Report on Impingement Studies Conducted at the Hanford Generating Project March and April 1975.* Prepared by Battelle, Pacific Northwest Laboratories for Washington Public Power Supply System, Richland, Washington.

Page, T.L., R.H. Gray, and D.A. Neitzel. 1976. *Fish Impingement Studies at the Hanford Generating Project (HGP), December 1975 through April 1976.* Prepared by Battelle, Pacific Northwest Laboratories for Washington Public Power Supply System, Richland, Washington.

Page, T.L., R.H. Gray, and D.A. Neitzel. 1977a. *Fish Impingement and Screen Passage Studies at Hanford Generating Project*. BN-SA-775, Pacific Northwest Laboratory, Richland, Washington.

Page, T.L., D.A. Neitzel, and R.H. Gray. 1977b. *Impingement Studies at the 100-N Reactor Water Intake*. BNWL-2401, Pacific Northwest Laboratory, Richland, Washington.

Page, T.L., D.A. Neitzel, and R.H. Gray. 1978. "Comparative Fish Impingement at Two Adjacent Water Intakes on the Mid-Columbia River." In: *Fourth National Workshop on Entrainment and Impingement*, ed. L. D. Jenson, pp. 251–266. Ecological Analysts Communications, Melvine, New York.

Raymond Keizer Engineers (RKE). 1982. *Outfall Diffuser Conceptual Design, Hanford Generating Plant, Final Report*. Report by RKE to Washington Public Power Supply System, Richland, Washington.

Sauer, R.H., and J.E. Leder. 1985. "The Status of Yellowcress in Washington." *Northwest Sci*. 59:191–203.

Schreiber, D.L., C.D. Becker, and J.J. Fuquay. 1973. *Appraisal of Water Intake Systems on the Central Columbia River*. Prepared by Battelle, Pacific Northwest Laboratories for Burns and Roe, Inc., for Washington Public Power Supply System, Richland, Washington.

Schreiber, D.L., C.D. Becker, J.J. Fuquay, and R.A. Chitwood. 1974. "Intake System Assessment for Central Columbia River." *J. Power Div*. American Society of Chemical Engineers 100(PO2): 131–155.

Singarella, P.N., and D.N. Brocard. 1982. *Hydrothermal Modelling of Cooling Water Diffuser Outfall Hanford Generating Station*. Prepared by Alden Research Laboratory for Washington Public Power Supply System, Richland, Washington.

Supply System. See Washington Public Power Supply System.

U.S. Atomic Energy Commission (AEC). 1972. *Final Environmental Statement Related to the Proposed Hanford Number Two Nuclear Power Plant, Washington Public Power Supply System, Docket No. 50–397*. Directorate of Licensing, U.S. Atomic Energy Commission, Washington, D.C.

U.S. Department of Energy (DOE). 1982. *316(a) Demonstration for Test of N Reactor in Plutonium-Only Mode of Operation*. Richland Operations Office, Richland, Washington.

U.S. Department of Energy (DOE). 1988. *316(a) Demonstration for Operation of N Reactor in Dual-Purpose Mode*. Richland Operations Office, Richland, Washington.

U.S. Nuclear Regulatory Commission (NRC). 1981. *Final Environmental Statement Related to the Operation of WPPSS Nuclear Project No. 2*. Docket No. 50–397, U.S. Nuclear Regulatory Commission, Washington, D.C.

Whelan, G., and C.A. Newbill. 1983. *Update of Columbia River Flow and Temperature Data Measured at Priest Rapids Dam and Vernita Bridge*. PNL-4868, Prepared by Pacific Northwest Laboratory for UNC Nuclear Industries, Inc., Richland, Washington.

Washington Public Power Supply System. 1977a. *Final Environmental Impact Statement on Continued Operation of the Hanford Generating Project*. Washington Public Power Supply System, Richland, Washington.

Washington Public Power Supply System. 1977b. *Environmental Report, WPPSS Nuclear Project No. 2, Operating License Stage, Docket No. 50–397*. Washington Public Power Supply System, Richland, Washington.

Washington Public Power Supply System. 1978. *Supplemental Information on the Hanford Generating Project in Support of a 316(a) Demonstration*. Washington Public Power Supply System, Richland, Washington.

Washington Public Power Supply System. 1982. *Technical Review of the Aquatic Monitoring Program of WNP-2*. Washington Public Power Supply System, Richland, Washington.

Washington Public Power Supply System. 1985. *Operational Ecological Monitoring Program for Nuclear Plant No. 2, 1984 Annual Report*. Washington Public Power Supply System, Richland, Washington.

Washington Public Power Supply System. 1986. *Operational Ecological Monitoring Program for Nuclear Plant 2, 1985 Annual Report*. Washington Public Power Supply System, Richland, Washington.

Washington Public Power Supply System. 1987. *Operational Ecological Monitoring Program for Nuclear Plant 2, 1987 Annual Report*. Washington Public Power Supply System, Richland, Washington.

INDEX